A. P. Fordy
J. C. Wood (Eds.)

**Harmonic Maps and
Integrable Systems**

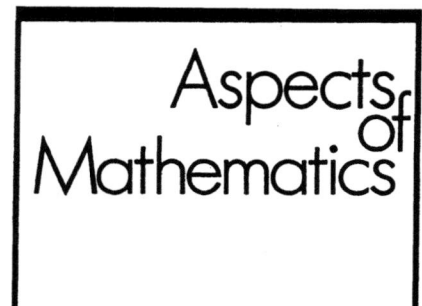

Edited by Klas Diederich

Vol. E 2: M. Knebusch/M. Kolster: Wittrings

Vol. E 3: G. Hector/U. Hirsch: Introduction to the Geometry of Foliations, Part B

Vol. E 5: P. Stiller: Automorphic Forms and the Picard Number of an Elliptic Surface

Vol. E 6: G. Faltings/G. Wüstholz et al.: Rational Points*

Vol. E 7: W. Stoll: Value Distribution Theory for Meromorphic Maps

Vol. E 9: A. Howard/P.-M. Wong (Eds.): Contribution to Several Complex Variables

Vol. E 10: A. J. Tromba (Ed.): Seminar of New Results in Nonlinear Partial Differential Equations*

Vol. E 13: Y. André: G-Functions and Geometry*

Vol. E 14: U. Cegrell: Capacities in Complex Analysis

Vol. E 15: J.-P. Serre: Lectures on the Mordell-Weil Theorem

Vol. E 16: K. Iwasaki/H. Kimura/S. Shimomura/M. Yoshida: From Gauss to Painlevé

Vol. E 17: K. Diederich (Ed.): Complex Analysis

Vol. E 18: W. W. J. Hulsbergen: Conjectures in Arithmetic Algebraic Geometry

Vol. E 19: R. Racke: Lectures on Nonlinear Evolution Equations

Vol. E 20: F. Hirzebruch, Th. Berger, R. Jung: Manifolds and Modular Forms*

Vol. E 21: H. Fujimoto: Value Distribution Theory of the Gauss Map of Minimal Surfaces in $\mathbf{R}^m$

Vol. E 22: D. V. Anosov/A. A. Bolibruch: The Riemann-Hilbert Problem

Vol. E 23: A. P. Fordy/J. C. Wood (Eds.): Harmonic Maps and Integrable Systems

Vol. E 24: D. S. Alexander: A History of Complex Dynamics

*A Publication of the Max-Planck-Institut für Mathematik, Bonn

Volumes of the German-language subseries "Aspekte der Mathematik" are listed at the end of the book.

Allan P. Fordy
John C. Wood
(Eds.)

Harmonic Maps and Integrable Systems

Mathematics Subject Classification: 58 E 20, 58 F 07, 58 F 39, 53 C 42

All rights reserved

© Springer Fachmedien Wiesbaden 1994
Ursprünglich erschienen bei Friedr. Vieweg & Sohn Verlagsgesellschaft mbH, Braunschweig/Wiesbaden, 1994
Softcover reprint of the hardcover 1st edition 1994

 No part of this publication may be reproduced, stored in a retrieval system or transmitted, mechanical, photocopying or otherwise without prior permission of the copyright holder.

Cover design: Wolfgang Nieger, Wiesbaden

ISSN 0179-2156
ISBN 978-3-528-06554-6   ISBN 978-3-663-14092-4 (eBook)
DOI 10.1007/978-3-663-14092-4

Preface

This book brings together experts in the field to explain the ideas involved in the application of the theory of integrable systems to finding harmonic maps and related geometric objects.

It had its genesis in a conference with the same title organised by the editors and held at Leeds in May 1992. However, it is *not* a conference proceedings, but rather a sequence of invited expositions by experts in the field which, we hope, together form a coherent account of the theory. The editors have added cross-references between articles and have written introductory articles in an effort to make the book self-contained. There are articles giving the points of view of both geometry and mathematical physics.

Leeds, England
October 1993

A. P. Fordy
J. C. Wood

Authors' addresses

J. Bolton, Dept. of Math. Sciences, Univ. of Durham, South Road, Durham, DH1 3LE, UK
A.I. Bobenko, FB Math., Technische Univ., Strasse des 17. Juni. 135, 10623 Berlin, Germany
M. Bordemann, Fak. für Physik, Albert-Ludwigs Univ., H.-Herder-Str. 3, 79104 Freiburg, Germany
F.E. Burstall, Dept. of Mathematics, Univ. of Bath, Claverton Down, Bath, BA7 7AY, UK
A.P. Fordy, School of Mathematics, Univ. of Leeds, Leeds, LS2 9JT, UK
M. Forger, Fak. für Physik, Albert-Ludwigs Univ., H.-Herder-Str. 3, 79104 Freiburg, Germany
M.A. Guest, Dept. of Mathematics, Univ. of Rochester, Rochester, NY 14627, USA
P.Z. Kobak, Math. Institute, Univ. of Oxford, 24-29 St. Giles, Oxford, OX1 3LB, UK
L. Laartz, Fak. für Physik, Albert-Ludwigs Univ., H.-Herder-Str. 3, 79104 Freiburg, Germany
M. Mañas, Depart. de Fisica Teorica I, Universidad Computense, 28040 Madrid, Spain
I. McIntosh, Dept. of Mathematics, Univ. of Bath, Claverton Down, Bath, BA7 7AY, UK
M. Melko, Dept. of Mathematics, Univ. of the West Indies, Mona Campus, Mona, Kingston 7, Jamaica
Y. Ohnita, Dept. of Mathematics, Tokyo Metropolitan Univ., Minami-Ohsawa 1-1, Hachioji-shi, Tokyo 192-03, Japan
F. Pedit, Dept. of Math. & Stats., Univ. of Massachussetts, Amherst, MA 01003, USA
U. Schäper, Fak. für Physik, Albert-Ludwigs Univ., H.-Herder-Str. 3, 79104 Freiburg, Germany
I. Sterling, Dept. of Math., Univ. of Toledo, 2801 W. Bancroft St., Toledo, OH 43606-3390, USA
R.S. Ward, Dept. of Math. Sciences, Univ. of Durham, South Road, Durham, DH1 3LE, UK
J.C. Wood, School of Mathematics, Univ. of Leeds, Leeds, LS2 9JT, UK
L. Woodward, Dept. of Math. Sciences, Univ. of Durham, South Road, Durham, DH1 3LE, UK

Contents

Introduction and background material

Introduction, *A.P. Fordy and J.C. Wood* 3
A historical introduction to solitons and Bäcklund tranformations, *A.P. Fordy* ... 7
Harmonic maps into symmetric spaces and integrable systems, *J.C. Wood* 29

The geometry of surfaces

The affine Toda equations and miminal surfaces, *J. Bolton and L. Woodward* ... 59
Surfaces in terms of 2 by 2 matrices: Old and new integrable cases,
A.I. Bobenko 83
Integrable systems, harmonic maps and the classical theory of solitons,
M. Melko and I. Sterling 129

Sigma and chiral models

The principal chiral model as an integrable system, *M. Mañas* 147
2-dimensional nonlinear sigma models: Zero curvature and Poisson structure,
M. Bordemann, M. Forger, J. Laartz and U. Schäper 175
Sigma models in $2+1$ dimensions, *R.S. Ward* 193

The algebraic approach

Infinite dimensional Lie groups and the two-dimensional Toda lattice,
I. McIntosh 205
Harmonic maps via Adler-Kostant-Symes theory, *F.E. Burstall and F. Pedit* .. 221
Loop group actions on harmonic maps and their applications,
M.A. Guest and Y. Ohnita 273

The twistor approach

Twistors, nilpotent orbits and harmonic maps, *P.Z. Kobak* 295

Index 323

INTRODUCTION AND BACKGROUND MATERIAL

Introduction

A.P. Fordy and J.C. Wood

Harmonic maps are mappings between Riemannian or pseudo-Riemannian manifolds which extremise a certain natural energy integral generalising Dirichlet's integral. Examples include geodesics, harmonic functions, complex analytic mappings between suitable (e.g. Kähler) manifolds, the Gauss maps of constant mean curvature surfaces, and harmonic morphisms, these last being maps which preserve Laplace's equation. The Euler-Lagrange equations for a harmonic map (the "harmonic equations" are a system of semi-linear equations, elliptic if the domain is Riemannian. If we impose sufficient symmetry or "equivariance", the harmonic equations can frequently be reduced to an ordinary or partial differential equation which can be interpreted as a Hamiltonian system; such reductions to ordinary and partial differential equations have been exploited to find many harmonic maps and morphisms. However, it is much more recently that the techniques of integrable systems used in soliton theory have been employed to find large families of harmonic maps, especially from a surface to a homogeneous space; in some cases these techniques have given *all* harmonic maps, for example from 2-tori to spheres or complex projective spaces. The method is to translate the harmonic equation into a Lax type differential equation for a map with values in a loop space, and to solve this either by finding commuting flows using r-matrices, or by using ideas of Kostant, Adler and Symes where the harmonic map appears as the projection of a complex geodesic. This book has a two-fold purpose: firstly, to explain the ideas and methods starting from an elementary level and secondly to bring the reader close to the current state of research in this area.

The book had its genesis in a conference with the same title organised by the editors and held at Leeds in May 1992. However, it is *not* a conference proceedings, but rather a sequence of invited expositions by experts in the field which, we hope, together form a coherent account of the theory. The editors have added cross-references between articles and have written introductory articles in an effort to make the book self-contained. There are articles giving the points of view of both geometry and mathematical physics. These are organized into five sections: A. Introduction and Background Material, B. The geometry of surfaces, C. Sigma and chiral Models, D. The algebraic approach, and E. The twistor approach.

The first section consists of this introduction together with two articles by the editors containing background material.

Firstly **A.P. Fordy** gives a historical introduction to some of the basic methods of integrable systems used in soliton theory, generalisations of which are used in later chapters to find harmonic maps. Specifically, soliton solutions, conservation laws, Miura maps and inverse scattering are briefly explained, mainly in the context of the KdV equation. The important idea of a *Lax hierarchy* is introduced. A short introduction to Bäcklund transformations is given, illustrated by the sine-Gordon and KdV equations. Finally, zero curvature representations of various integrable systems are given, including that for the 2D Toda lattice.

Secondly **J.C. Wood** gives an introduction to harmonic map theory and the basic ideas of applying integrable systems techniques to finding harmonic maps. First of all, harmonic maps are defined and examples given. Then the use of integrable systems techniques to find equivariant harmonic maps and morphisms is described. Next Bäcklund type transforms for harmonic maps are discussed including the Gauss (or ∂–)transform for harmonic maps from surfaces to complex projective space and K. Uhlenbeck's flag transform of "adding a uniton" for harmonic maps from a surface to, for example, the unitary group. Then, the key method of introducing a spectral parameter and the interpretation of the harmonic equations as zero curvature equations is discussed. It is then shown how solutions to these, and hence harmonic maps, can be found by integrating commuting flows on a loop space. The r-matrix formulation of the commuting flows is presented.

We then present a section on the geometry of surfaces.

This starts with a description by **J. Bolton** and **L.M. Woodward** of how the Toda equations arise naturally in the study of certain ("superconformal") harmonic maps of surfaces to complex projective spaces and spheres. In the case of the 6-sphere, the constuctions are related to the natural almost-complex structure on that manifold. The variables in the Toda system are interpreted as natural geometric invariants.

A.I. Bobenko continues by discussing the formulation of many important equations from differential geometry as zero curvature equations with spectral parameter, a key ingredient in applying integrable systems techniques to the harmonic and related equations; in particular he shows how to do this for minimal surfaces, surfaces of constant mean curvature or constant Gauss curvature, Bonnet and Bianchi surfaces, and surfaces with harmonic inverse mean curvature. The deformations corresponding to variations of the spectral parameter are described and the spin structure of these examples is discussed.

M. Melko and **I. Sterling** continue this theme, showing how to translate harmonicity of a map from the Euclidean plane $\mathbb{R}^2$ to the 2-sphere into zero curvature equations for a lift $\mathbb{R}^2 \to S^3$, and showing how to solve these by the integration of commuting flows. The resulting family of harmonic maps gives families of surfaces of constant Gauss or constant mean curvature. They also list some other types of surfaces with the corresponding soliton equations including Willmore surfaces and Hasimoto surfaces. There follow some historical remarks and computer generated pictures of surfaces of the types mentioned.

In the next section of the book we turn to the interpretation of these equations in mathematical physics. A harmonic map from $\mathbb{R}^2$ or the Minkowski plane $\mathbb{R}^{1,1}$ is called a *(two-dimensional)* σ*-model* — or a *(principal) chiral model* if the values of the map lie in a Lie group.

Firstly, **M. Mañas** discusses the chiral model, mainly in the Minkowski case, giving the zero curvature formulation and describing the Birkhoff factorisation leading to the dressing transformation of Zakharov and Shabat. This, in the limit, gives the flag transform of Uhlenbeck. Conservation laws are discussed as are soliton type solutions. These are contrasted with the uniton type solutions of Uhlenbeck in the Euclidean case.

Next **M. Bordemann, M. Forger, J. Laartz** and **U. Schäper** give general conditions under which a map from $\mathbb{R}^{1,1}$ to a homogeneous space has a zero curvature representation. They then give a historical outline of some of the key papers in the mathematical physics literature dealing with integrability of two-dimensional σ-models. Finally they show how to introduce a Poisson structure to give a Hamiltonian formulation of σ-models.

To finish this section, **R.S. Ward** introduces *time-dependent* σ*-models*, in particular discussing the dynamics of soliton-like solutions. The simplest way of introducing time dependence results in unstable solitons; he shows how to modify the Lagrangian so that the solitons become stable; then their behaviour under collisions is discussed.

In the next section we turn to *the algebraic approach*.

In the article by **I. McIntosh** we return to the Toda equations arising from harmonic maps of surfaces to complex projective space. McIntosh shows how a large class of solutions can be found explicitly in terms of Baker functions. He discusses the possibility of generalising this method to give explicit formulae for all harmonic tori in complex projective space.

F.E. Burstall and **F. Pedit** show how the Adler-Kostant-Symes scheme unifies the above soliton-theoretic construction of harmonic maps into various homogeneous spaces. In particular, they construct harmonic maps into Riemannian symmetric spaces and primitive maps into k-symmetric spaces; the latter include twistor lifts of the former and (affine or periodic) Toda fields. They use Symes' method of projecting geodesics to solve the Toda lattice in this context, thus relating integrable systems to the "extended solutions" of Uhlenbeck. They discuss the relation between the Adler-Kostant-Symes scheme and the r-matrix formalism and between finite type and finite uniton number. Among the results presented is the construction of all non-isotropic 2-tori in spheres.

M.A. Guest and **Y. Ohnita** finish this section with an exposition of finite and infinite dimensional group actions on the space of harmonic maps of a Riemann surface to a Lie group. These are then applied to study the deformations of harmonic maps. Some relations with twistor theory are described.

We turn finally to this *twistor theory*, a related powerful method of finding harmonic maps. **P.Z. Kobak** gives a description of the state of the art in the twistor theory of harmonic maps, in particular showing how well-known twistor constructions of harmonic maps into the 4-sphere can be generalised to compact quaternion-Kähler symmetric spaces, in particular giving large classes of harmonic maps from surfaces into $G_4(\mathbb{R}^2)$ and $G_2/SO(4)$.

We would like to thank F.E. Burstall for help in the final stages of producing this book. We are grateful to the London Mathematical Society and the European Community contract number SCI-0105-C(AM) for financial support of the conference which inspired this book.

A Historical Introduction to Solitons and Bäcklund Transformations

A.P.Fordy

1 An Introduction to the KdV Equation.

Soliton theory developed after the discovery by Gardner, Greene, Kruskal and Miura (GGKM)[11] of the Inverse Scattering Transform for the Korteweg de Vries (KdV) equation (see 1.1 below). They had been led to this by the earlier discovery of solitons by Kruskal and Zabusky [35], who were studying the Fermi-Pasta-Ulam problem of 1–dimensional lattices. Thus started the *modern* development of soliton theory.

However, with some historical perspective, a better starting point is 131 years earlier. In 1834 John Scott Russell, a nautical engineer, was riding on horseback by the side of the Union Canal near Edinburgh. He describes what he saw in [30]:

> I believe I shall best introduce this phenomenon by describing the circumstances of my own first acquaintance with it. I was observing the motion of a boat which was rapidly drawn along a narrow channel by a pair of horses, when the boat suddenly stopped - not so the mass of water in the channel which it had put in motion; it accumulated round the prow of the vessel in a state of violent agitation, then suddenly leaving it behind, rolled forward with great velocity, assuming the form of a large solitary elevation, a rounded, smooth and well defined heap of water, which continued its course along the channel apparently without change of form or diminution of speed. I followed it on horseback, and overtook it still rolling on at a rate of some eight or nine miles an hour, preserving its original figure some thirty feet long and a foot to a foot and a half in height. Its height gradually diminished, and after a chase of one or two miles I lost it in the windings of the channel. Such, in the month of August 1834, was my first chance interview with that singular

and beautiful phenomenon which I have called the Wave of Translation, a name which it now very generally bears.

He followed this observation with a number of experiments during which he determined the shape of a solitary wave to be that of a $sech^2()$ function. He also determined the relationship of the speed of the wave to its amplitude. At the time there was no equation describing such water waves and having such a solution. Thus John Scott Russell discovered the solution to an as yet unknown equation!

The equation describing the (unidirectional) propagation of waves on the surface of a shallow channel was derived by Korteweg and de Vries in 1895 [18]. After performing a Galilean and a variety of scaling transformations, the KdV equation can be written in simplified form:

$$u_t = u_{xxx} + 6uu_x . \tag{1.1}$$

The travelling wave solution (travelling to the left):

$$u(x,t) = \varphi(\xi), \quad \xi = x + ct, \tag{1.2}$$

is easily shown to satisfy the first order, nonlinear ODE:

$$\left(\frac{d\varphi}{d\xi}\right)^2 = b + a\varphi + c\varphi^2 - 2\varphi^3. \tag{1.3}$$

We can think of this as the energy equation for a simple Hamiltonian system with cubic potential:

$$E = \frac{1}{2}(\varphi')^2 + V(\varphi), \quad V = \varphi^3 - \frac{1}{2}\left(c\varphi^2 + a\varphi\right), \quad b = 2E, \tag{1.4}$$

generically having two equilibria. In the neighbourhood of the stable equilibrium, we have periodic solutions which can be expressed in terms of a Jacobi elliptic function, which corresponds to a periodic wave train of the KdV equation. Small amplitude oscillations correspond to the zero limit of the modulus of the elliptic function. The periodic solutions are bounded in the phase plane by the separatrix, which corresponds to the limit of infinite period and modulus 1. This is just the 1-soliton solution. We can, in fact, derive this directly, by imposing the boundary conditions that $\varphi, \frac{d\varphi}{d\xi}, \frac{d^2\varphi}{d\xi^2}$ vanish as $|\xi| \to \infty$. We then have $a = b = 0$ and the resulting equation has solution:

$$u = \varphi(\xi) = \frac{1}{2} c \, sech^2 \frac{1}{2}\sqrt{c}\,(x + ct + \delta), \tag{1.5}$$

where δ is the phase. This clearly represents the solitary wave observed by John Scott Russell and shows that the peak amplitude is exactly half the speed. Thus larger solitary waves have greater speeds. This suggests a numerical experiment: start with two solitary wave solutions, with centres well separated and the larger to the right. Initially, with negligible overlap, they will evolve *independently* as solitary wave solutions. However, the larger, faster one will start to overtake the smaller and the nonlinearity will play a significant role. For most dispersive evolution equations these solitary waves would scatter inelastically and lose 'energy' to radiation. Not so for the KdV equation: after a fully nonlinear interaction, the solitary waves re-emerge, retaining their identities (same speed and form), suffering nothing more than a phase shift (modified δ's, representing a displacement of their centres). It was after a similar numerical experiment that Kruskal and Zabusky coined the name 'soliton', to reflect the particle-like behaviour of the solitary waves under interaction.

Remark 1.1 *There are many equations which have solitary wave solutions, but these are not usually soliton solutions, in that they undergo inelastic scattering. For instance, the solitary wave solutions of the φ^4 Lagrangian field theory are NOT solitons, even though physicists often refer to them as such (see the discussion by Ward in [33]).*

1.1 Conservation Laws

It was this numerical evidence which prompted Kruskal and his co-workers in Princeton to analytically investigate the KdV equation. Initially, Miura investigated local conservation laws:

$$\partial_t \mathcal{T} + \partial_x \mathcal{F} = 0, \tag{1.6}$$

where $\mathcal{T}$ and $\mathcal{F}$ are polynomials in $u(x,t)$ and its x−derivatives and where ∂_t and ∂_x denote total derivatives. With appropriate boundary conditions this leads to a conserved quantity (constant of the motion). Integrating (1.6) with respect to x we get :

$$\partial_t \int_A^B \mathcal{T} + [\mathcal{F}]_A^B = 0. \tag{1.7}$$

Under periodic (in x) boundary conditions with $A - B$ an integer multiple of the period or with $u(x,t)$ rapidly decreasing as $x \to \pm\infty$ and $(A, B) = (-\infty, \infty)$, the square bracket in (1.7) vanishes and we have the constant of motion $\int_A^B \mathcal{T} dx$. The quantities $\mathcal{T}$ and $\mathcal{F}$ are respectively called conserved density and flux. Each $\mathcal{T}$ is, in fact, only determined up to an exact x−derivative, so defines an equivalence class of conserved densities:

$$\mathcal{T} \sim \mathcal{T} + \partial_x S, \tag{1.8}$$

since this only adds $\partial_t S$ to $\mathcal{F}$ and leaves the value of $\int \mathcal{T} dx$ unchanged. For the KdV equation the first three are:

$$\mathcal{T}_0 = u, \qquad \mathcal{F}_0 = -u_{xx} - 3u^2,$$

$$\mathcal{T}_1 = \tfrac{1}{2}u^2, \qquad \mathcal{F}_1 = -uu_{xx} + \tfrac{1}{2}u_x^2 - 2u^3, \tag{1.9}$$

$$\mathcal{T}_2 = u^3 - \tfrac{1}{2}u_x^2, \qquad \mathcal{F}_2 = u_x u_{xxx} - \tfrac{1}{2}u_{xx}^2 - 3u^2 u_{xx} + 6uu_x^2 - \tfrac{9}{2}u^4.$$

The first of these is just the equation itself. These three conservation laws have some physical interpretation, so it was no surprise that they exist. However, Miura discovered several more by direct calculation and was led to the conjecture that there should be *infinitely many*. I quote from a footnote in [24]:

> The author's own introduction to this field of research began here with the tedious work of deriving more explicit conservation laws. Four additional ones were discovered in rapid succession leading to the obvious conjecture that there were infinitely many. However, during the summer of 1966, there was a rumor that only nine polynomial conservation laws existed—exactly the number that had been found! Consequently, the author spent a week's vacation at a beautiful lake near Peterboro, Ontario, Canada, working out the tenth conservation law—which exists! An algorithm had been developed for computing the conserved densities and Donald Stevens of the Courant Institute of Mathematical Sciences devised a computer program for the AEC CDC 6600 computer which successfully computed the eleventh conserved density consisting of 45 terms. This was a significant accomplishment in view of the fact that a program using the FORMAC symbol manipulating language for an IBM 7094 computer written at the Los Alamos Scientific Laboratory for the same purpose only successfully computed up through the fifth conserved density before exceeding the available storage space.

In order to ascertain whether the KdV equation was the only such equation with so many conservation laws Miura investigated equations of the form:

$$u_t = u_{xxx} + 6u^n u_x, \tag{1.10}$$

and found that for $n = 1$ and $n = 2$ (and *only* these values) there existed many conservation laws. With a slight change in notation the second of these, called the modified KdV (MKdV) equation, can be written:

$$v_t = v_{xxx} - 6v^2 v_x = \left(v_{xx} - 2v^3\right)_x. \tag{1.11}$$

The first few conservation laws correspond to conserved densities and fluxes:

$$\tilde{\mathcal{T}}_{-1} = v, \qquad \tilde{\mathcal{F}}_{-1} = -v_{xxx} + 2v^3,$$

$$\tilde{\mathcal{T}}_0 = v^2, \qquad \tilde{\mathcal{F}}_0 = -2vv_{xx} + v_x^2 + 3v^4, \tag{1.12}$$

$$\tilde{\mathcal{T}}_1 = \tfrac{1}{2}(v_x^2 + v^4), \quad \tilde{\mathcal{F}}_1 = -v_x v_{xxx} + \tfrac{1}{2}v_{xx}^2 - 2v^3 v_{xx} + 6v^2 v_x^2 + 2v^6.$$

Miura [23] then noticed a remarkable fact: by substituting

$$u = -v_x - v^2 \tag{1.13}$$

into the conserved densities (1.9) of the KdV equation, they were transformed into those of the MKdV equation ($\tilde{\mathcal{T}}_{-1}$ is not included this way). For example,

A Historical Introduction to Solitons and Bäcklund Transformations

$$\mathcal{T}_0 = u = -v_x - v^2 \sim -\tilde{\mathcal{T}}_0,$$
$$\mathcal{T}_1 = \tfrac{1}{2}u^2 = \tfrac{1}{2}(v_x^2 + v^4 + 2v^2 v_x) \sim \tilde{\mathcal{T}}_1,$$
(1.14)

where '$\sim$' refers to the equivalence relation (1.8). The substitution (1.13) is now referred to as the Miura transformation (more properly, Miura *map*). Such maps play a very important role in the Hamiltonian theory of soliton equations. Furthermore, (1.13) gives a direct relation between equations (1.1) and (1.11) since

$$\begin{aligned} u_t &= -(\partial + 2v)\, v_t = -(\partial + 2v)(v_{xxx} - 6v^2 v_x) \\ &= (-v_x - v^2)_{xxx} + 6(-v_x - v^2)(-v_{xx} - 2vv_x) = u_{xxx} + 6uu_x, \end{aligned}$$
(1.15)

where $\partial \equiv \partial_x$. Thus if v satisfies the MKdV equation (1.11), then u satisfies the KdV equation (1.1). The remarkable occurrence in this calculation is that the mess of terms in v and its x-derivatives, resulting from the right-hand side of (1.15), should collect together to give an expression entirely in terms of the combination $(-v_x - v^2)$ and thus u.

Remark 1.2 *This is a one-way passage since we cannot deduce that if u satisfies the KdV equation then v satisfies the MKdV equation.*

We next come to the *proof* that there exist an infinite number of polynomial conservation laws for the KdV equation. In [24] Miura explains, *"Two existence proofs were found simultaneously by Gardner and by Kruskal and the author"*, with a foot note :

> We use the word "simultaneously" in the strict sense. While working late one afternoon, Martin Kruskal and the author found a proof for the existence of an infinite number of conservation laws using the WKB formalism. As we examined our result, Clifford Gardner called from his home and told us he had just obtained an existence proof—a different one!

Gardner's proof became one of the standard constructions and starts by adding a parameter to the Miura map [10], by exploiting the Galilean symmetry of the KdV equation (and the lack of such for the MKdV equation). Specifically, we add a constant onto the function u, so that

$$u = -v_x - v^2 + \lambda.$$
(1.16)

Whilst u still satisfies the KdV equation (1.1), v now satisfies:

$$v_t = \left(v_{xx} - 2v^3 - 6\lambda v\right)_x.$$
(1.17)

We introduce the asymptotic expansion:

$$v = k - \frac{1}{2}uk^{-1} + \sum_{i=2}^{\infty} v_i k^{-i}, \quad \lambda = k^2,$$
(1.18)

which starts as:
$$v_2 = \tfrac{1}{4}u_x, \qquad v_3 = -\tfrac{1}{8}\left(u^2 + u_{xx}\right),$$
$$v_4 = \tfrac{1}{16}\left(2u^2 + u_{xx}\right)_x, \quad v_5 = -\tfrac{1}{16}(u^3 - \tfrac{1}{2}u_x^2) - \tfrac{1}{32}\left(u_{xxx} + 6uu_x\right)_x,$$
(1.19)

and satisfies (for $i \geq 1$):
$$v_{i+1} = -\frac{1}{2}\Big(v_{ix} + \sum_{j=0}^{i} v_j v_{i-j}\Big). \tag{1.20}$$

Thus, at each level, v_{i+1} can be solved in terms of a *local* functional of $v_1, \cdots, v_i$. Since v satisfies the MKdV equation (1.17), it is itself conserved. Thus each term v_i of our series is individually conserved. However, it is easy to prove that the even elements of the expansion are exact derivatives and thus trivial as conserved densities. The odd elements give us our *infinite* number of (nontrivial) conserved densities.

1.2 The Inverse Spectral Transform

The Miura map (1.16) can be viewed as a Riccati equation for v and thus linearised by the substitution $v = \psi_x/\psi$, giving
$$L\psi \equiv \psi_{xx} + u\psi = \lambda\psi. \tag{1.21}$$

This is the time-independent Schrödinger equation (well known from quantum theory) with $u(x,t)$ playing the role of potential and λ the energy. It is important to realise that t *is not the time of the time-dependent Schrödinger equation*. We think of x as the spatial variable and t as a parameter. Considered as a Sturm-Liouville eigenvalue problem it is natural to ask how λ and ψ change with t as $u(x,t)$ evolves from some initial state according to the KdV equation. Gardner, Greene, Kruskal and Miura [11, 12] discovered the remarkable fact that the (discrete part of the) spectrum necessarily remains constant in 'time' while the corresponding wave functions ψ evolve according to a very simple *linear differential equation*.

Today (following Lax [19]) we usually take the opposite route. We postulate that ψ evolves through a linear differential equation:
$$\psi_t = P\psi. \tag{1.22}$$

Equations (1.21) and (1.22) form an overdetermined system, whose integrability conditions can be written:
$$L_t = [P, L] = PL - LP. \tag{1.23}$$

A consequence of (1.23) is that all eigenvalues corresponding to the function $u(x,t)$ remain constant. When P is given by :
$$P = 4\partial^3 + 6u\partial + 3u_x \tag{1.24}$$

A Historical Introduction to Solitons and Bäcklund Transformations 13

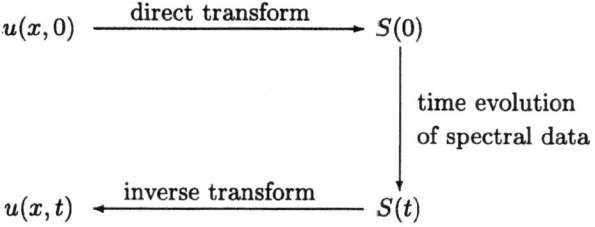

Figure 1 Inverse Scattering Transform

(1.23) reduces to the KdV equation (1.1). Since, under these conditions, the spectrum of L remains constant, the KdV equation is referred to as an *isospectral flow*. Equation (1.23) is called the *Lax representation* of the KdV equation.

The identification of the KdV equation as an isospectral flow of the Schrödinger operator enabled GGKM to devise a method of solving the KdV equation (with 'rapidly decreasing' boundary conditions), called the inverse scattering or inverse spectral transform (IST). This is a direct generalisation of the Fourier transform used to solve linear equations and can be represented by essentially the same scheme (Figure 1).

For these boundary conditions, the solutions of (1.21) are characterised by their asymptotic properties as $x \to \pm\infty$. For a given potential function, the continuous and discrete spectrum are treated separately. Corresponding to the continuous spectrum ($\lambda = -k^2$) the solutions are asymptotically oscillatory, characterised by two coefficients:

$$\psi \sim \begin{cases} T(k)\, e^{-ikx} & x \to -\infty \\ e^{-ikx} + R(k)e^{ikx} & x \to +\infty \end{cases} \quad (1.25)$$

subject to the condition $\mid R \mid^2 + \mid T \mid^2 = 1$. The constants $R(k)$ and $T(k)$ are respectively called the *reflection* and *transmission* coefficients, from their quantum mechanical interpretation. Under mild conditions on the potential function, the Schrödinger operator L has only a finite number of discrete eigenvalues $\{\kappa_n^2\}$. The corresponding eigenfunctions are square integrable and are the 'bound states' of quantum mechanics with asymptotic properties (for $\kappa_n > 0$):

$$\psi_n \sim \begin{cases} \tilde{c}_n e^{\kappa_n x} & x \to -\infty \\ c_n e^{-\kappa_n x} & x \to +\infty \end{cases}. \quad (1.26)$$

The *direct scattering transform* constructs the quantities $\{T(k), R(k), \kappa_n, c_n\}$ from a given potential function. The important inversion formulae were derived by Gel'fand and Levitan in 1955 [14]. These enable the potential u to be constructed out of the spectral or scattering data $S = \{R(k), \kappa_n, c_n\}$. This is considerably more complicated than the Inverse Fourier Transform, involving the solution of a nontrivial integral equation, whose kernel is built out of the scattering data (see [1, 7, 26, 28] for descriptions of this).

To solve the KdV equation we first construct the scattering data $S(0)$ from the initial condition $u(x,0)$. As a consequence of (1.22) (with an additional constant) with the given boundary conditions, the scattering data evolves in a very simple way. Indeed, we can give explicit formulae:

$$R(k,t) = R(k,0)e^{8ik^3 t}, \quad c_n(t) = c_n(0)e^{-4\kappa_n^3 t}. \tag{1.27}$$

Using the inverse scattering transform on the scattering date $S(t)$, we obtain the potential $u(x,t)$ and thus the solution to the initial value problem for the KdV equation.

Remark 1.3 *Although $T(k)$ plays no role in the IST it is responsible for the constants of motion, since $T(k)$ is independent of t.*

This process cannot be carried out *explicitly* for arbitrary initial data, although, in this case, it gives a great deal of information about the solution $u(x,t)$. However, whenever the reflection coefficient is zero, the kernel of Gel'fand-Levitan integral equation becomes separable and explicit solutions can be found. It is in this way that the N−soliton solution is constructed by IST from the initial condition:

$$u(x,0) = N(N+1)\operatorname{sech}^2 x. \tag{1.28}$$

The general formula for the multi-soliton solution is given by:

$$u(x,t) = 2\frac{\partial^2}{\partial x^2} \ln \det M, \tag{1.29}$$

where M is a matrix built out of the discrete scattering data. This determinant is in the form of a sum of exponentials.

1.3 The τ-Function.

The simple form of this solution inspired Hirota [15] to *directly* introduce the ansatz:

$$u = 2\frac{\partial^2}{\partial x^2} \ln \tau \tag{1.30}$$

into the KdV equation, giving:

$$\tau\tau_{xt} - \tau_x \tau_t = \tau\tau_{xxxx} - 4\tau_x \tau_{xxx} + 3\tau_{xx}^2. \tag{1.31}$$

Some apparently magical cancellations take place in order for this expression to be purely *quadratic* in τ. The 1− and 2−soliton solutions of the KdV equation correspond respectively to τ−functions:

$$\tau_1 = 1 + e^{\theta_1}, \tag{1.32}$$

$$\tau_2 = 1 + e^{\theta_1} + e^{\theta_2} + \left(\frac{k_1 - k_2}{k_1 + k_2}\right)^2 e^{\theta_1 + \theta_2}, \tag{1.33}$$

A Historical Introduction to Solitons and Bäcklund Transformations 15

where $\theta_i = k_i x - k_i^3 t + \alpha_i$. The coefficient of $e^{\theta_1+\theta_2}$ in τ_2 has been chosen so that the coefficient of $e^{\theta_1+\theta_2}$ in (1.31) vanishes. *Miraculously* the coefficients of $e^{2\theta_1+\theta_2}$, etc. in (1.31) also vanish. Analogous expansions correspond to the N-soliton solution for each N for which *all* necessary cancellations occur!

Hirota developed his 'direct method' with which he was able to build the multi-soliton solutions of many equations. The first step is to discover the correct substitution ((1.30) in the case of the KdV equation), such that the τ-function satisfies a quadratically nonlinear equation such as (1.31). As in (1.31) these formulae take the form of the 'Leibnitz formula' for the differentiation of a product, except that the signs alternate. For such cases, Hirota developed a calculus of 'bilinear derivatives' and wrote equations such as (1.31) in a compact notation, called 'Hirota's bilinear form'(see the review [27]). He proved that *all* such equations possess 1− and 2−soliton solutions. However, only for very special equations can the N-soliton solution, for $N \geq 3$, be constructed in this way. Requiring all the cancellations to take place puts very strong restrictions on the equation, thus selecting *integrable* equations.

Remark 1.4 *Periodic or finite gap solutions correspond to τ−functions which are the product of a simple exponential and a Riemann Θ−function. These are the solutions which are important for this book and their τ−function representations are discussed in other chapters [34, 20].*

As originally introduced, this method was rather ad-hoc (but *very* effective). However, it was later shown by the Kyoto group that the τ−function and Hirota's bilinear form naturally arise in the context of the representations of infinite dimensional Lie algebras (see the review [25] and the discussion in [26]).

1.4 The Lax Hierarchy

In [19] Lax reformulated GGKM's discovery [11, 12] of the isospectral nature of the KdV equation in algebraic form:

$$\left. \begin{array}{l} L\psi \equiv (\partial^2 + u)\psi = \lambda\psi \\ \psi_t = P\psi \equiv (4\partial^3 + 6u\partial + 3u_x)\psi \\ \lambda_t = 0 \end{array} \right\} \Rightarrow L_t = [P, L] \equiv PL - LP. \quad (1.34)$$

This led Lax to an interesting generalisation:

$$\left. \begin{array}{l} L\psi \equiv (\partial^2 + u)\psi = \lambda\psi \\ \psi_{t_m} = P_{(m)}\psi \equiv \partial^{2m+1} + \sum_{i=0}^{2m-1} b_i \partial^i \\ \lambda_t = 0 \end{array} \right\} \Rightarrow L_{t_m} = [P_{(m)}, L]. \quad (1.35)$$

The integrability condition $L_{t_m} = [P_{(m)}, L]$ is explicity written as:

$$u_{t_m} = ((2m+1)u_x - 2b_{2m-1x})\partial^{2m} + \cdots + (P_{(m)}u - Lb_0). \tag{1.36}$$

Equating coefficients of $\partial^i, i = 0, \cdots, 2m$, gives us $2m+1$ equations, from which we deduce the $2m$ coefficients $b_0, \cdots, b_{2m-1}$:

$$b_{2m-1} = \frac{1}{2}(2m+1)u, \quad b_{2m-2} = \frac{1}{4}(2m+1)(2m-1)u_x, \cdots \tag{1.37}$$

together with the isospectral flow:

$$u_{t_m} = P_{(m)}u - Lb_0. \tag{1.38}$$

Nontrivial flows only exist for odd-order equations, since the adjoint of the integrability condition implies $P^\dagger = -P$. There exists an infinite hierarchy of such isospectral flows, the first three of which are:

$$u_{t_0} = u_x \tag{1.39}$$

$$u_{t_1} = \frac{1}{4}(u_{xxx} + 6uu_x) \tag{1.40}$$

$$u_{t_2} = \frac{1}{16}\left(u_{xxxxx} + 10uu_{xxx} + 20u_x u_{xx} + 30u^2 u_x\right), \tag{1.41}$$

corresponding respectively to operators:

$$P_{(0)} = \partial, \tag{1.42}$$

$$P_{(1)} = \partial^3 + \frac{3}{4}(u\partial + \partial u), \tag{1.43}$$

$$P_{(2)} = \partial^5 + \frac{5}{4}(u\partial^3 + \partial^3 u) + \frac{5}{16}\left((3u^2 - u_{xx})\partial + \partial(3u^2 - u_{xx})\right). \tag{1.44}$$

Remark 1.5 *It was later shown by Gel'fand and Dikii [13] that these $P_{(m)}$ could be interpreted as fractional powers of L:*

$$P_{(m)} = (L^{m/2})_+, \tag{1.45}$$

where $()_+$ means the differential part of the operator. When m is even, $(L^{m/2})_+ = L^{m/2}$, which commutes with L, thus giving trivial equations. This is a general theory which also applies to higher order operators L.

Remark 1.6 *Zakharov and Shabat [38] gave a general scheme which enabled them to integrate by IST a number of equations of physical interest, including the Boussinesq equation (associated with a third order Lax operator) and the Kadomtsev-Petviashvili equation, which is an equation in $(2+1)$ dimensions (2 space and 1 time dimension). Taking soliton theory out of the purely $(1+1)$ regime was a very important step.*

1.5 Factorisation of the Schrödinger Operator

We have already seen that the KdV and MKdV equations are related through the Miura map (1.13) and that the spectral problem (1.21) arose through the linearisation of (1.16). We can, of course, reconstruct the Miura map through reversing the process and 'nonlinearising' equation (1.21).

However, there is an interesting alternative way which was, in a different context, introduced by Schrödinger [31] and more generally arises in the context of recurrence relations for special functions [32]. If we *factorise* the Schrödinger operator and write it as the produce of two first order operators we immediately obtain the Miura map:

$$L = \partial^2 + u = (\partial + v)(\partial - v) = \partial^2 - v_x - v^2. \tag{1.46}$$

We thus have an alternative way of writing (1.21) as a first order system:

$$\begin{pmatrix} \psi_1 \\ \psi_2 \end{pmatrix}_x = \begin{pmatrix} v & k \\ k & -v \end{pmatrix} \begin{pmatrix} \psi_1 \\ \psi_2 \end{pmatrix}, \quad \lambda = k^2, \tag{1.47}$$

where $\psi_1 = \psi$. This is a spectral problem for the MKdV hierarchy. A generalisation of this construction to the case of higher order scalar Lax operators was introduced in [8, 9].

2 Bäcklund Transformations

In 1875 A.V. Bäcklund discovered a transformation which enabled him to build new surfaces of constant negative curvature from old. In *asymptotic* co-ordinates the first fundamental form of such a surface (with Gaussian curvature $K = -1$) is given by:

$$ds^2 = du^2 + 2\cos\theta \, du \, dv + dv^2, \tag{2.1}$$

where $\theta(u, v)$ is the angle between the co-ordinate lines, and satisfies the sine-Gordon equation:

$$\theta_{uv} = \sin\theta. \tag{2.2}$$

A discussion of these classical surfaces can be found in [4, 22]. The Bäcklund transformation is actually a differential relation between two solutions θ and $\bar\theta$ of (2.2):

$$\begin{aligned} (\bar\theta - \theta)_u &= 2\alpha \sin\left(\frac{\bar\theta+\theta}{2}\right), \\ (\bar\theta + \theta)_v &= \frac{2}{\alpha} \sin\left(\frac{\bar\theta-\theta}{2}\right). \end{aligned} \tag{2.3}$$

Using the integrability conditions $\theta_{uv} = \theta_{vu}$, $\bar\theta_{uv} = \bar\theta_{vu}$ we can eliminate either θ or $\bar\theta$ to obtain equations (2.2) for respectively $\bar\theta$ and θ. The important feature is that the *second order* equation (2.2) arises as the integrability conditions of a pair of *first order* equations. Starting with a known solution $\theta(u,v)$ (known surface) we solve (2.3) to obtain a new solution (surface) $\bar\theta(u,v)$. Starting with the trivial solution $\theta \equiv 0$, we can explicitly solve (2.3) for $\bar\theta$:

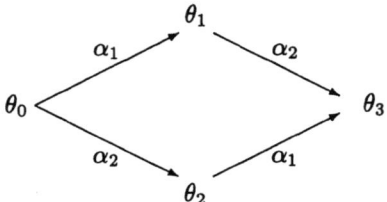

Figure 2 A commutative Bianchi diagram

$$\bar{\theta} = s_1 = 4tan^{-1}\left(exp\left(\alpha u + v/\alpha + \beta\right)\right), \qquad (2.4)$$

which is the 1-soliton (kink) solution of the sine-Gordon equation.

An important feature is that the Bäcklund transformation (2.3) possesses a parameter α (the Bäcklund parameter), which then enters the solution s_1, which we therefore label $s_1(\alpha)$. We supress the dependence upon β since this is not so important for the following argument. We could substitute $\theta = s_1(\alpha_1)$ into the (2.3) (with $\alpha = \alpha_2$) and attempt to solve for the new $\bar{\theta}$. Fortunately, it is possible to avoid this and use a purely algebraic construction. Consider Figure 2, which represents a commutative Bianchi diagram. Starting with solution θ_0 we use the Bäcklund transformation to obtain θ_1 and θ_2 (with the *same* functional form, but depending upon *different* Bäcklund parameters). To each of the solutions we apply the Bäcklund transformation again (with parameters as shown) and ask for the resulting solutions to be identical. We thus have:

$$(\theta_1 - \theta_0)_u = 2\alpha_1 \sin\left(\frac{\theta_1 + \theta_0}{2}\right), \quad (\theta_3 - \theta_1)_u = 2\alpha_2 \sin\left(\frac{\theta_3 + \theta_0}{2}\right),$$

$$(\theta_2 - \theta_0)_u = 2\alpha_2 \sin\left(\frac{\theta_2 + \theta_0}{2}\right), \quad (\theta_3 - \theta_2)_u = 2\alpha_1 \sin\left(\frac{\theta_3 + \theta_2}{2}\right), \qquad (2.5)$$

and similarly for the second part of (2.3). We can thus eliminate the derivatives to obtain:

$$\alpha_1 \sin\frac{1}{4}(\theta_0 - \theta_3 + \theta_1 - \theta_2) = \alpha_2 \sin\frac{1}{4}(\theta_0 - \theta_3 - \theta_1 + \theta_2), \qquad (2.6)$$

which can be written:

$$tan\frac{1}{4}(\theta_0 - \theta_3) = \frac{\alpha_2 + \alpha_1}{\alpha_2 - \alpha_1} tan\frac{1}{4}(\theta_1 - \theta_2). \qquad (2.7)$$

This is Bianchi's "Theorem of permutability" and is an example of a "nonlinear superposition formula".

Thus, starting with solution θ_0, we need to solve the differential equation (2.3) once to obtain θ_1 and θ_2, after which θ_3 is given by a purely algebraic operation.

A Historical Introduction to Solitons and Bäcklund Transformations

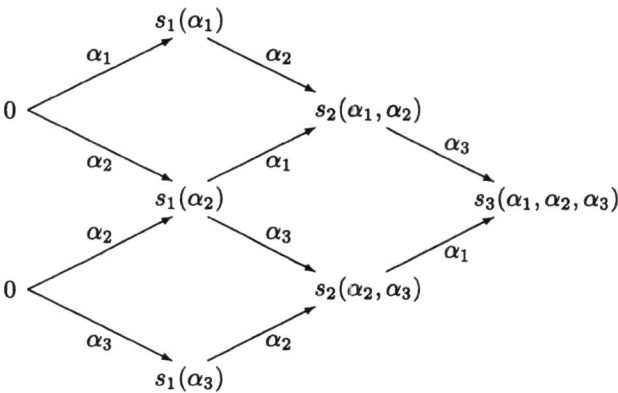

Figure 3 Bianchi diagram for constructing the 3-soliton solution

Knowing θ_2 and θ_3, we can continue the process to obtain θ_4 and so on, thus building an infinite sequence of solutions related through a chain of Bäcklund transformations. In particular, if we take $\theta_0 = 0, \theta_1 = s_1(\alpha_1), \theta_2 = s_1(\alpha_2)$ then:

$$\theta_3 = s_2(\alpha_1, \alpha_2) = 4\tan^{-1}\left[\frac{\alpha_1 + \alpha_2}{\alpha_1 - \alpha_2}\tan\frac{1}{4}(\theta_1 - \theta_2)\right], \tag{2.8}$$

which is the 2-soliton solution of the sine-Gordon equation. To obtain the 3-soliton solution we glue together 3 copies of Figure 2 (see Figure 3). The appearance of $\alpha_1 - \alpha_2$ in the denominator shows that we must have a different Bäcklund parameter at each step.

The sine-Gordon equation (2.2) is given here as the compatibility condition of a pair of *nonlinear* differential equations (2.3) whereas the KdV equation was previously given as the compatibility condition of a pair of *linear* differential equations. In fact it is an easy matter to produce the Bäcklund transformation for most equations with a linear spectral problem (see [29] for a review). For the KdV equation this is just:

$$\begin{aligned}\bar{w}_x + w_x &= \alpha - \tfrac{1}{2}(w - \bar{w})^2, \\ \bar{w}_t + w_t &= (\bar{w} - w)(w_{xx} - \bar{w}_{xx}) + 2(w_x^2 + w_x\bar{w}_x + \bar{w}_x^2),\end{aligned} \tag{2.9}$$

where $u = w_x$ and $\bar{u} = \bar{w}_x$.

Remark 2.1 *This can easily be derived from the Miura map (1.16). We note that the MKdV equation (1.17) is invariant under $v \to -v$ whilst the Miura map changes to:*

$$\hat{u} = v_x - v^2 + \lambda. \tag{2.10}$$

For a given solution v of the MKdV equation we get two different *solutions of the KdV equation, hence the $\hat{u}$. By considering the difference $\hat{u} - u = 2v_x$, we find*

$$v = \frac{1}{2}(\hat{w} - w). \tag{2.11}$$

Substitution of this into either of the Miura maps gives the x-part of the Bäcklund transformation (2.9). We use the MKdV equation (1.17), together with the potential KdV equation:

$$w_t = w_{xxx} + 3w_x^2, \tag{2.12}$$

and its hatted version, to obtain the second part of (2.9).

The sine-Gordon equation can also be written as the compatibility condition of a pair of *linear* differential equations which is the subject of the next section.

3 Zero Curvature

The linear equations for ψ, involving the higher order operators L and P of the Lax formulation, can always be replaced by a *system* of *first order* differential equations. This gives us a pair of linear equations for vector Ψ, whose number of components equals the order of the operator L. We can obviously consider a general system of this type, and the integrability condition $\Psi_{xt} = \Psi_{tx}$:

$$\left. \begin{array}{l} \Psi_x = U\Psi, \\ \Psi_t = V\Psi, \end{array} \right\} \Rightarrow U_t - V_x + [U, V] = 0. \tag{3.1}$$

This is equivalent to the *zero curvature* condition described in [21, Section 3] with connection form:

$$\omega = 2U\,dx + 2V\,dt. \tag{3.2}$$

An important ingredient is the dependence of U and V upon the spectral parameter λ. This is a crucial step employed by Zakharov and Mikhailov (Uhlenbeck) in order to represent the chiral field (harmonic map) equations in this way. It is then possible to employ the Zakharov-Shabat dressing and Riemann-Hilbert problem techniques to solve these equations. This is described in detail in [21], where full references can be found.

In this chapter we just explain some simple principles which underlie the typical construction of integrable equations associated with such zero curvature problems.

Consider a matrix U which is linear in λ:

$$U = \lambda A + Q, \tag{3.3}$$

where A is a constant diagonalisable matrix and Q is the matrix of potential functions. It is easily shown that $Q \in \text{Im ad}\,A$, where $(\text{ad}A)\,B = [A, B]$.

Remark 3.1 *A will not generally be regular (distinct eigenvalues) and not necessarily diagonalised.*

A Historical Introduction to Solitons and Bäcklund Transformations 21

Remark 3.2 *Case (3.3) is the simplest case to work with but is not essential. Zakharov and Mikhailov [36] use a pole expansion:* $U = \sum U_i(q)(\lambda - \lambda_i)^{-1}$, *while others [16, 17] favour polynomial expansions.*

Substitute (3.3) into (3.1), so that:

$$Q_t = V_x - [Q, V] - \lambda[A, V]. \tag{3.4}$$

We first seek solutions V which are polynomial in λ:

$$V^{(m)} = \sum_{i=0}^{m} V_{m-i}\lambda^i. \tag{3.5}$$

The coefficients of λ^k constitute $m+1$ equations, the solutions of which give a recursive definition of $V_0, \cdots, V_m$, together with the equations of motion as the coefficient of λ^0:

$$Q_{t_m} = V_{mx} - [Q, V_m]. \tag{3.6}$$

To achieve this for all $m \geq 0$ we seek an asymptotic solution:

$$\bar{V} = \sum_{i=0}^{\infty} V_i \lambda^{-1} \tag{3.7}$$

of the equation:

$$\bar{V}_x - [Q, \bar{V}] = \lambda[A, \bar{V}]. \tag{3.8}$$

This corresponds to the infinite recursion:

$$V_{kx} - [Q, V_k] = [A, V_{k+1}], \quad k \geq -1. \tag{3.9}$$

The solution (3.5) is then given by:

$$V^{(m)} = \left(\lambda^m \bar{V}\right)_+, \tag{3.10}$$

where $()_+$ is the truncation containing only non-negative powers of λ. Using (3.7) with $k = m$, the equations of motion (3.6) can be written:

$$Q_{t_m} = V_{mx} - [Q, V_m] = [A, V_{m+1}], \tag{3.11}$$

thus confirming that Q does indeed lie in Im ad A.

This is a recursive procedure for construction the *polynomial* flows. It is also possible to seek V, for the given U, which are rational. The sine-Gordon and 2D Toda lattices are such examples. Since these flows mutually commute they are often referred to as symmetries of each other.

3.1 Examples

The case of 2×2 matrices was considered in [37, 2]. This includes such familiar examples as the KdV and MKdV equations, the NLS equation and the sine-Gordon equation. Larger systems of integrable equations can be obtained by taking larger matrices (see the reviews [1, 6]) or by considering matrices with a higher degree dependence upon the spectral parameter [3].

Example 3.1 (Zakharov-Shabat/AKNS) The matrices:

$$U = \begin{pmatrix} \lambda & q \\ r & -\lambda \end{pmatrix}, \quad V = \begin{pmatrix} A & B \\ C & -A \end{pmatrix} \tag{3.12}$$

give rise to three scalar equations :

$$q_t = B_x - 2\lambda B + 2qA, \tag{3.13}$$
$$r_t = C_x + 2\lambda C - 2rA, \tag{3.14}$$
$$A_x = qC - rC. \tag{3.15}$$

The hierarchy of polynomial flows is obtained by the substitution :

$$V = \sum_{i=0}^{m} V_{m-i} \lambda^i. \tag{3.16}$$

We can recursively solve the resulting equations for $V_0, V_1, \ldots$ to obtain the isospectral flow:

$$q_{t_m} = B_{mx} + 2qA_m, \tag{3.17}$$
$$r_{t_m} = C_{mx} - 2rA_m. \tag{3.18}$$

The second-order expansion is

$$A = a_0 \left(\lambda^2 - \frac{1}{2} qr \right), \; B = a_0 \left(\lambda q + \frac{1}{2} q_x \right), \; C = a_0 \left(\lambda r - \frac{1}{2} r_x \right); \tag{3.19}$$

$$q_{t_2} = \frac{1}{2} a_0 \left(q_{xx} - 2q^2 r \right), \quad r_{t_2} = \frac{1}{2} a_0 \left(-r_{xx} + 2qr^2 \right). \tag{3.20}$$

With $a_0 = -2i$ and $r = -q*$ we get the NLS equation, which corresponds to the Hasimoto surface of [22, Table 2].

The third-order expansion is

$$\begin{aligned} A &= a_0 \left(\lambda^3 - \frac{1}{2} \lambda qr + \frac{1}{4} (qr_x - rq_x) \right), \\ B &= a_0 \left(\lambda^2 q + \frac{1}{2} \lambda q_x + \frac{1}{4} (q_{xx} - 2q^2 r) \right), \\ C &= a_0 \left(\lambda^2 r - \frac{1}{2} \lambda r_x + \frac{1}{4} (r_{xx} - 2qr^2) \right), \end{aligned} \tag{3.21}$$

giving the nonlinear evolution equation:

$$q_{t_3} = \tfrac{1}{4}a_0 \left(q_{xxx} - 6qrq_x\right),$$
$$r_{t_3} = \tfrac{1}{4}a_0 \left(r_{xxx} - 6qrr_x\right).$$
(3.22)

When $r = q$ (and $a_0 = 4$) we get the MKdV equation (1.11), while with $r = -1$ (and $a_0 = 4$) we get the KdV equation (1.1). This spectral problem for the MKdV hierarchy is related to (1.47) through a similarity transformation.

There are also isospectral flows for which V is not polynomial. If we choose

$$-r = q = \theta_x, \quad A = \tfrac{1}{2}\lambda^{-1}\cos\theta, \quad B = C = -\tfrac{1}{2}\lambda^{-1}\sin\theta,$$
(3.23)

then we obtain the sine-Gordon equation:

$$\theta_{xt} = \sin\theta.$$
(3.24)

Example 3.2 (The 2D Toda Lattice) We can generalise the sinh-Gordon equation by considering the following spectral problem:

$$\begin{pmatrix} \psi_1 \\ \vdots \\ \vdots \\ \psi_{N+1} \end{pmatrix}_x = \begin{pmatrix} v_1 & \lambda & \cdots & 0 \\ \vdots & \ddots & \ddots & \vdots \\ 0 & & \ddots & \lambda \\ \lambda & 0 & \cdots & v_{N+1} \end{pmatrix} \begin{pmatrix} \psi_1 \\ \vdots \\ \vdots \\ \psi_{N+1} \end{pmatrix} \equiv (\lambda A + v)\psi.$$
(3.25)

This spectral problem was first introduced in [9] and results from factorising the $(N+1)$-order scalar Lax operator in an analogous way to (1.46). Clearly (1.47) is a special case of (3.25). This factorisation gives rise to a generalised Miura map which relates the isospectral flows of (3.25) to those of the $(N+1)$-order Lax operator. Choosing V to be polynomial in λ leads to the infinite hierarchy of commuting *polynomial flows*.

To obtain a generalised sinh-Gordon flow we choose V to be proportional to λ^{-1}:

$$\begin{pmatrix} \psi_1 \\ \vdots \\ \vdots \\ \psi_{N+1} \end{pmatrix}_t = \frac{1}{\lambda}\begin{pmatrix} 0 & \cdots & \cdots & a_{1\,N+1} \\ a_{2\,1} & \ddots & & \vdots \\ \vdots & \ddots & \ddots & \vdots \\ 0 & \cdots & a_{N+1\,N} & 0 \end{pmatrix} \begin{pmatrix} \psi_1 \\ \vdots \\ \vdots \\ \psi_{N+1} \end{pmatrix},$$
(3.26)

where $a_{n\,n-1}$ and $a_{1\,N+1}$ are the only non-zero elements.

The integrability conditions of (3.25) and (3.26) are:

$$a_{n+1\,n\,x} = a_{n+1\,n}(v_{n+1} - v_n)$$
$$v_{n\,t} = a_{n\,n-1} - a_{n+1\,n},$$
(3.27)

where $n = 1, \cdots, N+1$ and indices are considered modulo $N+1$. To solve (3.27), define θ_i by $v_i = \theta_{ix}$ so that:

$$a_{n+1n} = A_{n+1n} \, exp\left(\theta_{n+1} - \theta_n\right), \tag{3.28}$$

where A_{n+1n} are constants of integration. The equations of motion then take the form of a 2D Toda lattice:

$$\theta_{nxt} = A_{nn-1} \, exp\left(\theta_n - \theta_{n-1}\right) - A_{n+1n} \, exp\left(\theta_{n+1} - \theta_n\right). \tag{3.29}$$

The polynomial flows are symmetries of these 2D Toda lattice equations.

In [8] a Bäcklund transformation is presented for the 2D Toda system (3.29) with $A_n = 1$. This itself is an integrable system, which is a 2D generalisation of the Kac-van Moerbeke system:

$$\partial_x \left(\theta_{n+1} - \bar{\theta}_n\right) = -\alpha \left(exp\left(\bar{\theta}_{n+1} - \theta_{n+1}\right) - exp\left(\bar{\theta}_n - \theta_n\right)\right),$$

$$\partial_t \left(\bar{\theta}_n - \theta_n\right) = \frac{1}{\alpha} \left(exp\left(\theta_{n+1} - \bar{\theta}_n\right) - exp\left(\theta_n - \bar{\theta}_{n-1}\right)\right). \tag{3.30}$$

There is also a nonlinear superposition law, given by:

$$exp\left(\theta_{n+1} + \tilde{\theta}_n\right) = \left(\frac{\hat{\alpha} e^{\hat{\theta}_{n+1}} - \bar{\alpha} e^{\bar{\theta}_{n+1}}}{\hat{\alpha} e^{\hat{\theta}_n} - \bar{\alpha} e^{\bar{\theta}_n}}\right) exp\left(\hat{\theta}_n + \bar{\theta}_n\right). \tag{3.31}$$

The two simplest examples (taking $A_{n+1n} = 1$, for all n) are:

1. $N = 1$:

$$-\theta_2 = \theta_1 = \theta, \quad \theta_{xt} = 2sinh \, 2\theta, \tag{3.32}$$

which is just the sinh-Gordon equation.

2. $N = 2$:

$$\theta_1 = \theta + \varphi, \quad \theta_2 = -2\varphi, \quad \theta_3 = -\theta + \varphi,$$
$$\theta_{xt} = e^{2\theta} - e^{-\theta} cosh \, 3\varphi, \quad \varphi_{xt} = e^{-\theta} sinh \, 3\varphi. \tag{3.33}$$

This corresponds to the Willmore surface listed in [22, Table 2]. A reduction of this system is:

$$\varphi \equiv 0, \quad \theta_{xt} = e^{2\theta} - e^{-\theta}, \tag{3.34}$$

which is the Dodd-Bullough equation and corresponds to the affine surface of [22, Table 2]. For $N = 2$ (3.30) is an auto-Bäcklund transformation for (3.33). However, the restriction $\varphi = 0$ is not invariant under this transformation, which transforms a solution of (3.34) onto a solution of (3.33).

More examples can be found in [6, 9]. This class of equation is also discussed in this book [5, 20]. The $2D$ Toda lattice (3.27) and its spectral problem (3.25) are associated with the root space A_N. This construction is easily generalised to other root spaces (see the review [6]).

3.2 Gauge Transformations

Equation (3.1) is invariant under gauge transformations :

$$\Psi \mapsto \bar{\Psi} = g\Psi \Rightarrow \begin{cases} U \mapsto \bar{U} = gUg^{-1} + g_x g^{-1} \\ V \mapsto \bar{V} = gVg^{-1} + g_t g^{-1} \end{cases}. \tag{3.35}$$

Without any constraint on g, this merely gives a *different* zero-curvature representation for the *same* nonlinear PDE. For instance :

$$U = \begin{pmatrix} v & 1 \\ \lambda & -v \end{pmatrix}, \quad g = \begin{pmatrix} 1 & 0 \\ v & 1 \end{pmatrix} \Rightarrow \bar{U} = \begin{pmatrix} 0 & 1 \\ \lambda + v_x + v^2 & 0 \end{pmatrix} \tag{3.36}$$

just gives two different representations of the MKdV hierarchy. The former is just the MKdV reduction of the hierarchy generated by (3.12) (see the remark following (3.22)). The $\bar{U}$ representation is less suitable for applying IST to the MKdV equation, since we would still need to solve a Riccati equation for v. However, we see that by defining u by (1.13), $\bar{U}$ corresponds to the Schrödinger equation written as a first order system. Thus, the Miura map naturally arises in the context of gauge transformations (see the appendix to [3]).

By choosing g to depend upon λ, and two sets of potential functions, we can construct Bäcklund transformations. For instance, if we choose

$$U = \begin{pmatrix} \theta_x & \lambda \\ \lambda & -\theta_x \end{pmatrix}, \quad \bar{U} = \begin{pmatrix} \bar{\theta}_x & \lambda \\ \lambda & -\bar{\theta}_x \end{pmatrix}, \quad g = g_0 + \lambda g_1, \tag{3.37}$$

we can construct the (x-part of the) Bäcklund transformation for the sinh-Gordon equation (3.32). It is convenient to multiply the $\bar{U}$ equation of (3.35) by g on the right, so that it is polynomial in λ :

$$(g_0 + \lambda g_1)(\lambda A + Q) + g_{0x} + g_{1x} = (\lambda A + \bar{Q})(g_0 + \lambda g_1), \tag{3.38}$$

where we assume that $\bar{U}$ and U have the same form. It is easy to show that there are two possible choices for g_1 (namely $\alpha^{-1}I$ or $\alpha^{-1}A$), each leading to a possible form of matrix g and each giving the usual Bäcklund transformation for the sinh-Gordon equation. For the first choice we have:

$$g = \begin{pmatrix} \alpha^{-1}\lambda & e^{\bar{\theta}+\theta} \\ e^{-\bar{\theta}-\theta} & \alpha^{-1}\lambda \end{pmatrix} \Rightarrow \bar{\theta}_x - \theta_x = 2\alpha \sinh(\bar{\theta} + \theta), \tag{3.39}$$

which just corresponds to the first part of (2.3). This type of gauge transformation is usually referred to as a *Darboux transformation*.

4 Conclusions

Most of the key ideas in soliton theory were formulated in the 3 years or so following the discovery of Kruskal and Zabusky [35]. Since then there have been an enormous number of examples of soliton equations found, verifying that the KdV is not just a freak equation. Many of these equations have important applications in the physical sciences and engineering. In particular, the NLS equation (3.20) models an optical pulse travelling down a fibre, which is currently being developed for Telecommunications. The sine-Gordon equation arises in the theory of Josephson junctions which arise in the theory of fast switching circuits, which are likely to be used in 'super computers' of the future. The 'Davidov soliton' describes the transmission of energy along protein chains.

Much of the development since the mid-seventies has been putting flesh on the early bones. Nevertheless, there have been some major advances in our fundamental understanding, such as the connection with Kač-Moody algebras (see [25]) and the more recent progress in the integration of equations in higher space-time dimensions (see [1]). This means that we now have quite a deep mathematical understanding of soliton equations. It should be noted that this has not merely been absorbing *known* mathematical ideas into soliton theory. Soliton theory has had a tremendous impact on mathematics and given rise to whole new branches (such as quantum groups) and stimulated others (such as Kač-Moody algebras). Indeed, the current volume is evidence of this, with geometers using soliton theory to solve important geometrical problems.

There are now many text books, monographs and conference proceedings on soliton theory. For general background I cite [1, 7]. Most of the important papers and books on soliton theory can be found in the references of these two books. The book by Newell [26] is very readable and has a good introduction to the τ-function. The book by Novikov et al [28] includes a review of the Zakharov-Shabat dressing method and of Zakharov and Mikhailov's work on chiral models.

Bibliography

[1] M.J. Ablowitz and P.A. Clarkson, *Solitons, Nonlinear Evolution Equations and Inverse Scattering*, London Math. Soc. **149**, CUP, Cambridge, 1991.

[2] M.J. Ablowitz, D.J. Kaup, A.C. Newell, and H. Segur, *The inverse scattering transform - Fourier analysis for nonlinear problems*, Stud.Appl.Math. **53** (1974), 249–315.

[3] M. Antonowicz and A.P. Fordy. *Hamiltonian structures of nonlinear evolution equations*, in: Fordy [7], 273–312.

[4] A.I. Bobenko. *Surfaces in terms of 2 by 2 matrices. Old and new integrable cases*, this volume.

[5] J. Bolton and L.M. Woodward. *The affine Toda equations and minimal surfaces*, this volume.

[6] A.P. Fordy. *Equations associated with simple Lie algebras and symmetric spaces*, in: Fordy [7], 315–337.

[7] A.P. Fordy, ed. *Soliton Theory : A Survey of Results*, MUP, Manchester, 1990.

[8] A.P. Fordy and J. Gibbons, *Factorization of operators I : Miura transformations*, J. Math. Phys **21** (1980), 2508–2510.

[9] A.P. Fordy and J. Gibbons. *Integrable nonlinear Klein-Gordon equations and Toda lattices*, Commun. Math. Phys. **77** (1980), 21–30.

[10] C.S. Gardner, *The Korteweg-de Vries equation and generalizations IV. The Korteweg-de Vries equation as a Hamiltonian system*, J. Math .Phys. **12** (1971), 1548–1551.

[11] C.S. Gardner, J.M. Greene, M.D. Kruskal, and R.M. Miura, *Method for solving the Korteweg-de-Vries equation*, Phys. Rev Lett. **19** (1967), 1095–1097.

[12] C.S. Gardner, J.M. Greene, M.D. Kruskal, and R.M. Miura, *The Korteweg-de Vries equation and generalizations VI. Methods for exact solution*, Commun. Pure Appl. Math. **27** (1974), 97–133.

[13] I.M. Gel'fand and L. Dikii, *Fractional powers of operators and Hamiltonian systems*, Funkts. Anal. Prilozh. **10** (1976), 13–29; English transl.: Funct. Anal. Appl. **10** (1976), 259–273.

[14] I.M. Gel'fand and B.M. Levitan. *On the determination of a differential equation from its spectral function*, Izv. Akad. Nauk. SSR Ser.Math **15** (1951), 309–360; English transl.: Amer. Math. Soc. Transl. Ser. 2, **1** (1955), 259–309.

[15] R. Hirota, *Exact solutions of the Korteweg-de Vries equation for multiple collisions of solitons*, Phys. Rev. Letts. **27** (1972), 1192–1194.

[16] D.J. Kaup and A.C. Newell, *An exact solution for a derivative nonlinear Schrödinger equation*, J. Math .Phys. **19** (1978), 798–801.

[17] B.G. Konopelchenko, *Nonlinear Integrable Equations*, Springer, Berlin, 1987.

[18] D.J. Korteweg and G. de Vries, *On the change of form of long waves advancing in a rectangular canal, and on a new type of long stationary waves*, Philos. Mag. Ser. 5, **39** (1895), 422–443.

[19] P.D. Lax. *Integrals of nonlinear equations of evolution and solitary waves*, Commun. Pure Appl. Math **21** (1968), 467–490.

[20] I. McIntosh. *Infinite dimensional Lie groups and the two-dimensional Toda lattice*, this volume.

[21] M. Mañas, *The principal chiral model as an integrable system*, this volume.

[22] M. Melko and I. Sterling, *Integrable systems, harmonic maps and the classical theory of surfaces*, this volume.

[23] R.M. Miura, *Korteweg - de Vries equation and generalizations I. A remarkable explicit nonlinear transformation*, J. Math. Phys. **9** (1968), 1202–1204.

[24] R.M. Miura, *The Korteweg-de Vries equation: a survey of results*, SIAM Review, **18** (1976), 412–559.

[25] T. Miwa, *Infinite-dimensional Lie algebras of hidden symmetries of soliton equations*, in: Fordy [7], 338–353.

[26] A.C. Newell, *Solitons in Mathematics and Physics*, SIAM, Philadelphia, 1985.

[27] J.J.C. Nimmo. *Hirota's method*, in Fordy [7], 75–96.

[28] S.P. Novikov, S.V. Manakov, L.P. Pitaevskii and V.E. Zakharov, *Theory of Solitons*, Plenum, NY, 1984.

[29] C. Rogers, *Bäcklund transformations in soliton theory*, in: Fordy [7], 97–130.

[30] J. Scott Russell. *Report on waves*, Fourteenth meeting of the British Association for the Advancement of Science, 1844.

[31] E. Schrödinger, *A method of determining quantum mechanical eigenvalues and eigenfunctions*, Proc. R. Ir. Acad. **46A** (1940), 9–16.

[32] N.J. Vilenkin. *Special Functions and the Theory of Group Representations*, Amer. Math. Soc., Providence, R.I. 1968.

[33] R.S. Ward. *Sigma models in $2+1$ dimensions*, this volume.

[34] J.C. Wood, *Harmonic maps into symmetric spaces and integrable systems*, this volume.

[35] N.J. Zabusky and M.D. Kruskal, *Interaction of solitons in a collisionless plasma and the recurrence of initial states*, Phys. Rev. Lett. **15** (1965), 240–243.

[36] V.E. Zakharov and A.V. Mikhailov. *Relativistically invariant two-dimensional model of field theory which is integrable by means of the inverse scattering problem method*, Zh. Eksp. Teor .Fiz. **74** (1978), 1953–1973; English transl.: Sov. Phys. JETP **47** (1978) 1017–1027.

[37] V.E. Zakharov and A.B. Shabat, *Exact theory of two-dimensional self focusing and one-dimensional self modulation of waves in nonlinear media*, Zh. Eksp. Teor. Fiz. **61** (1971)118–134; English transl.: Sov. Phys. JETP **34** (1972), 62–69.

[38] V.E. Zakharov and A.B. Shabat., *A scheme for integrating the nonlinear equations of mathematical physics by the method of the inverse scattering problem I*, Funkts. Anal. Prilozh. **8** (1974), 43–53; English transl.: Func. Anal. Appl. **8** (1974), 226–235.

Harmonic maps into symmetric spaces and integrable systems

J.C. Wood

1 Introduction

Let $M = (M^m, g)$, $N = (N^n, h)$ be C^∞ Riemannian manifolds of dimensions m, n respectively and let $\phi : M^m \to N^n$ be a C^∞ mapping between them. We define the *energy (integral) of ϕ* over a compact domain D of M^m to be the non-negative number:

$$E(\phi, D) = \tfrac{1}{2} \int_D e(\phi) v_M \tag{1.1}$$

where v_M is the volume element of M and $e(\phi)$, the *energy density*, is defined by $e(\phi) = \tfrac{1}{2}|d\phi|^2$ where $|d\phi|$ is the Hilbert–Schmidt norm of the differential $d\phi : T_p M \to T_{\phi(p)} N$ of ϕ at a point p of D. If M is compact we may take $D = M$ and write $E(\phi) = E(\phi, D)$. In local coordinates $(x^1, \ldots, x^m)$ on M, $(u^1, \ldots, u^n)$ on N writing the metrics as $g = g_{ij} dx^i dx^j$, $h = h_{\alpha\beta} du^\alpha du^\beta$ the energy integral (1.1) reads

$$E(\phi, D) = \tfrac{1}{2} \int_D g^{ij} h_{\alpha\beta} \frac{\partial \phi^\alpha}{\partial x^i} \frac{\partial \phi^\beta}{\partial x^j} \sqrt{g} dx^1 \ldots dx^n \tag{1.2}$$

where $g = \det(g_{ij})$. Here, and throughout the paper, we employ the double summation convention. A *harmonic map* is a C^∞ mapping which is a critical point of $E(\phi, D)$ with respect to variations of ϕ supported in D, i.e.

$$\left. \frac{d}{dt} \right|_{t=0} E(\phi_t) = 0$$

for all smooth variations ϕ_t of ϕ with $d\phi_t/dt$ supported in D. The Euler–Lagrange equations for this variational problem are written

$$\tau(\phi) = 0 \tag{1.3}$$

where $\tau(\phi)$ is a section of the pull-back bundle $\phi^{-1}TN$ called the *tension field* of ϕ; $\tau(\phi)$ is the (generalized) divergence of the differential $d\phi$, see [33]; at a point x of M, the components of $\tau(\phi)(x) \in T_{\phi(x)}N$ in local coordinates are given by

$$\tau(\phi)^\gamma = g^{ij}\left(\frac{\partial^2 \phi^\gamma}{\partial x^i \partial x^j} - \Gamma_{ij}^k \frac{\partial \phi^\gamma}{\partial x^k} + L_{\alpha\beta}^\gamma \frac{\partial \phi^\alpha}{\partial x^i}\frac{\partial \phi^\beta}{\partial x^j}\right)$$
$$= \Delta \phi^\gamma + g^{ij} L_{\alpha\beta}^\gamma \frac{\partial \phi^\alpha}{\partial x^i}\frac{\partial \phi^\beta}{\partial x^j}, \qquad (\gamma = 1, \ldots, n)$$

where Γ_{ij}^k (resp. $L_{\alpha\beta}^\gamma$) are the Christoffel symbols of M (resp. N) and Δ is the Laplace–Beltrami operator on M. Thus a harmonic map is a (C^∞) solution of (1.3). Equivalently, $\tau(\phi)$ is the trace of the second fundamental form $\beta_\phi = \nabla d\phi$ of ϕ — this is the section of $T^*M \otimes T^*M \otimes \phi^{-1}TN$ defined by

$$\beta_\phi(X,Y) = D_X^{\phi^{-1}TN} d\phi(Y) - d\phi(D_X^M Y), \qquad (X,Y \in C^\infty(TM)) \tag{1.4}$$

where $D^{\phi^{-1}TN}$ (resp. D^M) is the pull-back connection on the bundle $\phi^{-1}TN$ (resp. the Levi–Civita connection on TM). Thus (1.3) can be written

$$\sum_{i=1}^m \left\{D_{e_i}^{\phi^{-1}TN} d\phi(e_i) - d\phi(D_{e_i}^M e_i)\right\} = 0 \tag{1.5}$$

where $\{e_i\}$ is an orthonormal frame on M. If, further, the frame $\{e_i\}$ comes from normal coordinates centred at a point p, then at that point (and, unless M is flat, at that point only), the second term on the left-hand side vanishes.

The system (1.3) is a system of n semi-linear (but not, in general, linear) elliptic partial differential equations. If g is a pseudo-Riemannian metric, the system is no longer elliptic. In this case, harmonic maps are sometimes called *pseudoharmonic*. See, for example [46, 64, 13] for some results in this case. Harmonic maps from $\mathbb{R}^2$ or $\mathbb{R}^{1,1}$ are called *(two-dimensional non-linear) σ-models* in particle physics (see [13]). It is possible to consider weak solutions to (1.3) and ask whether such solutions are *regular*, i.e. C^∞; [34] and, for a recent result, [49]. However, throughout this article we shall suppose that all manifolds, structures on them and maps between them are smooth, i.e. C^∞.

Harmonic maps include many well-known examples, some of which we list:

1. A map from $\mathbb{R}$ (resp. S^1) to N is harmonic if and only if it is a geodesic (resp. closed geodesic) linearly parametrised. In this case the energy (1.1) is the usual energy or action integral of the curve.

2. More generally, a *totally geodesic map* is a map $\phi : M \to N$ with vanishing second fundamental form. Equivalently, by (1.4), ϕ maps linearly parametrised geodesics to linearly parametrised geodesics. Such a map is harmonic; however, totally geodesic maps are rather rare.

3. A map $\phi : M \to \mathbb{R}$ is a harmonic map if and only if it is a harmonic function in the usual sense. In this case the energy is just the Dirichlet integral and (1.3) is the linear partial differential equation (Laplace's equation):

Harmonic maps into symmetric spaces and integrable systems 31

$$\Delta \phi = 0.$$

4. Holomorphic, i.e. complex analytic, and antiholomorphic maps between Kähler manifolds are harmonic, see [36], and [60, 68, 87] for generalizations.

5. Harmonic maps from surfaces have many nice properties, for example the equations (1.3) are invariant under conformal changes of the domain metric so that methods of complex analysis may be applied; indeed if M^2 is a Riemann surface, then $\phi : M^2 \to N$ is harmonic (with respect to any Hermitian metric) if and only if, in a local complex coordinate z it satisfies the equation

$$\frac{D^{\phi^{-1}TN}}{\partial \bar{z}} \frac{\partial^c \phi}{\partial z} = 0. \tag{1.6}$$

or, equivalently, the conjugate equation

$$\frac{D^{\phi^{-1}TN}}{\partial z} \frac{\partial^c \phi}{\partial \bar{z}} = 0. \tag{1.7}$$

Here

$$\frac{\partial^c \phi}{\partial z} = \tfrac{1}{2}\left(\frac{\partial \phi}{\partial x} - i\frac{\partial \phi}{\partial y}\right) \quad \text{and} \quad \frac{\partial^c \phi}{\partial \bar{z}} = \tfrac{1}{2}\left(\frac{\partial \phi}{\partial x} + i\frac{\partial \phi}{\partial y}\right).$$

(1.6) can be interpreted as saying that $\frac{\partial^c \phi}{\partial z}$ is holomorphic with respect to the Koszul–Malgrange holomorphic structure on $\phi^{-1}TN \to M$ (see, for example ([33], §9.22) and this is responsible for many nice properties. For example, for any map $\phi : M^2 \to N$ from a Riemann surface to a Riemannian manifold we define its *quadratic differential* by

$$\eta = \left\langle \frac{\partial^c \phi}{\partial z}, \frac{\partial^c \phi}{\partial z} \right\rangle^c dz^2. \tag{1.8}$$

Here $\langle \, , \, \rangle^c$ denotes the complex bilinear extension of the inner product on TN to $T^cN = TN \otimes \mathbb{C}$. Note that η is well-defined, and is zero if and only if ϕ is weakly conformal. Further, if ϕ is harmonic, (1.6) implies that η is holomorphic. Since any holomorphic differential on the 2-sphere is zero, we conclude that any harmonic map from the 2-sphere is weakly conformal; this and related "vanishing theorems" account for the special place of the 2-sphere in the theory of harmonic maps — see §3. On the other hand, any holomorphic differential on the 2-torus is of the form constant.dz^2, see §4 for the application of this remark. The holomorphicity of η is an example of a *conservation law* for harmonic maps: as for any system of equations which are the Euler–Lagrange equations for the critical points of a functional

$$\int L(j^k\phi)v_M$$

there is a stress-energy tensor S_ϕ which is conservative, i.e. has zero divergence, at the critical point ϕ, see [50, 39]. If X is a Killing vector field on the domain, then $\text{div}(i(X)S_\phi) = 0$, this giving conservation of momentum and energy when the domain is a Lorentzian manifold. For the functional (1.1), whose critical points are harmonic maps, the stress-energy tensor reads

$$S_\phi = e(\phi)g - \phi^*h. \tag{1.9}$$

If ϕ is harmonic then $\text{div} S_\phi = 0$, the converse is true if ϕ is submersive almost everywhere; furthermore, if the domain M is two-dimensional, this last condition is equivalent to the holomorphicity of η. See [4, 69] for an exposition of the applications of this tensor.

6. A map $\phi : M \to N$ is called *weakly conformal* if, for each point $p \in M$, $d\phi_p$ is either zero or maps T_pM conformally into $T_{\phi(p)}N$. In the first case, p is called a *branch point*. Given a weakly conformal map $\phi : M^2 \to N^n$ from a surface, ϕ is harmonic if and only if its image is minimal away from branch points. Such a map is a *minimal branched immersion* in the sense of [48], that is, its branch points are isolated and in suitable normal charts have the form

$$\begin{aligned}\phi^1(z) &= c\text{Re}(z^k) + o(|z|^k) \\ \phi^2(z) &= c\text{Im}(z^k) + o(|z|^k) \\ \phi^\alpha(z) &= o(|z|^k), \quad (3 \leq \alpha \leq n),\end{aligned}$$

for some $c \neq 0$, see [41]. Given a weakly conformal map $\phi : M^2 \to \mathbb{R}^n$ of a Riemann surface, its *Gauss map* $\gamma_\phi : M^2 \to G_2^0(\mathbb{R}^n)$ to the Grassmannian of oriented 2-planes in $\mathbb{R}^n$ is defined away from the branch points by $\gamma_\phi(x) = T_{\phi(x)}M^2 \subset \mathbb{R}^n$. When the Grassmannian $G_2^0(\mathbb{R}^n)$ is identified with the complex hyperquadric $Q_{n-2} = \{[z_1, \ldots, z_n] \in \mathbb{C}P^{n-1} : z_1^2 + \cdots z_n^2 = 0\}$ the Gauss map $\gamma_\phi(z)$ is given by the span of $\frac{\partial^c \phi}{\partial \bar{z}} = \frac{1}{2}\left(\frac{\partial \phi}{\partial x} + i\frac{\partial \phi}{\partial y}\right)$ and S.S. Chern's result [26] that the Gauss map extends over the branch points and is antiholomorphic if and only if ϕ is minimal (equivalently harmonic)) quickly follows from (1.6). This and its generalizations are the starting point for many twistorial constructions of harmonic maps , see [83, 15, 56].

7. The above Gauss map is *harmonic* if and only if ϕ has parallel mean curvature, see [72] for a generalization to higher dimensions and codimensions. In particular, if M^2 is simply-connected, every harmonic map $M^2 \to S^2$ arises as the Gauss map of a constant mean curvature surface M'^2 in $\mathbb{R}^3$ composed with a weakly conformal map $M^2 \to M'^2$, see [55] and [52, Theorem 3.9].

8. A *harmonic morphism* is a map $\phi : M^m \to N^n$ which preserves Laplace's equation in the sense that if $f : V \to \mathbb{R}$ is a harmonic function where $V \subset N$ is an open set, then so is $f \circ \phi : \phi^{-1}(V) \to \mathbb{R}$ provided $\phi^{-1}(V)$ is non-empty. Harmonic morphisms can be characterized as harmonic maps ϕ which satisfy a quadratic constraint on their first derivatives called *horizontal weak conformality*:

$$g^{ij}\frac{\partial \phi^\alpha}{\partial x^i}\frac{\partial \phi^\beta}{\partial x^j} = \lambda h^{\alpha\beta} \quad \text{for some } \lambda : M^m \to [0,\infty)$$

which expresses the fact that, away from points where $d\phi$ is zero, ϕ is a submersion and $d\phi$ is conformal on the horizontal spaces $(\ker d\phi)^\perp$ with conformality factor λ. Harmonic morphisms *to* surfaces behave in many ways dual to (weakly conformal) harmonic maps *of* surfaces; for example the equations for the former are conformally invariant on the *codomain*. Examples of harmonic morphisms include Riemannian submersions with minimal fibres, for example Hopf maps, and there are some relations with twistors in 4 dimensions, see [3, 6, 7, 8, 9, 88, 87].

9. The tension field of the composition of two maps $\phi : M^m \to N^n$ and $\psi : N^n \to P^p$ is given [36] by

$$\tau(\psi \circ \phi) = d\psi \circ \tau(\phi) + \operatorname{tr}\nabla d\psi(d\phi, d\phi) \tag{1.10}$$

where the last term means the sum $\sum_{i=1}^{m} \nabla d\psi(d\phi(e_i), d\phi(e_i))$ over an orthomormal frame $\{e_i\}$ on M. From this we see that

(a) If $\phi : M^m \to N^n$ is harmonic and $\psi : N^n \to P^p$ is totally geodesic, then $\psi \circ \phi : M^m \to P^p$ is harmonic. Note that the composition of two harmonic maps is not, in general, a harmonic map; however the composition of two harmonic morphisms is a harmonic morphism, as is clear from the definition.

(b) If $\phi : M^2 \to N^2$ is a weakly conformal map between surfaces and $\psi : N^2 \to P^p$ is harmonic, then $\psi \circ \phi$ is harmonic; this confirms the conformal invariance noted in item 5 above. More generally, if $\phi : M^m \to N^n$ is a harmonic *morphism* and $\psi : N^n \to P^p$ is a harmonic map, then $\psi \circ \phi$ is a harmonic map.

(c) If $\psi : N^n \to P^p$ is the inclusion map of a submanifold (or, more generally, an isometric immersion) then

$$\tau(\phi) = \text{the component of } \tau(\psi \circ \phi) \text{ tangential to } N^n. \tag{1.11}$$

The fundamental question in the theory of harmonic maps is the *existence problem*: given two Riemannian manifolds (M^m, g), (N^n, h) and a continuous map $\phi_0 : M^m \to N^n$, when can we deform ϕ_0 to a harmonic map, i.e. when is there a harmonic map in the homotopy class of ϕ_0? If M^m and N^n are compact and (N^n, h) has non-positive sectional curvatures, it was established by J. Eells and J.H. Sampson [36] that any homotopy class contains a harmonic map; their method was to deform the map ϕ_0 under the negative gradient flow of the energy integral (1.1) by solving the associated non-linear heat equation and showing that a smooth limiting map exists and is harmonic; in fact it is of absolute minimum energy in its homotopy class. If (N^n, h) does not have non-positive sectional curvatures, this flow may blow up in finite time or may fail to have a smooth limit, see for example [25, 28, 45]. In this case there may be no solution to the existence problem. For example consider the case of maps to a sphere. Let $i : S^n \to \mathbb{R}^{n+1}$ be the standard inclusion map, then from (1.10) or by Lagrange multipliers we can show that a map $\phi : (M^m, g) \to S^n$ is harmonic if and only if $\Phi = i \circ \phi$ satisfies

$$\Delta \Phi = \lambda \Phi \tag{1.12}$$

for some function $\lambda : M^m \to \mathbb{R}$. Further, $\lambda = -|d\phi|^2$. For example, a map $\phi : \mathbb{R}^m \to S^n$ is harmonic if and only if Φ satisfies the equation

$$\sum_{i=1}^{m} \frac{\partial^2 \Phi}{\partial x^{i^2}} = \lambda \Phi; \tag{1.13}$$

further $\lambda = \sum_{i=1}^{m} \left|\frac{\partial \Phi}{\partial x^i}\right|^2$ where now $|\ |$ denotes the standard Euclidean norm on $\mathbb{R}^n$. A map of a torus $T^m = \mathbb{R}^m/\Gamma$ may be regarded as a suitably periodic solution of this equation. Then [37] *there is no harmonic map from a 2-torus T^2 to the 2-sphere S^2 of (Brouwer) degree* ± 1. On the other hand we shall see in §3 that there are many interesting harmonic maps from S^2, T^2, other surfaces and higher dimensional manifolds to S^n, $\mathbb{C}P^n$ and other symmetric spaces. In the rest of this article we shall discuss some methods from the theory of integrable systems that have been applied to find harmonic maps in these cases; we shall start with a discussion of equivariant harmonic maps where the reduction to an integrable system is most transparent.

The author thanks F.E. Burstall and J. Eells for useful comments on a draft of this article.

2 Equivariant harmonic maps and integrable systems

One way of finding harmonic maps is to impose enough symmetry to reduce the harmonic map equation (1.3) to an ordinary differential equation and show that the latter has suitably behaved solutions. This was done first by R.T. Smith [73] who looked for harmonic maps from a rectangular 2-torus $S^1 \times S^1$ to the 2-sphere S^2 as follows: Regard the torus as $S^1 \times \mathbb{R}/\mathbb{Z}a$ where $a > 0$ and look for maps ϕ from the tube $S^1 \times \mathbb{R} = \{(\theta, s) : 0 \leq \theta \leq 2\pi, s \in \mathbb{R}\}$ to S^2 of the S^1-equivariant form:

$$\phi(\theta, s) = \bigl(\cos\theta \sin\alpha(s), \sin\theta \sin\alpha(s), \cos\alpha(s)\bigr) \tag{2.1}$$

for some function $\alpha : \mathbb{R} \to \mathbb{R}$. The equation for harmonicity (1.12) reduces to

$$\alpha''(s) = \tfrac{1}{2}\sin 2\alpha(s). \tag{2.2}$$

This is the equation for a simple pendulum (or 1-dimensional sine–Gordon equation), a well-known Hamiltonian system. It has periodic solutions which can be expressed in terms of elliptic functions. The resulting maps (2.1) give harmonic maps of degree zero from a rectangular torus $S^1 \times \mathbb{R}/\mathbb{Z}a$ for some $a > 0$ which are either surjective — corresponding to circular motions of the pendulum, or have image a narrow band about the equator — corresponding to oscillatory motion. As pointed out by E. Calabi (see [31] for the full tale), these harmonic maps $\phi : T^2 \to S^2$ are all Gauss maps of Delaunay surfaces. More generally, K. Uhlenbeck [76] looked for harmonic maps $\phi : S^1 \times \mathbb{R} \to S^n$ of the form $\phi(\theta, s) = e^{B\theta}x(s)$ where $B \in so(n+1)$ and $x : \mathbb{R} \to \mathbb{R}^{n+1}$ has image in S^n. Such maps are equivariant with respect to the representation $\rho : S^1 \to SO(n+1)$ given by $\rho(\theta) = e^{B\theta}$. Then ϕ is harmonic if and only if

$$x''(s) + B^2 x(s) + \bigl(|Bx(s)|^2 + |x'(s)|^2\bigr)x(s) = 0. \tag{2.3}$$

This equation is the Neumann equation describing the motion of a particle moving on a sphere under the influence of a quadratic potential $V = \tfrac{1}{2}\langle Ax, x\rangle$ where $A = B^2$, see [65]. The conservation law for harmonic maps (see §1) gives conservation of energy and angular momentum for the Neumann equation, see [5].

Writing $A = B^2$ and diagonalizing B, P. Baird and A. Ratto [5] note that $x : \mathbb{R} \to \mathbb{R}^{n+1}$ satisfies (2.3) if and only if its components are eigenfunctions of the Hill's operator

$$-\frac{d^2}{ds^2} + u(s). \tag{2.4}$$

Exploiting this and a higher-dimensional analogue for maps $x : \mathbb{R}^m \to \mathbb{R}^n$ where (2.4) is replaced by the Schrödinger operator

$$-\Delta + u(s) \tag{2.5}$$

and using knowledge of the eigenfunctions of (2.5) for some special choices of $u(s)$ together with the *remarkable identity* [62], they find many harmonic maps from tori to spheres including linearly full ones $T^n \to S^n$ for $n = 1, 2, \ldots, 5$ (all of degree 0). Related methods give harmonic morphisms from S^4 to S^2 where S^4 is given a metric conformally equivalent to the standard one, the harmonic morphism representing the non-trivial element of $\pi_4(S^2) = Z_2$.

For reduction theorems and their applications to finding harmonic maps, morphisms and minimal immersions see also [3, 77, 54, 35, 53].

3 The Bäcklund transform and adding a uniton

A feature of many integrable systems is the existence of Bäcklund transforms which give new solutions to a system of partial differential equations from old ones by solving ordinary differential equations. The best known is the Bäcklund transform for the sine–Gordon equation

$$u_{xt} = \sin u \,, \tag{3.1}$$

namely, if $u(x,t)$ is a solution to (3.1), then $\tilde{u}(x,t)$ is another solution if and only if

$$\tilde{u}_x = u_x - 2\beta \sin \frac{u+\tilde{u}}{2}$$
$$\tilde{u}_t = -u_t + \frac{2}{\beta} \sin \frac{u-\tilde{u}}{2}.$$

There is a similar version for the sinh–Gordon equation

$$u_{xx} + u_{yy} = \sinh u \tag{3.2}$$

see [43, 74].

Now let $F : U \to \mathbb{R}^3$ be a conformal map from an open set U of $\mathbb{R}^2 = \mathbb{C}$. The induced metric on U can be written as $ds^2 = 4e^u dz d\bar{z}$ for some function $u : U \to \mathbb{R}$. Then the orthogonal frame $\sigma = (F_z, F_{\bar{z}}, N)^T$ satisfies the Gauss–Weingarten equations

$$\sigma_z = U\sigma, \qquad \sigma_{\bar{z}} = V\sigma \tag{3.3}$$

with

$$U = \begin{pmatrix} u_z & 0 & A \\ 0 & 0 & B \\ -e^{-u}B/2 & -e^{-u}A/2 & 0 \end{pmatrix}, \quad V = \begin{pmatrix} 0 & 0 & B \\ 0 & u_{\bar{z}} & \bar{A} \\ -e^u \bar{A}/2 & -e^u B/2 & 0 \end{pmatrix}$$

where $A = \langle F_{zz}, N \rangle$ and $B = \langle F_{z\bar{z}}, N \rangle$. Note that the mean curvature of $F(U)$ is $H = \frac{1}{2} e^{-u} B$. The compatibility condition for existence of a solution to (3.3) is

$$U_{\bar{z}} - V_z + [U, V] = 0. \tag{3.4}$$

The components of this give the Gauss and Codazzi–Mainardi equations of the surface. In the case of a surface of constant mean curvature H these reduce to

$$u_{z\bar{z}} + 2H^2 e^u - \frac{1}{2} A\bar{A} e^{-u} = 0 \tag{3.5}$$
$$A_{\bar{z}} = 0. \tag{3.6}$$

In particular A is a holomorphic function and so, if F covers a map from a 2-torus $T^2 = \mathbb{R}^2/\Gamma$, A must be constant. Furthermore, by scaling the coordinates on the domain and range we can assume that $A\bar{A} = 1$ and $H = \frac{1}{2}$, then (3.5) reduces to the sinh–Gordon equation

$$u_{z\bar{z}} + \sinh u = 0. \tag{3.7}$$

Thus a constant mean curvature torus gives rise to a solution to the sinh–Gordon equation (3.7) and conversely, given a solution to (3.7) and a choice of A, we may integrate (3.3) to find a constant mean curvature torus. Thus the Bäcklund transform for the sinh–Gordon equation transforms constant mean curvature tori to constant mean curvature tori and this induces a transformation of their harmonic Gauss maps $T^2 \to S^2$.

Remarks

1. From (3.3) we see that the Gauss map $N : U \to S^2$ satisfies

$$N_z = -HF_z - \tfrac{1}{2}e^{-u}AF_{\bar{z}};$$

Differentiating this shows that

$$N_{z\bar{z}} = H_{\bar{z}}F_z + \text{normal component}$$

showing that N is harmonic if and only if the surface $F(U)$ has constant mean curvature, see Examples 1 item 7.

2. If the surface has constant mean curvature H the differential $\eta = 2HAdz^2$ is holomorphic; it is simply the holomorphic differential (see Examples 1 item 5) of the Gauss map N.

3.1 A Bäcklund-type transform for harmonic maps to complex projective spaces.

Let $\phi : M^2 \to \mathbb{C}P^n$ be a harmonic map from a Riemann surface. Assume that ϕ is not antiholomorphic. Let $\Phi : U \to \mathbb{C}^{n+1} \setminus \{0\}$ be a lift of ϕ over a coordinate neighbourhood (U, z), i.e a representation of ϕ in homogeneous coordinates. Define $\Phi_1 : U \to \mathbb{C}^{n+1}$ by $\Phi_1 = \pi_0^\perp(\partial \Phi/\partial z)$ where π_0 denotes the orthogonal projection onto the line defined by ϕ, explicitly

$$\Phi_1(z) = \frac{\partial \Phi}{\partial z} - \langle \frac{\partial \Phi}{\partial z}, \Phi \rangle \Phi/|\Phi|^2$$

where $\langle \, , \, \rangle$ denotes the standard Hermitian inner product on $\mathbb{C}^{n+1}$. Then Φ_1 may vanish at isolated points, but because of (1.6), it can be shown that it defines a holomorphic section of the subbundle of $\mathbb{C}^{n+1}$ defined by $\phi^\perp$ equipped with its Koszul–Malgrange holomorphic structure (see Examples 1 item 5 and [22, Proposition 2.2]), in particular its zeros are all of holomorphic type and so Φ_1 gives a well-defined map $G'(\phi) : M^2 \to \mathbb{C}P^n$ called the ∂'-*Gauss map* (or ∂'-*Gauss transform*) [22] or ∂-*transform* [27] of ϕ. Assuming instead that φ is not holomorphic and replacing z by $\bar{z}$ gives the ∂''-*Gauss map* (or *transform*) or $\bar{\partial}$-*transform* $G''(\phi) : M \to \mathbb{C}P^n$ of ϕ. These two transforms are inverses to each other, viz. $G''(G'(\phi)) = \phi$ if ϕ is not antiholomorphic and $G'(G''(\phi)) = \phi$ if ϕ is not holomorphic. We then have the following

Proposition 1 *Let $\phi : M^2 \to \mathbb{C}P^n$ be harmonic. Then so are $G'(\phi)$ and $G''(\phi)$.*

This can be established by (i) direct computation in local frames (see [27]), (ii) the diagram method of F. Burstall, S. Salamon and the author, see [22, Propostion 2.3], [21, Proposition 3], (iii) an easy direct computation for weakly conformal harmonic maps (recall all harmonic maps are weakly conformal if $M^2 = S^2$), see [47], (iv) as a special case of a general result of K. Uhlenbeck [78], see below. There is a similar construction for harmonic maps from surfaces into complex Grassmannians, see [82, 40, 85], and various related transforms for other symmetric spaces, see [1, 2, 81], which give all harmonic maps in these cases; all these transforms are particular cases of the transform of Uhlenbeck discussed in §3.3.

The sequence of harmonic maps

$$\cdots \longrightarrow G''\big(G''(\phi)\big) \longrightarrow G''(\phi) \longrightarrow \phi \longrightarrow G'(\phi) \longrightarrow G'\big(G'(\phi)\big) \longrightarrow \cdots$$

(together with natural maps between them, see [27, 22, 82]) is called the *harmonic sequence* of ϕ. A harmonic map is called *complex isotropic* (or *pseudo-holomorphic* or *superminimal*) if the members of its harmonic sequence are mutually orthogonal. In this case, for dimension reasons, the harmonic sequence must terminate in each direction and its first (resp. last) element is holomorphic (resp. antiholomorphic). Hence any complex isotropic harmonic map $\phi : M^2 \to \mathbb{C}P^n$ can be obtained from a holomorphic map by successive applications of the ∂'–Gauss transform. If ϕ is not complex isotropic, defining $G^{(i)}(\phi)$ inductively by $G^{(1)}(\phi) = G'(\phi)$ and $G^{(i)}(\phi) = G'(G^{(i-1)}(\phi))$, the largest $r \in \{1, 2, \ldots\}$ such that $\phi \perp G^{(i)}(\phi)$ for all i, $1 \leq i \leq r$ is called the *isotropy order* of ϕ. The isotropy order is measured by the vanishing of a sequence of differentials η_k with $\eta_1 = \eta$, each one holomorphic if the previous ones vanish (see [84, 22]. Since any holomorphic differential on S^2 is a holomorphic section of a line bundle of negative degree, it vanishes, hence any harmonic map $\phi : S^2 \to \mathbb{C}P^n$ is complex isotropic; we thus obtain

Theorem 1 ([38], Theorem 6.9) *Any harmonic map from the 2-sphere to $\mathbb{C}P^n$ can be obtained from a holomorphic one by repeated application of the ∂'–Gauss transform.*

For the history of this result, see [38]. We have a similar result for harmonic maps to the n-sphere: A harmonic map $\phi : M^2 \to S^n$ is called *(real) isotropic* (or *pseudo-holomorphic* or *superminimal*) if the harmonic map

$$\Phi : M^2 \stackrel{\phi}{\to} S^n \stackrel{\pi}{\to} \mathbb{R}P^n \stackrel{i}{\to} \mathbb{C}P^n$$

is (complex) isotropic. (Here π is the standard double cover and i is the totally geodesic inclusion.) Equivalently, ϕ is real isotropic if the *infinite order ∂' osculating space* $\theta'_\infty(z) = \text{span}\left\{\partial'^\alpha \phi(z) : \alpha = 1, 2, \ldots\right\}$ is isotropic for each $z \in M^2$ in the usual sense that $\langle v, \bar{v}\rangle = 0$ for all $v \in \theta'_\infty(z)$. Without loss of generality we can assume ϕ is full, then n must be even, say $n = 2r$. Any harmonic map $\phi : S^2 \to S^{2r}$ is isotropic. The corresponding maps Φ are precisely those obtained as the r'th ∂'–Gauss transform of a full holomorphic map satisfying the additional condition of

Harmonic maps into symmetric spaces and integrable systems

total isotropy [38, Definition 3.13] which can be stated as the requirement that the $(r-1)$-st "augmented" osculating space of any representation F of f in homogeneous coordinates: span$\{\partial'^\alpha F(z) : \alpha = 0, \ldots, r-1\}$ is isotropic for all z (no higher osculating space can be), see [23], [38, Corollary 6.11], [12]. The isotropy order of a non-isotropic harmonic map $\phi : S^2 \to S^n$ is defined to be the isotropy order of the corresponding map $\Phi : M^2 \to \mathbb{C}P^n$.

3.2 Harmonic maps to Lie groups

We first recall some Lie group theory. Let G be a Lie group equipped with a biinvariant Riemannian metric. Let $\mathfrak{g}$ be its Lie algebra identified in the standard way with the tangent space at the identity T_eG. The (left) Maurer–Cartan form is a $\mathfrak{g}$-valued 1-form ω on G which gives an isomorphism $\omega_\gamma : T_\gamma G \to \mathfrak{g}$ for each $\gamma \in G$, in fact ω_γ is simply the differential of left translation by γ^{-1}. Note that ω gives a trivialisation (or gauge) for the tangent bundle

$$\omega : TG \cong G \times \mathfrak{g}. \tag{3.8}$$

There is a connection D^+ on G called the $+$-*connection* in which the left translations are parallel; in the gauge (3.8) this has the formula $D^+ = d + \omega$, i.e. $D_X^+(s) = ds + [\omega, s]$ for $X \in TM$, $s : M \to \mathfrak{g}$. The Maurer–Cartan equation

$$d\omega + \tfrac{1}{2}[\omega \wedge \omega] = 0 \tag{3.9}$$

expresses the fact that this connection is flat. (Here $[\omega \wedge \omega]$ denotes the $\mathfrak{g}$-valued 2-form on G defined by $[\omega \wedge \omega](X,Y) = 2[\omega(X), \omega(Y)]$ for $X, Y \in T_\gamma G, \gamma \in G$.) On the other hand, the Levi–Civita connection is given by $D = d + \tfrac{1}{2}\omega$ and is not flat (unless G is Abelian).

Now let $\phi : M^m \to G$ be a smooth map from a Riemannian manifold to the Lie group. Let α be a 1-form on M with values in $\mathfrak{g}$ given by

$$\alpha = \phi^*(\omega). \tag{3.10}$$

(note that Uhlenbeck [78] and the author in [86] sets $A = \tfrac{1}{2}\alpha$; we prefer to follow the notations of [18]). If G is a matrix group we have

$$\alpha = \phi^{-1}d\phi. \tag{3.11}$$

Note that α represents the differential $d\phi \in C^\infty(T^*M \otimes \phi^{-1}TG)$ in the gauge

$$\phi^{-1}TG \cong M \times \mathfrak{g} \tag{3.12}$$

given by pulling back the gauge (3.8). Then pulling back (3.9) shows that α satisfies

$$d\alpha + \tfrac{1}{2}[\alpha \wedge \alpha] = 0 \tag{3.13}$$

which expresses the fact that the pull-back $d+\alpha$ of the connection D^+ to $\phi^{-1}TG \to M$ is flat. It is an integrability condition: if M^m is simply connected and α is a given $\mathfrak{g}$-valued 1-form on M, then we can solve (3.10) (or (3.11)) for ϕ if and only if α satisfies (3.13). Further, such a solution is unique up to left translation by some element of G. The map ϕ can be interpreted as a gauge transformation which gauges the flat connection $d + \alpha$ to the trivial connection.

Now, in the gauge (3.12), the pull-back of the Levi-Civita connection on G to the bundle $\phi^{-1}TG \to M$ is

$$D^{\phi^{-1}TG} = d + \tfrac{1}{2}\alpha$$

that is,

$$D^{\phi^{-1}TG}_X(s) = X(s) + \tfrac{1}{2}[\alpha, s] \qquad (X \in M, \; s : M \to \mathfrak{g}).$$

To calculate the tension field at a point $p \in M$, let $\{e_i\}$ be a frame coming from normal coordinates at p, then by (1.5),

$$\tau(\phi) = \sum_{i=1}^{m} D^{\phi^{-1}TG}_{e_i}(d\phi(e_i)) = \sum_{i=1}^{m}\{e_i(\alpha(e_i)) + \tfrac{1}{2}[\alpha(e_i), \alpha(e_i)]\}$$
$$= d^*\alpha.$$

Hence ϕ is harmonic if and only if

$$d^*\alpha = 0. \tag{3.14}$$

Remarks

We can generalise this to any Riemannian homogeneous space $N = G/H$ equipped with a left-invariant metric h as follows: For $\xi \in \mathfrak{g}$ let $\tilde{\xi}$ be the corresponding Killing field ("fundamental field") defined at each point $y \in N$ by $\tilde{\xi}_y = \frac{d}{dt}(\exp t\xi) \cdot y\big|_{t=0}$ where . denotes the action of G on N. Then $\xi \mapsto \tilde{\xi}$ defines a surjective bundle endomorphism a from the trivial bundle $N \times \mathfrak{g}$ to TN. This has kernel the bundle $[\mathfrak{h}] = G \times_H \mathfrak{h}$ where H acts on $\mathfrak{h}$ by Ad. Assume that N is *naturally reductive* so that $\mathfrak{g} = \mathfrak{h} \oplus \mathfrak{m}$ with $[\mathfrak{h}, \mathfrak{m}] \subset \mathfrak{m}$ and the metric h corresponds to an Ad(H)-invariant inner product B on $\mathfrak{m}$ such that $B(X, [Y, Z]_\mathfrak{m}) + B([Z, X]_\mathfrak{m}, Y) = 0$ for all $X, Y, Z \in \mathfrak{m}$, see [57], then $[\mathfrak{h}]$ has a complement $[\mathfrak{m}] = G \times_H \mathfrak{m}$. Note that the fibre of this bundle at $y = gH \in N$ is $y \times \mathrm{Ad}_\mathfrak{g}(g)(\mathfrak{m}) \subset N \times \mathfrak{g}$. The map a above restricts to a bundle isomorphism $[\mathfrak{m}] \to TN$. Its inverse followed by the inclusion of $[\mathfrak{m}]$ in $N \times \mathfrak{g}$ and the natural projection $N \times \mathfrak{g} \to \mathfrak{g}$ defines a $\mathfrak{g}$-valued one-form ω on N again called the Maurer–Cartan form. Given a map $M \to N$ we can pull this back to M to define a $\mathfrak{g}$-valued 1-form $\alpha = \phi^*(\omega)$ on M which again represents the differential $d\phi$. Similar considerations to those above show that this form satisfies (3.14) if and only if ϕ is harmonic.

If N is not naturally reductive, then we must instead consider the *moment* or *momentum map* $\mu : TN \to \mathfrak{g}^*$ defined on each fibre as the composition of the isomorphism $\flat : T_yN \to T^*_yN$ defined by the metric and the dual $T^*_yN \to \mathfrak{g}^*$ of the map $\mathfrak{g} \to T_yN, \xi \mapsto \tilde{\xi}_y$ defined above. Then again pulling this back to a 1-form $\alpha = \phi^*\mu$ on M, but this time with values in $\mathfrak{g}^*$, a calculation shows that ϕ is harmonic if and only if $d^*\alpha = 0$, cf. [13],see [69] for an alternative treatment using Noether's Theorem, and see [64] and [19, §3.2] for an alternative approach based on lifting to maps into G.

Harmonic maps into symmetric spaces and integrable systems 41

Returning to the case of a Lie group we established that if ϕ is harmonic, α satisfies (3.13) and (3.14). Conversely, if M is simply-connected, then given a $\mathfrak{g}$-valued 1-form on M satisfying (3.13), we can solve (3.10) for ϕ; then, if (3.14) is also satisfied, this map is harmonic. Thus we have translated the harmonic equation (1.3) into the pair of equations (3.13, 3.14).

Now suppose that $M = M^2$ is a Riemann surface with complex coordinate chart (U, z). Write $\alpha = \alpha_z dz + \alpha_{\bar{z}} d\bar{z}$ where α_z and $\alpha_{\bar{z}}$ are maps from U with values in $\mathfrak{g}^c = \mathfrak{g} \otimes \mathbb{C}$, the complexified Lie algebra, then the pair of equations (3.13, 3.14) is equivalent to

$$\frac{\partial \alpha_z}{\partial \bar{z}} + \frac{\partial \alpha_{\bar{z}}}{\partial z} = 0$$

$$\frac{\partial \alpha_z}{\partial \bar{z}} - \frac{\partial \alpha_{\bar{z}}}{\partial z} + [\alpha_{\bar{z}}, \alpha_z] = 0;$$

this pair of equations can be written more symmetrically as

$$\frac{\partial \alpha_z}{\partial \bar{z}} + \tfrac{1}{2}[\alpha_{\bar{z}}, \alpha_z] = 0 \qquad (3.15)$$

$$\frac{\partial \alpha_{\bar{z}}}{\partial z} + \tfrac{1}{2}[\alpha_z, \alpha_{\bar{z}}] = 0. \qquad (3.16)$$

Note that, if ϕ is a harmonic map satisfying (3.10), these equations reduce to the equations (1.6) and (1.7).

3.3 Flag transforms

We now show how a harmonic map from a Riemann surface to $U(n)$ can be transformed to another by multiplying by a suitable map into a Grassmannian. A point of the Grassmannian $G_k(\mathbb{C}^n)$ represents a k-plane and may be identified with the orthogonal projection $\pi : \mathbb{C}^n \to \mathbb{C}^n$ onto that k-plane. We then have the Cartan embedding [24] of $G_k(\mathbb{C}^n)$ in $U(n)$ given by $\pi \mapsto \pi - \pi^\perp$. Since this is totally geodesic, it follows from the composition law (1.10) for the tension field that a map is harmonic into $G_k(\mathbb{C}^n)$ if and only if its composition with the Cartan embedding is harmonic into $U(n)$. We have the following result of Uhlenbeck [78]:

Proposition 1 Let $\phi : M^2 \to U(n)$ be a harmonic map from a Riemann surface. Write $\alpha = \phi^{-1} d\phi = \alpha_z dz + \alpha_{\bar{z}} d\bar{z}$. Suppose that $\pi : M^2 \to G_k(\mathbb{C}^n)$ is a map such that the following two equations hold:

$$\pi^\perp \alpha_z \pi = 0 \qquad (3.17)$$

$$\pi^\perp \left(\frac{\partial}{\partial \bar{z}} + \tfrac{1}{2}\alpha_{\bar{z}} \right) \pi = 0. \qquad (3.18)$$

Then the product

$$\tilde{\phi} = \phi(\pi - \pi^\perp) : M^2 \to U(n) \qquad (3.19)$$

is harmonic.

Following [20] we shall call the transform $\phi \mapsto \tilde{\phi}$ a *flag transform*. The Proposition can be established by direct calculation using equations (3.15, 3.16) but an easier way is possible after introduction of a spectral parameter, see [78]. The equations can be interpreted as a degenerate Bäcklund transform, in fact a degenerate version of the dressing transform of Zakharov et al. [89, 90], see [78, 61], and the conditions can be interpreted geometrically, see [86]. The ∂'-Gauss transform of §3.1 above is the case where ϕ has image in $\mathbb{C}P^{n-1} = G_1(\mathbb{C}^n) \subset U(n)$ and π is defined by setting $\pi^\perp = \phi \oplus G'(\phi)$ in the notation of §3.1.

Notes. Uhlenbeck shows that *any* harmonic map $\phi : S^2 \to U(n)$ can be obtained by means of a finite number of flag transforms from a constant map ϕ_0 giving a *factorisation:* $\phi = \phi_0(\pi_1 - \pi_1^\perp) \cdots (\pi_s - \pi_s^\perp)$ of the harmonic map (here the π_i are maps of S^2 to Grassmannians $G_{k_i}(\mathbb{C}^n)$); for simple proofs of this (giving different factorisations to that of Uhlenbeck) see [80, 86]. This is generalized to any compact simple Lie group which admits a Hermitian symmetric space as quotient (all compact simple Lie groups except E_8, F_4 and G_2), or any symmetric space of such a group, by F.E. Burstall and J.H. Rawnsley in [20]. For various symmetric spaces explicit methods of obtaining all harmonic maps based on using suitable flag transforms (cf. §3.1) are given in [1, 2].

4 Introduction of a spectral parameter

4.1 Two elementary examples

We start by giving two elementary examples of the introduction of a parameter before discussing how it is introduced into the theory of harmonic maps into a Lie group. For more examples, see [19].

(A) **Loops of constant mean curvature surfaces in $\mathbb{R}^3$**. Let U be an open subset of $\mathbb{R}^2 = \mathbb{C}$. Recall the integrability conditions (3.4) for a surface of constant mean curvature $F : U \to \mathbb{R}^3$. If A is replaced by λA, and so $\bar{A}$ by $(1/\lambda)\bar{A}$ where $\lambda \in S^1$ the equations are still satisfied so that we may integrate (3.3) to get a 1-parameter family or *loop* of isometric constant mean curvature surfaces: $F_\lambda : U \to \mathbb{R}^3$; note that we require $\lambda \in S^1$ to ensure that these surfaces are real, i.e. lie in $\mathbb{R}^3$. Taking Gauss maps, this procedure gives a loop of harmonic maps $\phi_\lambda : U \to S^2$. Note that if F is actually (conformal and) minimal ($H = 0$), then all the maps F_λ are conformal and minimal and form the well-known loop of associate minimal surfaces.

(B) **Harmonic conjugates** Again let U be an open subset of $\mathbb{R}^2$. If $\phi : U \to \mathbb{R}^n$, $x + iy = z \mapsto \phi(z)$ is harmonic then $\alpha = d\phi$ satisfies

$$d\alpha = 0, \quad d * \alpha = 0. \tag{4.1}$$

Explicitly, writing $\alpha = \alpha_x dx + \alpha_y dy$, (4.1) reads

$$\frac{\partial \alpha_y}{\partial x} - \frac{\partial \alpha_x}{\partial y} = 0, \quad \frac{\partial \alpha_x}{\partial x} + \frac{\partial \alpha_y}{\partial y} = 0. \tag{4.2}$$

These two equations can be combined into a single equation as follows: Write $\alpha = \alpha_z dz + \alpha_{\bar z} d\bar z$. Then equations (4.2) can be written:

$$\frac{\partial \alpha_z}{\partial \bar z} = 0, \quad \frac{\partial \alpha_{\bar z}}{\partial z} = 0. \tag{4.3}$$

Now introduce a parameter $\lambda \in S^1$ and write $A_\lambda = \lambda \alpha_z dz + \frac{1}{\lambda} \alpha_{\bar z} d\bar z$ so that $dA_\lambda = \lambda \bar\partial \alpha_z + \frac{1}{\lambda} \partial \alpha_{\bar z}$. Then

$$dA_\lambda = 0 \quad \text{for all } \lambda \in S^1 \tag{4.4}$$

if and only if (4.3) holds, i.e. the two equations of (4.1) have been encoded into a single equation (4.4). Note further that the transformation $\alpha_z \mapsto \lambda \alpha_z$, $\alpha_{\bar z} \mapsto \frac{1}{\lambda} \alpha_{\bar z}$ preserves the equations (4.3) so that we obtain a loop of harmonic maps $\phi_\lambda : U \to \mathbb{R}^3$, $(\lambda \in S^1)$.

Remark. If ϕ is conformal, so are all the maps ϕ_λ in the loop, and if $n = 3$ we obtain the loop of isometric minimal surfaces of Example (A). In this case, all the Gauss maps are antiholomorphic and don't change with λ.

4.2 Harmonic maps into a Lie group

Now we return to the case of a harmonic map from a simply-connected domain U of $\mathbb{R}^2$ to a Lie group G. Recall the pair of equations (3.13, 3.14) for $\alpha = \phi^*\omega$ ($= \phi^{-1}d\phi$ for a matrix group) is equivalent to harmonicity of ϕ. Now we introduce a loop of $\mathfrak{g}^c$-valued 1-forms

$$A_\lambda = \tfrac{1}{2}(1-\lambda)\alpha_z dz + \tfrac{1}{2}(1-\lambda^{-1})\alpha_{\bar z} d\bar z \tag{4.5}$$

Note that $A_{-1} = A$ and $A_1 = 0$ and that A_λ agrees with Uhlenbeck [78] except that, following [18], our λ is the reciprocal of hers. Then a short calculation shows [78] that A satisfies (3.13,3.14) if and only if

$$dA_\lambda + \tfrac{1}{2}[A_\lambda \wedge A_\lambda] = 0 \quad \text{for all} \quad \lambda \in S^1. \tag{4.6}$$

So we have encoded the equations (3.13,3.14) into the single equation (4.6). This equation can be interpreted as saying that for each $\lambda \in S^1$, $d + A_\lambda$ is a flat connection and is an example of a *zero curvature equation*. We get a loop of maps $\phi_\lambda : M \to G$, $(\lambda \in S^1)$ defined by $\phi_\lambda^*(\omega) = A_\lambda$. Indeed, for each λ, ϕ_λ is the map which gauges $d + A_\lambda$ to the trivial connection. However, unless G is Abelian, the maps ϕ_λ are not harmonic but rather extremise certain functionals of Novikov, see [75]. In the case $G = U(n)$ a map $\Phi : M \to \Omega G$ with $\Phi(p)(\lambda) = \phi_\lambda(p)$ is called *an extended solution corresponding to ϕ* [78]. It is holomorphic with respect to a suitable complex structure on ΩG and we have a diagram common in twistor theory, see [20, §8F]:

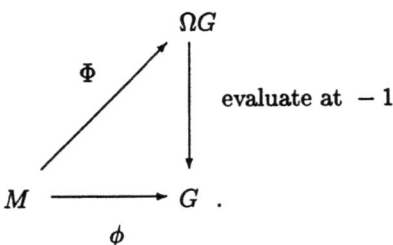

where Φ is holomorphic and "pseudo-horizontal".

Remarks

1. The idea of introducing a spectral parameter to get a zero curvature equation of the form (4.6) is credited to Pohlmeyer [67], see [13].

2. There is way of defining a loop of maps $\tilde{\phi}_\mu$, $(\mu \in S^1)$ which *are* harmonic, namely, set $\tilde{\phi}_\mu = \phi_{-\mu}\phi_\mu^{-1}$. This defines an S^1-action on the space of harmonic maps $U \to G$ due to C.-L. Terng, see [78, §7] and [64].

3. The flag transforms of §3 can be explained in terms of the extended solution. Indeed the transform (3.19) corresponds to multiplying the extended solution by $\pi + \lambda\pi^\perp$, see [78], and see [20] for other Lie groups.

Now let $\Omega\mathfrak{g}$ be the space of based loops in $\mathfrak{g}$, viz.

$$\Omega\mathfrak{g} = \{\xi : S^1 \to \mathfrak{g} : \xi(1) = 0\};$$

we also need its complexification

$$\Omega\mathfrak{g}^c = \{\xi : S^1 \to \mathfrak{g}^c : \xi(1) = 0\}.$$

We write any loop $\xi \in \Omega\mathfrak{g}^c$ as a Fourier series, which, since $\xi(1) = 0$, takes the form

$$\xi(\lambda) = \sum_{n \neq 0} \xi_n(1 - \lambda^n)$$

where $\xi_n \in \mathfrak{g}^c$. Note that ξ is real, i.e. lies in $\Omega\mathfrak{g}$, if and only if the following *reality condition* is satisfied:

$$\xi_{-n} = \overline{\xi_n} \quad \forall n. \tag{4.7}$$

Now $\Omega\mathfrak{g}$ has a filtration $\Omega_1 \subset \Omega_2 \subset \cdots \subset \Omega\mathfrak{g}$ by finite dimensional subspaces $\Omega_d = \{\xi \in \Omega\mathfrak{g} : \xi_n = 0 \text{ for } |n| > d\}$. $\Omega\mathfrak{g}^c$ contains two subalgebras consisting of the positive and negative frequencies:

Harmonic maps into symmetric spaces and integrable systems 45

$$\Omega^+ = \{\xi : \xi_n = 0 \text{ for } n < 0\}, \quad \Omega^- = \{\xi : \xi_n = 0 \text{ for } n > 0\}.$$

We have a conjugation on $\Omega\mathfrak{g}^c$ given by $\xi \mapsto \bar{\xi} = \sum_{n \neq 0} \overline{\xi_n}(1 - 1/\lambda^n)$ under which Ω^+ corresponds to Ω^-, and $\Omega\mathfrak{g}^c = \Omega^+ \oplus \Omega^-$. Then a family of 1-forms A_λ of the type (4.5) is the same as an $\Omega\mathfrak{g}$-valued 1-form A such that

(i) $A^{(1,0)}$ takes values in $\Omega^+ \cap \Omega_1^c$,

(ii) the connection $d + A$ is flat, i.e. (4.6) is satisfied.

Given any harmonic map $\phi : \mathbb{R}^2 \to G$ we define such an A by (4.5). Conversely, given such an A, there is a harmonic map $\phi : \mathbb{R}^2 \to G$ given by integrating the equation (3.10) (or (3.11)). Thus to find harmonic maps we must find $\Omega\mathfrak{g}$-valued 1-forms A which satisfy (i) and (ii).

Now let $\xi : \mathbb{R}^2 \to \Omega_d$ be a smooth mapping. If we do the ansatz:

$$A_\lambda = 2\mathrm{i}(1-\lambda)\xi_d dz - 2\mathrm{i}(1-\lambda^{-1})\xi_{-d} d\bar{z}, \tag{4.8}$$

then condition (i) above is satisfied and, by equating coefficients we see that condition (ii) is equivalent to

$$\frac{\partial \xi_{-d}}{\partial z} = 2\mathrm{i}[\xi_{-d}, \xi_d],$$

$$\frac{\partial \xi_d}{\partial \bar{z}} = -2\mathrm{i}[\xi_d, \xi_{-d}].$$

Alternatively, simply note that these are equations (3.15, 3.16) with the ansatz $\alpha_z = 4\mathrm{i}\xi_d$, $\alpha_{\bar{z}} = -4\mathrm{i}\xi_{-d}$. However, the key observation is that these equations are simply the $1 - \lambda^{-d}$ coefficient of

$$\frac{\partial \xi}{\partial z} = [\xi, 2\mathrm{i}(1-\lambda)\xi_d] \tag{4.9}$$

and the $1 - \lambda^d$ coefficient of the conjugate equation

$$\frac{\partial \xi}{\partial \bar{z}} = -[\xi, 2\mathrm{i}(1-\lambda^{-1})\xi_{-d}]. \tag{4.10}$$

To interpret these, writing

$$\tfrac{1}{2}(X_1 - \mathrm{i}X_2) = [\xi, 2\mathrm{i}(1-\lambda)\xi_d] \tag{4.11}$$

defines vector fields X_1, X_2 on Ω_d; we are trying to find $\xi : \mathbb{R}^2 \to \Omega_d$ such that

$$\frac{\partial \xi}{\partial x} = X_1 \text{ and } \frac{\partial \xi}{\partial y} = X_2.$$

The integrability condition for this is that the Lie bracket $[X_1, X_2] = 0$, equivalently, writing $Z = \tfrac{1}{2}(X_1 - \mathrm{i}X_2)$, $[Z, \bar{Z}] = 0$. A direct calculation of $[Z, \bar{Z}] = d_Z \bar{Z} - d_{\bar{Z}} Z$ shows it is zero (see [18]). Furthermore, it can be shown that the flows evolve on spheres in Ω_d and so are complete. Thus we can solve (4.9,4.10) for $\xi : \mathbb{R}^2 \to \Omega_d$, define an $\Omega\mathfrak{g}$-valued 1-form A by (4.8) and find a harmonic map $\phi : \mathbb{R} \to \mathfrak{g}$ by integrating (4.6).

To interpret all this, recall that a *Poisson structure* on a manifold M is a skew bilinear map $C^\infty(M) \times C^\infty(M) \to C^\infty(M)$, $f, g \mapsto \{f, g\}$ satisfying the derivation property

$$\{f, gh\} = \{f, g\}h + g\{f, h\} \tag{4.12}$$

and the Jacobi identity

$$\{f, \{g, h\}\} + \{g, \{h, f\}\} + \{h, \{f, g\}\} = 0. \tag{4.13}$$

In particular, $(C^\infty(M), \{,\})$ is a Lie algebra. Given such a structure we can associate to any $f \in C^\infty(M)$ a vector field X_f on M defined as a derivation by $X_f h = \{f, h\}$, $(h \in C^\infty(M))$. (That this is a derivation follows from (4.12).) Now let $[\,,\,]$ be the standard Lie bracket of vector fields on M, then from (4.13) it quickly follows that

$$(C^\infty(M), \{\,,\,\}) \to (C^\infty(TM), [\,,\,]), \quad f \mapsto X_f$$

is a Lie algebra homomorphism, i.e.

$$[X_f, X_g] = X_{\{f, g\}}, \quad (f, g \in C^\infty(M)).$$

Indeed, we have, for $f, g, h \in C^\infty(M)$,

$$\begin{aligned}
[X_f, X_g]h &= X_f X_g h - X_g X_f h \\
&= \{f, \{g, h\}\} - \{g, \{f, h\}\} \\
&= \{\{f, g\}, h\} \\
&= X_{\{f, g\}} h.
\end{aligned}$$

X_f is called the *Hamiltonian vector field* of f. The classic example of a Poisson structure due to Lie is the following: Let $\mathbf{G}$ be a Lie group with Lie algebra $(\mathbf{g}, [,]^0)$. Then its dual $M = \mathbf{g}^*$ has a canonical Poisson structure given by

$$\{f, g\}^0(\xi) = \langle \xi, [df_\xi, dg_\xi]^0 \rangle, \quad (f, g \in C^\infty(\mathbf{g}^*), \xi \in \mathbf{g}^*).$$

Here $\langle\,,\,\rangle$ denotes the canonical pairing of $\mathbf{g}^*$ and $\mathbf{g}$ and df_ξ lies in $T^*\mathbf{g}^*$ which we can identify with $\mathbf{g}$. The *co-adjoint action* of $\mathbf{g}$ on $\mathbf{g}^*$ is defined by

$$\langle \mathrm{Ad}_x^*(\xi), Y \rangle = \langle \xi, \mathrm{Ad}_x Y \rangle, \quad (x \in G,\ Y \in \mathbf{g},\ \xi \in \mathbf{g}^*)$$

where $\mathrm{Ad}_x Y$ is the adjoint action of $\mathbf{G}$ on $\mathbf{g}$ given for matrix groups by $\mathrm{Ad}_x Y = xYx^{-1}$. Its differential is the co-adjoint action of $\mathbf{g}$ on $\mathbf{g}^*$ given by

$$\langle \mathrm{ad}_X^*(\xi), Y \rangle = \langle \xi, \mathrm{ad}_X Y \rangle = \langle \xi, [X, Y]^0 \rangle, \quad (X, Y \in \mathbf{g},\ \xi \in \mathbf{g}^*).$$

Call a function $f \in C^\infty(\mathbf{g}^*)$ *invariant* if it is invariant under the action Ad^*. Equivalently, in terms of the action ad^*,

$$\langle \mathrm{ad}_X^* \xi, df_\xi \rangle = 0 \text{ for all } \xi \in \mathbf{g}^*, \mathbf{X} \in \mathbf{g} \tag{4.14}$$

Harmonic maps into symmetric spaces and integrable systems 47

since $\{\mathrm{ad}_X^*\xi : X \in \mathbf{g}\}$ gives the tangent space to the orbit of Ad^* through ξ. Equation (4.14) can be written as $\langle \xi, \mathrm{ad}_X(df_\xi)\rangle = 0$; this is equivalent to $\langle \xi, \mathrm{ad}_{df_\xi} X\rangle = 0$, which in turn is equivalent to $\langle \mathrm{ad}_{df_\xi}^* \xi, X\rangle = 0$ for all $\xi \in \mathbf{g}^*, X \in \mathbf{g}$, and so f is invariant if and only if

$$\mathrm{ad}_{df_\xi}^* \xi = 0 \text{ for all } \xi \in \mathbf{g}^*. \tag{4.15}$$

The basic problem in Hamiltonian theory is the following: Given a function $H \in C^\infty(M)$ on a Poisson manifold $(M, \{\,,\,\})$ find functions f_i in involution with themselves and with H, i.e. with $\{f_i, H\} = \{f_i, f_j\} = 0$ for all i. Such functions are integrals of the flow of X_H, i.e. constant along the solution curves $t \mapsto \sigma(t)$ of the ordinary differential equation $\frac{d\sigma}{dt} = X_H$; indeed $X_H(f_i) = \{H, f_i\} = 0$. Now in the case $M = \mathbf{g}^*$ any invariant function f is in involution with an arbitrary $H \in C^\infty(\mathbf{g}^*)$ for

$$\begin{aligned}\{f, H\}^0(\xi) &= \langle \xi, [df_\xi, dH_\xi]^0\rangle \\ &= \langle \xi, \mathrm{ad}_{df_\xi} dH_\xi\rangle \\ &= \langle \mathrm{ad}_{df_\xi}^* \xi, dH_\xi\rangle \\ &= 0 \quad \text{by (4.15)}.\end{aligned}$$

Thus, although invariant functions provide integrals in involution they cannot distinguish one flow from another. So we refine this method by introducing r-matrices:

An *r-matrix* on $\mathbf{g}$ is a linear map $R : \mathbf{g} \to \mathbf{g}$ such that the map $[\,,\,]^R : \mathbf{g} \times \mathbf{g} \to \mathbf{g}$ defined by

$$[X, Y]^R = [RX, Y]^0 + [X, RY]^0, \quad (X, Y \in \mathbf{g})$$

is a Lie bracket. This new Lie bracket defines a new Poisson bracket $\{\,,\,\}^R$ on $C^\infty(\mathbf{g}^*)$ by

$$\{f, g\}^R(\xi) = \langle \xi, [df_\xi, dg_\xi]^R\rangle, \quad (\xi \in \mathbf{g}^*,\ f, g \in C^\infty(\mathbf{g}^*))$$

and a new Lie algebra homomorphism $C^\infty(\mathbf{g}^*) \to (\mathbf{C}^\infty(\mathbf{Tg}^*), [\,,\,])$, $f \mapsto X_f^R$ where the vector field X_f^R is defined as the derivation $X_f^R(h) = \{f, h\}^R$, $(h \in C^\infty(\mathbf{g}^*))$. Here $[\,,\,]$ is unchanged, it is the standard Lie bracket of vector fields on $M = \mathbf{g}^*$. This time the invariant functions (no change of meaning) give interesting Hamiltonian vector fields; indeed, let $f, g \in C^\infty(\mathbf{g}^*)$ with f invariant, then we calculate at $\xi \in \mathbf{g}^*$,

$$\begin{aligned}(X_f^R)_\xi(h) &= \{f, h\}^R(\xi) \\ &= \langle \xi, [df_\xi, dh_\xi]^R\rangle \\ &= \langle \xi, [Rdf_\xi, dh_\xi]^0\rangle + \langle \xi, [df_\xi, Rdh_\xi]^0\rangle \\ &= \langle \mathrm{ad}_{Rdf_\xi}^* \xi, dh_\xi\rangle + \langle \mathrm{ad}_{df_\xi}^* \xi, Rdh_\xi\rangle \\ &= (\mathrm{ad}_{Rdf_\xi}^* \xi)(h) + 0\end{aligned}$$

by (4.15), so that

$$(X_f^R)_\xi = \mathrm{ad}_{Rdf_\xi}^* \xi. \tag{4.16}$$

This is all clearer if **g** has an ad-invariant inner product allowing us to identify **g*** with **g**. Then the ad* action just becomes the adjoint action of **g** on itself, df is identified with the gradient ∇f, and, for an invariant function f, (4.16) reads

$$(X_f^R)_x = [x, R\nabla f_x], \quad (x \in \mathbf{g}). \tag{4.17}$$

Thus the flow of X_f^R is the curve $\sigma(t)$ in **g** which satisfies

$$\frac{d\sigma}{dt} = [\sigma, R\nabla f_\sigma], \tag{4.18}$$

an equation in Lax form. Furthermore, the invariant functions are in involution with respect to $\{\,,\,\}^R$ for, if f, h are invariant functions on **g*** (or on **g** if this has an ad-invariant inner product),

$$\begin{aligned}
\{f, h\}^R(\xi) &= \langle \xi, [df_\xi, dh_\xi]^R \rangle \\
&= \langle \xi, [Rdf_\xi, dh_\xi]^0 \rangle + \langle \xi, [df_\xi, Rdh_\xi]^0 \rangle \\
&= -\langle \mathrm{ad}^*_{dh_\xi} \xi, Rdf_\xi \rangle + \langle \mathrm{ad}^*_{df_\xi} \xi, Rdh_\xi \rangle \\
&= 0.
\end{aligned}$$

by (4.15). In particular, invariant functions give integrals for (4.18) and provide a powerful tool for finding such integrals. More relevant for our purposes is the fact that a set of invariant functions $\{f_1, \ldots, f_k\}$ together with an r-matrix gives a set of commuting vector fields $\{X_{f_i}^R\}$ defined by (4.17) on **g**.

To return to the problem at hand, note that if we set $\mathbf{g} = \Omega\mathfrak{g}$, where $\mathfrak{g}$ is the Lie algebra of a compact group, we can define an r-matrix on **g** by

$$R = \begin{cases} i & \text{on } \Omega^+ \\ -i & \text{on } \Omega^- \end{cases}.$$

Now $\Omega\mathfrak{g}$ has an inner product

$$(\xi_1, \xi_2)_{L^2} = \int_{S^1} (\xi_1, \xi_2)_\mathfrak{g} = \int_{S^1} (\xi_1(\lambda), \xi_2(\lambda))_\mathfrak{g} |d\lambda|$$

where $(\,,\,)_\mathfrak{g}$ is an ad-invariant inner product on $\mathfrak{g}$. This is invariant under the pointwise adjoint action of $\Omega\mathfrak{g}$ on itself. Define f_1, f_2 by

$$(f_1 + if_2)(\xi) = \int_{S^1} \lambda^{1-d} (\xi, \xi)_\mathfrak{g}. \tag{4.19}$$

Then f_1 and f_2 are clearly invariant functions and so the corresponding vector fields $X_1 = X_{f_1}^R$ and $X_2 = X_{f_2}^R$ given by (4.17) commute. We claim that these vector fields restricted to Ω_d are just the X_1 and X_2 respectively of (4.11). To see this, note that the flow $\xi(x+iy)$ of these vector fields is the mapping $\xi: \mathbb{R}^2 \to \Omega_d$ satisfying the equations

$$\frac{\partial \xi}{\partial x} = [\xi, R\nabla df_1] \tag{4.20}$$

$$\frac{\partial \xi}{\partial y} = [\xi, R\nabla df_2] \tag{4.21}$$

Harmonic maps into symmetric spaces and integrable systems

and, clearly from (4.19), $(\nabla f_1 + \nabla f_2)_\xi = \lambda^{1-d}\xi$. Thus (4.20, 4.21) can be combined into the single equation

$$\frac{\partial \xi}{\partial z} = [\xi, R\lambda^{1-d}\xi]$$
$$= [\xi, (R+\mathrm{i})\lambda^{1-d}\xi]$$

since $[\xi, \xi] = 0$

$$= [\xi, 2\mathrm{i}(1-\lambda)\xi_d],$$

since $R + \mathrm{i}$ annihilates Ω^-, which is just (4.9).

Remarks

1. Equations of type for $\xi : \mathbb{R} \to \mathfrak{g}$:

$$\frac{d\xi}{dt} = [\xi, B(\xi)],$$

have many nice properties. For example, if $P : \mathfrak{g} \to \mathbb{R}$ is any (ad-)invariant function on $\mathfrak{g}$ then $P(\xi)$ is a constant of the motion. Thus, if $\mathfrak{g}$ is a matrix group, $\operatorname{tr}\xi$, $\det \xi$ and the other coefficients of the characteristic polynomial of ξ are all constant. In particular, the eigenvalues η of ξ are constant. If $\mathfrak{g}$ is infinite dimensional the same applies to the discrete eigenvalues and is of great use, for example, in the study of the KdV equation see for example [43]. Now let $\mathfrak{g} = \Omega\mathfrak{g}$ and let $\xi \in \Omega_d$, then evaluating the equation (4.9) at λ gives

$$\frac{\partial \xi(\lambda)}{\partial z} = [\xi(\lambda), 2\mathrm{i}(1-\lambda)\xi_d]. \tag{4.22}$$

For each λ, $\xi(\lambda) \in u(n)$ has eigenvalues η which do not change with z so we get an affine curve

$$\{(\lambda, \eta) \in \mathbb{C}^2 : \det\big(\xi(\lambda) - \eta\operatorname{Id}\big) = 0\}$$

which does not vary with z. The normalization of the compactification of this is called the *spectral curve* $\mathcal{C}$, cf. [51]. Now suppose that the initial condition $\xi(0)$ is chosen such that the eigenvalues η are simple for generic λ. Then the eigenspaces of η are subspaces of $\mathbb{C}^n$ which vary with η and thus form a line bundle over $\mathcal{C}$. This line bundle *does* vary with z: its evolution is determined by (4.22). Thus (4.22) corresponds to a dynamical system on the Jacobian variety of all line bundles over the complex curve $\mathcal{C}$. Following ideas of [70, 44], Burstall [16] shows that this dynamical system is linear.

2. By finding more commuting integrals, one can construct a large class of *pluriharmonic* maps from a higher dimensional Euclidean space or torus to a Lie group, see [18].

3. In the case of constant mean curvature tori in $\mathbb{R}^3$ (also S^3 and $\mathbb{H}^3$) (cf. §1), all the solutions can be expressed in terms of θ functions [10], see [30] for a general exposition and [63] for such constructions for the case of certain harmonic tori in $\mathbb{C}P^n$. This programme should be possible in the present case but has not yet been carried out.

4. Let G/H be a symmetric space and let σ be the involutive isometry of G/H with fixed point the identity coset. Then $\tau : g \mapsto g^\tau = \sigma g \sigma$ defines an involution of G and the map $i : G/H \to G$ defined by $gH \mapsto g^\tau = g^\tau g^{-1}$ is well-defined and is a totally geodesic immersion (cf. [24]) called the Cartan immersion. By the composition law (1.10) a map $\phi : M \to G/H$ is harmonic if and only if $i \circ \phi : M \to G$ is harmonic; thus a harmonic map into G/H is just a harmonic map into G with image in G/H. In [18, Theorem 1.12], it is shown that if we choose a suitable initial condition ξ_0 the solution to (4.9) will give a harmonic map $\mathbb{R}^2 \to G/H$.

5. We have constructed a large class of harmonic maps $\phi : \mathbb{R}^2 \to G$ by solving (4.9); such maps are said to be of *finite type*. Some of these maps descend to maps of a 2-torus, Conversely, ([18, Corollary 3.3]), any non-conformal harmonic map of a 2-torus to the 4-sphere S^4 is of finite type and is therefore constructed by the methods above. As for the *conformal* harmonic maps, these are **either** superminimal (= real-isotropic = pseudo-holomorphic) in which case they can all be constructed from holomorphic maps [23, 56, 14, 59] as described in §3.1 **or** they are not superminimal, in which case they are all constructed in [42] by a forerunner of the integrable systems methods of [18] discussed above. Thus all harmonic maps from the 2-torus to the 4-sphere have been constructed.

6. A harmonic map $M \to \mathbb{C}P^n$ is said to be *superconformal* if its harmonic sequence is periodic, or, equivalently, its isotropy order is the maximum possible finite value, i.e. $2m + 1$ (cf. §3.1). The Gauss–Codazzi–Mainardi–Ricci equations lead to 2-dimensional periodic Toda systems of $SU(n+1)$ type, and thus such harmonic maps can be found by solving this system, see [29, 11, 12]. Indeed the sequence of osculating curves of a harmonic map $M^2 \to \mathbb{C}P^n$ provides a lift to a holomorphic map into the full flag manifold $SU(n + 1)/S(U(1) \times \cdots U(1))$; this is an $(n+1)$-symmetric space (see [58]), and it is this lift which can be constructed from the Toda equations. All such harmonic maps are then expressible in terms of τ functions. The story is similar for maps into S^{2n}, the flag manifold being replaced by $SO(2n + 1)/SO(2) \times \cdots \times SO(2)$.

7. Burstall takes this further [17] to construct all non-isotropic harmonic tori in spheres and complex projective spaces (for isotropic ones see §3.1); indeed given a harmonic map of a torus to $\mathbb{C}P^n$ of isotropy order $k \leq n - 1$, the flag of ∂'-osculating spaces gives a lift of ϕ into a $(k + 2)$-symmetric space $U(n + 1)/U(1) \times \cdots U(1) \times U(n - k)$ which is of finite type, i.e. can be constructed by the above methods. Similarly, all non-isotropic harmonic tori in

S^{2n} are constructed replacing the symmetric space by $SO(2n+1)/SO(2) \times \cdots \times SO(2) \times SO(2n-2k)$.

8. For an alternative powerful method of constructing harmonic maps from surfaces, see [19].

The author would like to thank F.E. Burstall and J. Eells for useful comments on a draft version of this article.

Bibliography

[1] A. Bahy-El-Dien and J.C. Wood, *The explicit construction of all harmonic 2-spheres in $G_2(\mathbb{R}^n)$*, J. Reine u. Angew. Math. **398** (1989), 36–66.

[2] A. Bahy-El-Dien and J.C. Wood, *The explicit construction of all harmonic two-spheres in quaternionic projective space*, Proc. London Math. Soc. **62** (1991), 202–224.

[3] P. Baird, *Harmonic maps with symmetry, harmonic morphisms and deformations of metrics*, Research Notes in Math. 87, Pitman, London (1983).

[4] P. Baird and J. Eells, *A conservation law for harmonic maps*, in: Geometry Symposium, Utrecht 1980, ed. E. Looijenga, D. Siersma and F. Takens, Lecture Notes in Math. 894 , Springer, (1981), 1–25.

[5] P. Baird and A. Ratto, *Conservation laws, equivariant harmonic maps and harmonic morphisms*, Proc. London Math. Soc. (3) **64** (1992), 197–224.

[6] P. Baird and J.C. Wood, *Bernstein theorems for harmonic morphisms from $\mathbb{R}^3$ and S^3*, Math. Ann. **280** (1988), 579–603.

[7] P. Baird and J.C. Wood, *Harmonic morphisms and conformal foliations by geodesics of three-dimensional space forms*, J. Austral. Math. Soc. (A)**51** (1991), 118–153.

[8] P. Baird and J.C. Wood, *Harmonic morphisms, Seifert fibre spaces and conformal foliations*, Proc. London Math. Soc. (3) **64** (1992), 170-196.

[9] P. Baird and J.C. Wood, *The geometry of a pair of Riemannian foliations by geodesics and asociated harmonic morphisms*, Bull. Soc. Math. Belg. Ser. B **44** (1992), 115–139.

[10] A.I. Bobenko, *All constant mean curvature tori in R^3, S^3, H^3 in terms of theta-functions*, Math. Ann. **290** (1991), 209-245.

[11] J. Bolton, F. Pedit and L.M. Woodward, *Minimal surfaces and the Toda field model*, preprint, Universities of Durham and Massachussetts, Amherst.

[12] J. Bolton and L.M. Woodward, *The affine Toda equations and minimal surfaces*, this volume.

[13] M. Bordemann, M. Forger, J. Laartz, U. Schäper, *2-dimensional nonlinear sigma models: zero curvature and Poisson structure*, this volume.

[14] R. Bryant, *Conformal and minimal immersions of compact surfaces into the 4-sphere*, J. Diff. Geom **17** (1982), 455–473.

[15] F.E. Burstall, *Twistor methods for harmonic maps*, in: Differential Geometry (Proceedings, Lyngby 1985), ed. V.L. Hansen, Lecture Notes in Math. 1263, Springer, Berlin (1987), 55–96.

[16] F.E. Burstall, *Harmonic maps and soliton theory*, Mathematica Contemporanea **2** (1992), 1–18.

[17] F.E. Burstall, *Harmonic tori in spheres and complex projective spaces*, in preparation.

[18] F.E. Burstall, D. Ferus, F. Pedit and U. Pinkall, *Harmonic tori in symmetric spaces and commuting Hamiltonian systems on loop algebras*, Ann. of Math. **138** (1993), 173–212.

[19] F.E. Burstall and F. Pedit, *Harmonic maps via Adler-Kostant-Symes theory*, this volume.

[20] F.E. Burstall and J.H. Rawnsley, *Twistor theory for Riemannian symmetric spaces with applications to harmonic maps of Riemann surfaces*, Lecture Notes in Math. 1424, Springer, Berlin 1990.

[21] F.E. Burstall and S.M Salamon, *Tournaments, flags and harmonic maps*, Math. Ann. **277** (1987), 249–266.

[22] F. Burstall and J.C. Wood, *The construction of harmonic maps into complex Grassmannians*, J. Diff. Geom. **23** (1986), 255–298.

[23] E. Calabi, *Quelques applications de l'analyse complexe aux surfaces d'aire minima* in: Topics in Complex Manifolds, Université de Montréal (1967), 59–81.

[24] J. Cheeger and D. Ebin, *Comparison Theorems in Riemannian Geometry*, North Holland, Amsterdam 1975.

[25] Y-M. Chen and W-Y. Ding, *Blow-up and global existence for heat flows of harmonic maps*, Invent. Math. **99** (1990), 567–578.

[26] S.S. Chern, *Minimal surfaces in Euclidean space of N dimensions*, in: Symposium in honor of Marston Morse, Princeton University Press 1965, 187–198.

[27] S.S. Chern and J. Wolfson, *Harmonic maps of the two-sphere into a complex Grassmannian manifold II*, Ann. of Math. 125 (1987), 301–335.

[28] J.-M. Coron and J.-M. Ghidaglia, *Explosion en temps fini pour le flot des applications harmoniques*, C.R. Acad. Sci. Paris Série 1, Math. **308**(1989), 339–344.

[29] A. Doliwa and A. Sym, *The non-linear σ-model in spheres*, preprint, Warsaw University, 1992.

[30] B.A. Dubrovin, *Theta functions and non-linear equations*, Uspekhi Mat. Nauk **36:2** (1981), 11–80, English transl. Russian Math, Surveys **36:2** (1981), 11–92.

[31] J. Eells, *On the surfaces of Delaunay and their Gauss maps*, Proc. 4th Inter. Colloq. Diff. Geom., Santiago de Compostela (1978), Univ. de Santiago de Compostela 1979, 97–116; reprinted as *The surfaces of Delauney*, Math. Intell. **9** (1978), 53–57.

[32] J. Eells and L. Lemaire, *A report on harmonic maps*, Bull. London Math. Soc. **10** (1978), 1–68.

[33] J. Eells and L. Lemaire, *Selected topics in harmonic maps*, C.B.M.S. Regional Conference Series **50**, Amer. Math. Soc. 1983.

[34] J. Eells and L. Lemaire, *Another report on harmonic maps*, Bull. London Math. Soc. **20** (1988), 385–524.

[35] J. Eells and A. Ratto, *Harmonic maps and minimal immersions with symmetries, Methods of ordinary diferential equations applied to elliptic variational problems*, Annals of Math. Studies, 130, Princeton Univ. Press 1993.

[36] J. Eells and J.H. Sampson, *Harmonic mappings of Riemannian manifolds*, Amer. J. Math. **86** (1964), 109–160.

[37] J. Eells and J.C. Wood, *Restrictions on harmonic maps of surfaces*, Topology **15** (1976), 263–266.

[38] J. Eells and J.C. Wood, *Harmonic maps from surfaces to complex projective spaces*, Adv. in Math. **49** (1983), 217–263.

[39] G.F.R. Ellis and S.W. Hawking, *The large scale structure of space time*, Cambridge Monographs on Mathematical Physics, Cambridge Univ. Press (1973).

[40] S. Erdem and J.C. Wood, *On the construction of harmonic maps into a Grassmannian*, J. London Math. Soc. (2) **28** (1983), 161–174.

[41] J. Eschenburg and R. Tribuzy, *Branch points of conformal mappings of surfaces*, Math. Ann. **278** (1988), 621-633.

[42] D. Ferus, F. Pedit, U. Pinkall and I. Sterling, *Minimal tori in S^4*, J. Reine Angew. Math. **429** (1992), 1–47.

[43] A.P. Fordy (ed.) *Soliton theory: a survey of results*, Manchester Univ. Press 1990.

[44] P.A. Griffiths, *Linearizing flows and a cohomological interpretation of Lax equations*, Amer. J. Math. **107** (1985), 1445–1483.

[45] J.F. Growtowski, *Harmonic map heat flow for axially symmetric data*, Man. Math. **73** (1991), 207–228.

[46] C.-H. Gu, *On the harmonic maps from $R^{1,1}$ to $S^{1,1}$*, J. Reine u. Angew. Math. **346** (1984), 101–109.

[47] M. Guest, *Harmonic 2-spheres in complex projective space and some open problems*, Expos. Math. **10** (1992), 61–87.

[48] R. Gulliver, R. Osserman and H.L. Royden, *A theory of branched immersions of surfaces*, Amer. J. Math. **95** (1973), 750–812.

[49] F. Hélein, *Régularité des applications faiblement harmoniques entre une surface et une variété riemannienne*, C.R. Acad. Sci. Paris **312** (1991), 591–596.

[50] D. Hilbert, *Die Grundlagen der Physik*, Nachr. Ges. Wiss. Göttingen (1915), 395–407, (1917), 53–76.

[51] N.J. Hitchin, *Harmonic maps from a 2-torus to the 3-sphere*, J. Diff. Geom. **31** (1990), 627–710.

[52] D.A. Hoffman and R. Osserman, *The Gauss map of surfaces in $\mathbb{R}^3$ and $\mathbb{R}^4$*, Proc. London Math. Soc. **50** (1985), 27–57.

[53] W-Y. Hsiang and H.B. Lawson, *Minimal submanifolds of low cohomogeneity*, J. Diff. Geom. **5** (1971), 1–38.

[54] H. Karcher and J.C. Wood, *Non-existence results and growth properties for harmonic maps and forms*, J. Reine u. Angew. Math. **353** (1984), 165–180.

[55] K. Kenmotsu, *Weierstrass formula for surfaces of prescribed mean curvature*, Math. Ann. **245** (1979), 89–99.

[56] P.Z. Kobak, *Twistors, nilpotent orbits and harmonic maps*, this volume.

[57] S. Kobayashi and K. Nomizu, *Foundations of Differential Geometry*, vol II, Interscience, Wiley, New York 1969.

[58] A.O. Kowalski, *Generalized symmetric spaces*, Lecture Notes in Math. **805**, Springer-Verlag, Berlin (1980).

[59] B. Loo, *The space of harmonic maps of S^2 into S^4*, Trans. Amer. Math. Soc. **313** (1989), 81–102.

[60] A. Lichnerowicz, *Applications harmoniques et variétés Kählériennes*, Symp. Math. III (Bologna 1970), 341–402.

[61] M. Mañas, *The principal chiral model as an integrable system*, this volume.

[62] H. McKean and E. Trubowitz, *Hill's equation and hyperelliptic function theory in the presence of infinitely many branch points*, Comm. Pure Appl. Math. **29** (1976), 143-226.

[63] I. McIntosh, *Infinite dimensional Lie groups and the two-dimensional Toda lattice*, this volume.

[64] M. Melko and I. Sterling, *Integrable systems and the classical theory of surfaces*, this volume.

[65] J. Moser, *Various aspects of integrable Hamiltonian systems*, Progress in Math. **8**, Birkhäuser, Boston 1980.

[66] Y. Ohnita and G. Valli, *Pluriharmonic maps into compact Lie groups and factorization into unitons*, Proc. London Math. Soc. (3) **61** (1990), 546–570.

[67] K. Pohlmeyer, *Integrable Hamiltonian systems and interactions through quadratic constraints*, Comm. Math. Phys. **46** (1976), 207–221.

[68] J.H. Rawnsley, *f-structures, f-twistor spaces and harmonic maps*, Sem. Geom. L. Bianchi II, 1984, Lecture Notes in Math. 1164, Springer, Berlin 1985, 85–159.

[69] J.H. Rawnsley, *Noether's theorem for harmonic maps*, in: Diff. Geom. Methods in Math. Phys., ed. S. Sternberg, Reidel, Dodrecht 1984, 197–202.

[70] A.G. Reyman and M.A. Semenov-Tian-Shansky, *Reduction of Hamiltonian systems, affine Lie algebras and Lax equations*, II, Invent. Math. **63** (1981), 423–432.

[71] C. Rogers and W.F. Shadwick, *Bäcklund transforms and their applications*, Academic Press, New York 1982.

[72] E.A. Ruh and J. Vilms, *The tension field of the Gauss map*, Trans. Amer. Math. Soc. **149** (1970), 569–573.

[73] R.T. Smith, *Harmonic maps of spheres*, Thesis, University of Warwick, 1972.

[74] I. Sterling and H.C. Wente, *Existence and classification of constant mean curvature multibubbletons of finite and infinite type*, preprint, Univ. Toledo (1992).

[75] A.V. Tyrin, *Harmonic spheres in compact Lie groups and extremals of a multivalued Novikov functional*, Uspekhi Mat. Nauk. **46:3** (1991), 197–198; English transl.: Russ. Math. Surveys **46:3** (1991), 235–236.

[76] K. Uhlenbeck, *Minimal 2-spheres and tori in S^k*, preprint (1975).

[77] K. Uhlenbeck, *Equivariant harmonic maps into spheres*, Harmonic maps, Proceedings, New Orleans 1980, ed. R.J. Knill, M. Kalka and H.C.J. Sealey, Lecture Notes in Math. 949, Springer, Berlin 1982, 146–158.

[78] K. Uhlenbeck, *Harmonic maps into Lie groups (classical solutions of the chiral model)*, J. Diff. Geom. **30** (1989), 1–50.

[79] M. Umehara and K. Yamada, *Harmonic non-holomorphic mappings of 2-tori into the 2-sphere*, in: Geometry of Manifolds (Matsumoto 1988), ed. K. Shiohama, Perspectives in Math. 8, Academic Press, Boston, London 1989.

[80] G. Valli, *On the energy spectrum of harmonic 2-spheres in unitary groups*, Topology **27** (1988), 129–136.

[81] J. Wolfson, *Harmonic maps of the two-sphere into the complex hyperquadric*, J. Diff. Geom. **24** (1986), 141–152.

[82] J. Wolfson, *Harmonic sequences and harmonic maps of surfaces into complex Grassmannian manifolds*, J. Diff. Geom. **27** (1988), 161–178.

[83] J.C. Wood, *Twistor constructions for harmonic maps*, in: Differential Geometry and Differential Equations (Proceedings, Shanghai, 1985), ed. Gu Chaohao, M. Berger and R.L. Bryant, Lecture Notes in Math. 1255, Springer, Berlin (1987), 130–159.

[84] J.C. Wood, *Holomorphic differentials and classification theorems for harmonic maps and minimal immersions*, in: Global Riemannian Geometry, ed. T.J. Willmore and N.J. Hitchin, Ellis Horwood, Chichester 1984, 168–175.

[85] J.C. Wood, *The explicit construction and parametrization of all harmonic maps from the two-sphere to a complex Grassmannian*, J. Reine u. Angew. Math. **386** (1988), 1–31.

[86] J.C. Wood, *Explicit construction and parametrization of harmonic 2-spheres in the unitary group*, Proc. London Math. Soc. (3) **58** (1989), 608–624.

[87] J.C. Wood, *Harmonic morphisms and Hermitian structures on Einstein 4-manifolds*, Internat. J. Math. **3** (1992), 415–439.

[88] J.C. Wood, *Harmonic maps and harmonic morphisms*, Lobachevshy Semester, Euler International Math. Inst. (to appear).

[89] V.E. Zakharov and A.V. Mikhailov, *Relativistically invariant two-dimensional model of field theory which is integrable by means of the inverse scattering problem method*, Zh. Eksp. Teor .Fiz. **74** (1978), 1953–1973; English transl.: Sov. Phys. JETP **47** (1978) 1017–1027.

[90] V.E. Zakharov and A.V. Shabat *Integration of nonlinear equations of Mathematical-Physics by inverse scattering II*, Funkts. Anal. Prilozh. **13** (1978), 13–22; English transl.: Func. Anal. Appl. **13** (1979), 166–174.

THE GEOMETRY OF SURFACES

The affine Toda equations and minimal surfaces

J. Bolton and L.M. Woodward

1 Introduction

In this article we consider geometrical interpretations of the two-dimensional affine Toda equations for a compact simple Lie group G. These equations originated from the work of Toda [33],[34] over 25 years ago on vibrations of lattices, and they have received considerable attention from both pure and applied mathematicians particularly over the last 15 years. (For the original context of the ideas the reader is referred to [35], [27] and [1] and for a survey of recent work to [29] and [21]).

Our purpose is to explain how solutions of these equations correspond to certain types of harmonic maps of surfaces into certain homogeneous spaces of G. We elucidate this relationship in the cases where G is $SU(n+1)$, $SO(2m+1)$ or the exceptional Lie group G_2 and, in particular, explain how the unknowns in the Toda equations correspond to natural geometric invariants for minimal immersions into $\mathbb{C}P^n$, S^{2m} and S^6 respectively. For each such G there are several possible geometrical situations which can arise, corresponding to different homogeneous spaces of G. These other geometrical models, together with the cases of other simple Lie groups, will be considered elsewhere.

We do not deal here with the question of how explicit solutions to the Toda equations may be found, although the formulation of the ideas we give leads naturally to the use of loop algebra calculations along the lines of [20] and [13] for the case of solutions of finite type. These methods, in principle, give a description of all doubly periodic solutions of the affine Toda equations on $\mathbb{C}$, which enables us to make some brief remarks concerning minimal tori. The exposition given here is derived in large measure from that of [6] to which the reader is referred for further details.

The affine Toda equations for a compact simple Lie group G of rank ℓ may be written (c.f. [21]) as

$$2\frac{\partial^2 \Omega}{\partial z \partial \bar{z}} + \sum_{p=0}^{\ell} m_p e^{2\alpha_p(\Omega)} \alpha_p^{\#} = 0, \tag{1.1}$$

where $\Omega(z, \bar{z})$ is a function defined on an open subset of the complex numbers taking values in it, (t being the Cartan subalgebra of G), $\alpha_1, \ldots, \alpha_\ell \in it^*$ are a set of positive simple roots of G, $m_0 = 1$, $-\alpha_0 = \sum_{p=1}^{\ell} m_p \alpha_p$ is the highest root and for each $p = 0, \ldots, \ell$, $\alpha_p^{\#} \in it$ denotes the dual of α_p under the Killing form $(\,,\,)$ defined by

$$(\alpha_p^{\#}, v) = \alpha_p(v), \qquad v \in t.$$

In the case where $G = SU(n+1)$ and we write $\Omega = \mathrm{diag}(\omega_0, \ldots, \omega_n)$, with $\omega_0, \ldots, \omega_n \in \mathbb{R}$, (1.1) can be written as

$$2\frac{\partial^2 \omega_p}{\partial z \partial \bar{z}} + e^{2(\omega_p - \omega_{p-1})} - e^{2(\omega_{p+1} - \omega_p)} = 0, \tag{1.2}$$

where

$$\omega_{n+p+1} = \omega_p, \qquad \omega_0 + \ldots + \omega_n = 0. \tag{1.3}$$

The equations (1.2) may be more naturally written as

$$2\frac{\partial^2 \alpha_p}{\partial z \partial \bar{z}} - e^{2\alpha_{p-1}} + 2e^{2\alpha_p} - e^{2\alpha_{p+1}} = 0, \tag{1.4}$$

where $\alpha_p(\Omega) = \omega_p - \omega_{p-1}$. In the case where (1.3) are satisfied $\alpha_1, \ldots, \alpha_n$ are the positive simple roots of $SU(n+1)$, $-\alpha_0 = \alpha_1 + \ldots + \alpha_n$ is the highest root, and the coefficients of the exponentials in (1.4) are the non-zero entries in the Cartan matrix for $SU(n+1)$. As we shall see in Section 2, this formulation also arises naturally in a geometrical context.

In particular, in the case $n = 1$, if (1.3) holds then writing $\omega = \omega_1 - \omega_0$, Equation (1.4) reduces to the sinh-Gordon equation

$$\frac{\partial^2 \omega}{\partial z \partial \bar{z}} + 2\sinh 2\omega = 0. \tag{1.5}$$

As is well-known (see [37],[4], [32], and [31] for a brief history with references) this equation arises geometrically in the study of surfaces of constant mean curvature 2 in $\mathbb{R}^3$ where, in terms of a local isothermal coordinate system (x, y) whose coordinate curves are lines of curvature, the metric on the surface is given by

$$ds^2 = 4e^{2\omega}|dz|^2, \qquad z = x + iy. \tag{1.6}$$

The relationship with harmonic maps comes from the fact that the Gauss map of an oriented surface in $\mathbb{R}^3$ is harmonic (but not holomorphic or antiholomorphic) if and only if the surface has constant mean curvature (but is neither part of a sphere nor a minimal surface). Equation (1.5) is then the condition for the Gauss map to be harmonic so that, at least locally, solutions correspond to surfaces of constant mean curvature with metric given by (1.6). The crucial observation in the above is that there is a 1-1 correspondence between solutions of the affine Toda equations for $SU(2)$ (or equivalently $SO(3)$ and harmonic maps into S^2, with the local complex coordinate z and the unknown ω in (1.5) having geometrical significance. Using this and soliton theoretic methods for solving (1.5), Pinkall-Sterling [32] were able to describe all CMC-tori in $\mathbb{R}^3$.

These ideas were extended in [20] to the case of conformal minimal (but not pseudo-holomorphic) surfaces in S^4. In this case the choice of complex coordinate is dictated by conformality and a condition on the second fundamental form, and it turns out that there is a correspondence between the local theory of such immersions and the affine Toda equations for $SO(5)$. We note that $SO(5)$ is of rank 2 and the Toda equations in this case involve two functions, namely ω which corresponds to the induced metric and η which measures the eccentricity of the ellipse of curvature. Again, using soliton theoretic methods it is possible to describe all conformal minimal (but not pseudo-holomorphic) tori in S^4.

The strategy we adopt for arriving at a geometrical context in which the affine Toda equations for an arbitrary compact simple Lie group G arise has two steps:

Step(A) Show that solutions of (1.1) on a simply connected domain in $\mathbb{C}$ correspond to τ-primitive maps (defined in Section 3) of a suitable open subset of a Riemann surface into G/T, where T is the maximal torus of G. These are maps which are harmonic with respect to any G-invariant metric on G/T and are discussed in detail in [2], where it is shown that τ-primitive maps into G/T project to harmonic maps into G/K, where K is a subgroup of G of maximal rank.

Step(B) Identify geometrically the type of harmonic map into G/K arising in Step(A) and relate the unknowns in (1.1) to geometric invariants.

We indicate how to achieve Step(A) for general G, and Step(B) for $G = SU(n+1)$, $SO(2m+1)$ and G_2, where solutions of the affine G-Toda equations correspond respectively to superconformal harmonic maps into $\mathbb{C}P^n$ and S^{2m} and almost complex curves in the nearly Kähler S^6 which are not pseudo-holomorphic.

We begin our discussion in Section 2 with the case of $G = SU(n+1)$ which is particularly simple and illuminating. In this case, using harmonic sequences, one can see clearly the ways in which the Toda equations, τ-primitive maps into $SU(n+1)/T^n$ and orthogonally periodic harmonic sequences (whose elements are, by definition, superconformal harmonic maps into $\mathbb{C}P^n$) are interrelated.

Then in Section 3 we describe Step(A) in the general case.

In Section 4 we adapt the ideas of Section 2 to deal with Step(B) for the case of $G = SO(2m+1)$ and also explain how the m unknowns in the affine Toda equations are directly related to invariants of the higher order fundamental forms for the corresponding minimal immersion in S^{2m}.

Finally in Section 5 we make some brief remarks on how the ideas of Section 4 may be further adapted to relate the solutions of the affine Toda equations for G_2 to almost complex curves in S^6.

2 The $SU(n+1)$-case

After some introductory remarks, we discuss the relation between solutions of the affine Toda equations for $SU(n+1)$ and certain harmonic maps, called *superconformal* harmonic maps, of a connected Riemann surface S into $\mathbb{C}P^n$. This relationship is summarised in Theorems 2.2 and 2.3.

We then discuss superconformal harmonic maps from a twistorial point of view and, in Theorem 2.5, show how they correspond to maps of a certain type, called τ-*primitive*, into the flag manifold $SU(n+1)/T^n$. Finally, we complete the picture by producing in Theorem 2.6 a correspondence between τ-primitive maps into $SU(n+1)/T^n$ and solutions of the affine Toda equations for $SU(n+1)$. It is this latter correspondence which we discuss in a more general setting in Section 3. We wish to emphasize in the present section the way in which a superconformal harmonic map into $\mathbb{C}P^n$ (and hence a τ-primitive map into $SU(n+1)/T^n$) determines special complex coordinate systems on the Riemann surface S. This enables us to obtain a natural correspondence between solutions of the affine $SU(n+1)$-Toda equations on an open subset of $\mathbb{C}$ and certain harmonic maps on open subsets of S.

The simple group $SU(n+1)$ has rank n and, as maximal torus, we take

$$T^n = \{\operatorname{diag}(e^{i\omega_0},\ldots,e^{i\omega_n}) \,|\, \omega_0,\ldots,\omega_n \in \mathbb{R}, \ \omega_0+\ldots+\omega_n = 0\}.$$

The Lie algebra t^n of T^n is then given by

$$t^n = \{\operatorname{diag}(i\omega_0,\ldots,i\omega_n) \,|\, \omega_0,\ldots,\omega_n \in \mathbb{R}, \ \omega_0+\ldots+\omega_n = 0\}.$$

and the affine Toda equations (1.1) may be written in terms of $\omega_0,\ldots,\omega_n$ as (1.2) subject to (1.3).

We will discuss the relationship between solutions of these equations and certain harmonic maps into $\mathbb{C}P^n$. This is achieved using the harmonic sequence, which we now describe very briefly, see also [37]. For simplicity we will give a local description, and we will ignore problems arising from singularities. However, a more detailed account of harmonic sequences and their properties may be found in [9].

Let $\phi: S \to \mathbb{C}P^n$ be a harmonic map. Then the *harmonic sequence* $\{\phi_p\}$ associated to ϕ consists of harmonic maps $\phi_p: S \to \mathbb{C}P^n$ with $\phi_0 = \phi$, where $\phi_p = [f_p]$ with f_p a locally defined $\mathbb{C}^{n+1}$-valued function, satisfying

$$\begin{aligned}
\langle f_{p+1}, f_p \rangle &= 0, \\
\frac{\partial f_p}{\partial z} &= f_{p+1} + \frac{\partial}{\partial z}\log|f_p|^2 \, f_p, \\
\frac{\partial f_p}{\partial \bar{z}} &= -\frac{|f_p|^2}{|f_{p-1}|^2} f_{p-1}.
\end{aligned} \qquad (2.1)$$

In fact for a given local coordinate z on S the sequence $\{f_p\}$ is characterized up to multiplication by a meromorphic function by (2.1) and $\phi = [f_0]$. For later use we note that each map ϕ_p determines a complex line subbundle L_p of the trivial bundle $S \times \mathbb{C}^{n+1}$, where

The affine Toda equations and minimal surfaces 63

$$L_p = \{(x,v) \in S \times \mathbb{C}^{n+1} \mid v \in \phi_p(x)\}.$$

Then each L_p inherits the structure of a holomorphic vector bundle from its inclusion in $S \times \mathbb{C}^{n+1}$ and f_p is a local meromorphic section of L_p.

It is easy to check [9] that the integrability conditions

$$\frac{\partial^2 f_p}{\partial z \partial \bar{z}} = \frac{\partial^2 f_p}{\partial \bar{z} \partial z}$$

correspond to

$$\frac{\partial^2}{\partial z \partial \bar{z}} \log |f_p|^2 = \frac{|f_{p+1}|^2}{|f_p|^2} - \frac{|f_p|^2}{|f_{p-1}|^2}. \tag{2.2}$$

Writing

$$|f_p| = e^{\omega_p}, \tag{2.3}$$

we see that (2.2) and (1.2) are equivalent.

Since [5] the forms $(|f_p|^2/|f_{p-1}|^2)|dz|^2$ are globally defined, it is more natural to rewrite (2.2) in terms of the quotients $|f_p|^2/|f_{p-1}|^2$, in which case using (2.3) we obtain (1.4).

The simplest case is that in which the harmonic sequence has finite length and thus reduces to the Frenet frame of a holomorphic curve. The harmonic map is then said to be *pseudo-holomorphic* [16],[9], *superminimal* [36] or *complex isotropic* [19]. Such maps may be studied using holomorphic data, and they are relatively well-understood (see for example [19],[5]). It is well known [17],[19] that all harmonic maps from S^2 to $\mathbb{C}P^n$ are pseudo-holomorphic.

Assuming ϕ is not pseudo-holomorphic, the harmonic sequence is infinite, with ϕ_p being defined for all $p \in \mathbb{Z}$. In this case the corresponding equations (1.2) are precisely the Toda equations associated to the infinite dimensional Lie algebra A_∞. Even in the special case where the sequence of functions $\{\omega_p\}$ satisfies (1.3), it turns out that the harmonic map ϕ may not be recovered by just solving (1.2) subject to the constraint (1.3) since [9] other invariants are needed. However, corresponding to solutions of the pair of equations (1.2) and (1.3), it turns out that there is a natural choice of harmonic sequence, namely that for which any $n+1$ consecutive elements are mutually orthogonal. Here two maps from S into $\mathbb{C}P^n$ will be called *orthogonal* if the lines they determine in $\mathbb{C}^{n+1}$ are everywhere orthogonal. In order to describe this situation we recall an orthogonality property of harmonic sequences, whose proof may be found in [15] or [9].

Lemma 2.1. *If some k consecutive maps in a harmonic sequence are mutually orthogonal then every k consecutive maps of this sequence are mutually orthogonal.*

We will say that a harmonic map $\phi: S \to \mathbb{C}P^n$ is *k-orthogonal* (or *of isotropy order* $\geq k - 1$ in the terminology of [15]) if k (or more) consecutive maps in the harmonic sequence of ϕ are mutually orthogonal. In particular [9], ϕ is conformal if and only if ϕ is 3-orthogonal and, in all cases, ϕ is at most $(n + 1)$-orthogonal. There are two possible ways in which this latter case can arise. If ϕ is pseudo-holomorphic and linearly full then ϕ is $(n + 1)$-orthogonal since the harmonic sequence is just the Frenet frame of a linearly full holomorphic curve. If ϕ is $(n+1)$-orthogonal but not pseudo-holomorphic then it follows from Lemma 2.1 that the harmonic sequence of ϕ is orthogonally periodic with period $n + 1$, that is to say that $\phi_0, \ldots, \phi_n$ are mutually orthogonal and $\phi_{p+n+1} = \phi_p$ for all $p \in \mathbb{Z}$. In this case we say that ϕ is *superconformal*.

We note that every harmonic map $\phi: S \to \mathbb{C}P^1$ and every conformal harmonic map $\phi: S \to \mathbb{C}P^2$ is either pseudo-holomorphic or superconformal.

We now show how a superconformal harmonic map $\phi: S \to \mathbb{C}P^n$ determines, in a natural way, local coordinate systems and sequences of functions $\{f_p\}$ such that the corresponding functions $\{\omega_p\}$ given by (2.3) are solutions of the A_∞-Toda equations (1.2) which also satisfy (1.3). It then follows that $\omega_0, \ldots, \omega_n$ give solutions of the affine $SU(n+1)$-Toda equations. To do this we first pick any local complex coordinate z on a simply connected open subset U of S and let $\{f_p\}$ be a sequence of functions on U satisfying (2.1) with $\phi = [f_0]$. Then f_0 and f_{n+1} are both meromorphic sections over U of the subbundle L_0 of $S \times \mathbb{C}^{n+1}$ determined by $\phi = \phi_0$. Thus

$$f_{n+1} = \alpha f_0$$

for some meromorphic function α, which is easily seen to be independent of the choice of local meromorphic section f_0 of L_0. The non-removable singularities of α occur exactly at the higher order singularities [9] of ϕ, and away from these isolated points we may change coordinates so that $\alpha = 1$. In this case, the sequence $\{f_p\}$ is itself orthogonally periodic in the sense that $f_0, \ldots, f_n$ are mutually orthogonal and $f_{p+n+1} = f_p$ for all $p \in \mathbb{Z}$. Further, assuming that the domain of our new complex coordinate is also simply connected, it follows from (2.1) that $\det(f_0, \ldots, f_n)$ is a holomorphic function which we may take to be identically 1 by multiplying f_0 by a suitable holomorphic function. This gives the following theorem.

Theorem 2.2. *Let $\phi: S \to \mathbb{C}P^n$ be a superconformal harmonic map. Then the higher order singularities of ϕ occur at isolated points. If ϕ has no higher order singularities at $x \in S$ then there exists a local complex coordinate z around x such that for any meromorphic local section f_0 of L_0,*

$$f_{p+n+1} = f_p, \qquad p \in \mathbb{Z}.$$

Moreover, this coordinate system is unique up to arbitrary translations and rotation by an $(n + 1)$st root of unity. Further, f_0 may be chosen so that if $|f_p| = e^{\omega_p}$ then $\omega_0, \ldots, \omega_n$ give a solution to the affine $SU(n + 1)$-Toda equations (1.2) and (1.3).

A local complex coordinate z on a simply connected open subset U of S with the above property will be said to be *special*. We note, in particular, that special coordinates are essentially unique. For such a coordinate system z, if we pick a holomorphic nowhere vanishing section f_0 of L_0 over U then each f_p is a holomorphic nowhere vanishing section of L_p over U. Hence, if we write $F_p = e^{-\omega_p} f_p$, where $|f_p| = e^{\omega_p}$, then (2.1) gives

$$\frac{\partial F_p}{\partial z} = e^{\omega_{p+1} - \omega_p} F_{p+1} + \frac{\partial \omega_p}{\partial z} F_p, \qquad (2.4)$$

$$\frac{\partial F_p}{\partial \bar{z}} = -e^{\omega_p - \omega_{p-1}} F_{p-1} - \frac{\partial \omega_p}{\partial \bar{z}} F_p, \qquad (2.5)$$

$$F_{n+p+1} = F_p. \qquad (2.6)$$

We observe that, if $F_0, \ldots, F_n$ satisfy (2.4), (2.5) and (2.6) then, writing $F = (F_0, \ldots, F_n)$, we have

$$F^\dagger F = \text{constant}. \qquad (2.7)$$

Indeed, given (2.6), we see that any two of (2.4), (2.5) and (2.7) imply the third. If we choose f_0 so that $\det(f_0, \ldots, f_n) = 1$, we then have a map

$$F = (F_0, \ldots, F_n) : U \to SU(n+1), \qquad (2.8)$$

and (2.4) may be written in terms of the $su(n+1)^\mathbb{C}$-valued function $F^{-1} \frac{\partial F}{\partial z}$ as

$$F^{-1} \frac{\partial F}{\partial z} = \begin{pmatrix} \frac{\partial \omega_0}{\partial z} & & & & e^{\omega_0 - \omega_n} \\ e^{\omega_1 - \omega_0} & \frac{\partial \omega_1}{\partial z} & & & \\ & e^{\omega_2 - \omega_1} & \ddots & & \\ & & \ddots & \frac{\partial \omega_{n-1}}{\partial z} & \\ & & & e^{\omega_n - \omega_{n-1}} & \frac{\partial \omega_n}{\partial z} \end{pmatrix}. \qquad (2.9)$$

Alternatively, we may write

$$F^{-1} \frac{\partial F}{\partial z} = \frac{\partial \Omega}{\partial z} + e^{\Omega} B e^{-\Omega}, \qquad (2.10)$$

where $\Omega = \text{diag}(\omega_0, \ldots, \omega_n)$ and B is the permutation matrix corresponding to the $(n+1)$-cycle $(0 \ldots n)$, so that B has entry 1 in the $(i+1, i)$th place for $i = 0, \ldots, n-1$, entry 1 in the $(0, n)$th place, and zeros elsewhere.

The integrability condition

$$\frac{\partial^2 F}{\partial z \partial \bar{z}} = \frac{\partial^2 F}{\partial \bar{z} \partial z}$$

for the system (2.4), (2.5) to have a solution (or equivalently for (2.9) to have a solution arising from an $SU(n+1)$-valued map F) is easily seen to be the condition that $\omega_0, \ldots, \omega_n$ should be the set of solutions of the affine $SU(n+1)$-Toda system (1.2), (1.3).

Hence we have the following converse to Theorem 2.2. The last statement is a consequence of the congruence result of [9].

Theorem 2.3. *Let z be a complex coordinate on a simply connected open subset U of a Riemann surface S and let $\omega_0, \ldots, \omega_n$ be a set of solutions of the affine Toda equations (1.2) and (1.3) for $SU(n+1)$. Then there exists a superconformal harmonic map $\phi: U \to \mathbb{C}P^n$ for which z is a special coordinate and $\omega_0, \ldots, \omega_n$ are the solutions determined by ϕ and z as in Theorem 2.2. Moreover, any two such harmonic maps ϕ are $PSU(n+1)$-congruent.*

We now consider superconformal harmonic maps from a twistorial point of view. Recall that the full flag manifold consisting of $(n+1)$-tuples $(X_0, \ldots, X_n)$ of mutually orthogonal complex lines through the origin in $\mathbb{C}^{n+1}$ is a homogeneous Kähler manifold which may be identified with $SU(n+1)/T^n$, and that each of the projections $\pi_k: SU(n+1)/T^n \to \mathbb{C}P^n$ given by $\pi_k(X_0, \ldots, X_n) = X_k, k = 0, \ldots n$, is a Riemannian submersion. If $\phi: S \to \mathbb{C}P^n$ is a harmonic map which is linearly full and pseudo-holomorphic then there is a lift $\tilde{\phi}: S \to SU(n+1)/T^n$ given by the corresponding Frenet frame, such that if $\phi = \phi_k$ is the $(k+1)$-st element of the Frenet frame then $\phi_k = \pi_k \tilde{\phi}$. Moreover, $\tilde{\phi}$ is holomorphic (but not horizontal).

Similarly, if $\phi: S \to \mathbb{C}P^n$ is a superconformal harmonic map then there is again a lift $\tilde{\phi} = (L_0, \ldots, L_n): S \to SU(n+1)/T^n$ given by the corresponding orthogonally periodic harmonic sequence, so that if $\phi = \phi_0$ then $\phi = \pi\tilde{\phi}$, where from now on we write π in place of π_0. In this case, however, $\tilde{\phi}$ is neither holomorphic nor horizontal, but does have nice properties of a twistorial nature as we now show.

In order to discuss these properties we first define an $SU(n+1)$-invariant subbundle of the complexified tangent bundle $T(SU(n+1)/T^n)^{\mathbb{C}}$. The complexification of $su(n+1)$ is equal to $sl(n+1, \mathbb{C})$ and we let $\mathcal{M}_1$ be the subspace of $sl(n+1, \mathbb{C})$ given by

$$\mathcal{M}_1 = \left\{ \begin{pmatrix} & & & a_{0n} \\ a_{10} & & & \\ & \ddots & & \\ & & a_{nn-1} & \end{pmatrix} \mid a_{10}, \ldots, a_{nn-1}, a_{0n} \in \mathbb{C} \right\}. \quad (2.11)$$

Then $\mathcal{M}_1$ is AdT^n-invariant and determines, in a natural way, an $SU(n+1)$-invariant subbundle $[\mathcal{M}_1]$ of $T(SU(n+1)/T^n)^{\mathbb{C}}$. An element (a_{ij}) of $\mathcal{M}_1$ is said to be *cyclic* if $a_{10}, \ldots, a_{0n}$ are all non-zero. Observe that the cyclic elements form an AdT^n-invariant subset of $\mathcal{M}_1$. An element of $[\mathcal{M}_1]$ is thus said to be *cyclic* if it is obtained from a cyclic element of $\mathcal{M}_1$ via the action of $SU(n+1)$.

A map $\psi: S \to SU(n+1)/T^n$ is said to be τ-*primitive* (or just *primitive* [14]) if

(a) $d\psi(\partial/\partial z) \in [\mathcal{M}_1]$;

(b) $d\psi(\partial/\partial z)$ is cyclic except at an isolated set of points (referred to henceforth as the set of *singular points* of ψ).

The notation is chosen to accord with that of the general theory discussed in Section 3. These maps are also discussed in [12],[30].

The affine Toda equations and minimal surfaces   67

Lemma 2.4. *Let $\hat{\pi}: SU(n + 1) \to SU(n + 1)/T^n$ be the projection map and let $F = (F_0, \ldots, F_n): S \to SU(n+1)$. If $\psi = \hat{\pi} F$ then $d\psi(\partial/\partial z) \in [\mathcal{M}_1]$ if and only if $F^{-1}\partial F/\partial z \in \mathfrak{t}^\mathbb{C} \oplus \mathcal{M}_1$, in which case $d\psi(\partial/\partial z)$ is cyclic if and only if, for each $p = 0, \ldots, n$, $\partial F_p/\partial z$ is not a scalar multiple of F_p.*

The following theorem describes the correspondence between superconformal harmonic maps and τ-primitive maps.

Theorem 2.5. *Suppose that $\phi: S \to \mathbb{C}P^n$ is a superconformal harmonic map. Then the lift $\widetilde{\phi}: S \to SU(n + 1)/T^n$ given by*

$$\widetilde{\phi} = (L_0, \ldots, L_n)$$

is τ-primitive. Conversely, if $\psi: S \to SU(n+1)/T^n$ is τ-primitive then $\phi = \pi\psi: S \to \mathbb{C}P^n$ is a superconformal harmonic map with corresponding lift $\widetilde{\phi}$ equal to ψ.

Let $\phi: S \to \mathbb{C}P^n$ be a superconformal harmonic map and let $F: U \to SU(n+1)$ be the map given by (2.8) which is defined by ϕ and a special coordinate z on U. Then $\widetilde{\phi} = \hat{\pi} F$, so the fact that $\widetilde{\phi}$ is τ-primitive follows from (2.9) and Lemma 2.4.

Conversely, if $\psi: S \to SU(n+1)/T^n$ is τ-primitive, it follows from general results of [2] that $\phi = \pi\psi$ is harmonic (although this may also be proved in an elementary manner in this special case [6]). The final statement of the theorem now follows from the method of construction of the harmonic sequence of ϕ. □

We note that, in the correspondence of Theorem 2.5, $x_0 \in S$ is a singular point of $\widetilde{\phi}$ (in the sense that $d\widetilde{\phi}_{x_0}(\partial/\partial z)$ is not a cyclic element) if and only if x_0 is a higher order singularity of the superconformal harmonic map ϕ.

We now explicitly write down the relation between τ-primitive maps into $SU(n+1)/T^n$ and solutions of the affine Toda equations for $SU(n+1)$. It is this relationship which we generalize in the next section to the case of an arbitrary simple Lie group G, but in the special case of $SU(n+1)$ the result follows immediately from Theorem 2.5 and the discussion following Theorem 2.2.

Theorem 2.6. *Let $\psi: S \to SU(n + 1)/T^n$ be τ-primitive and assume that $x_0 \in S$ is not a singular point of ψ. Then there is a local complex coordinate z about x_0, a local lift $F = (F_0, \ldots, F_n)$ of ψ into $SU(n+1)$ and smooth functions $\omega_0, \ldots, \omega_n$ such that $\omega_0 + \ldots + \omega_n = 0$ and F satisfies (2.10). Moreover, the integrability conditions for the existence of $SU(n+1)$-valued solutions of (2.10) are precisely the affine Toda equations (1.2) and (1.3) for $SU(n+1)$.*

We end this section by making some remarks concerning tori. Let T be a 2-torus with universal cover $\widetilde{\pi}: \mathbb{C} \to T$. It may be shown [36] that if $\phi: T \to \mathbb{C}P^n$ is a superconformal harmonic map then ϕ has no higher order singularities. Moreover [6], there is a linear complex coordinate on $\mathbb{C}$ and a globally defined lift $F: \mathbb{C} \to SU(n + 1)$ of $\widetilde{\phi}\widetilde{\pi}$ satisfying (2.10) for some solution $\omega_0, \ldots, \omega_n$ of the affine Toda equations for $SU(n+1)$ defined on $\mathbb{C}$. In fact, F factors through a finite cover $\tilde{T}$ of T while the corresponding solution of the affine Toda equations factors through T. It is known [6] that all such solutions are of *finite type*, that is to say that they may be obtained by integrating a particular pair of commuting Hamiltonian vector fields on a finite dimensional subspace of a twisted loop algebra of the Lie algebra of G (where $G = SU(n+1)$ in this case). (See [32], [20],[6],[13] for further details). This leads to the following generalization of a result of [32].

Corollary 2.7. *Every superconformal harmonic torus in $\mathbb{C}P^n$ is of finite type. In particular, every harmonic torus in $\mathbb{C}P^1$ and every conformal harmonic torus in $\mathbb{C}P^2$ is either pseudo-holomorphic or superconformal of finite type.*

We remark that the functions $\omega_0, \ldots, \omega_n$ arising from a solution of the affine Toda equations for $SU(n+1)$ have a natural interpretation in terms of geometric invariants of the harmonic sequence $\{\phi_p\}$ of the corresponding superconformal harmonic map $\phi = \phi_0$ into $\mathbb{C}P^n$. Indeed, as shown in [9], the metric ds_p^2 induced on S by ϕ_p is given by

$$ds_p^2 = (e^{2(\omega_p - \omega_{p-1})} + e^{2(\omega_{p+1} - \omega_p)})|dz|^2,$$

while the Kähler angle θ_p of ϕ_p is given by

$$\tan^2(\theta_p/2) = e^{-2\omega_{p-1} + 4\omega_p - 2\omega_{p+1}}.$$

Finally, we remark (cf. [25],[10]) that the superconformal harmonic maps of $\mathbb{C}$ into $\mathbb{C}P^n$ corresponding to the zero solution of (1.2) and (1.3), which are given (up to holomorphic isometries of $\mathbb{C}P^n$) by

$$\phi(z) = [e^{z-\bar{z}}, \ldots, e^{\zeta^n z - \overline{\zeta^n z}}], \qquad \zeta = e^{\frac{2\pi i}{n+1}}, \qquad (2.12)$$

are flat and have constant Kähler angle $\pi/2$. Furthermore, it is not hard to show that ϕ factors through a torus if and only if $n = 1, 2, 3$, or 5. In the cases $n = 2, 3$, and 5 these are isometric embeddings of tori whose corresponding lattice in $\mathbb{C}$ is generated by $2\pi/\sqrt{3}$ and $2\pi i$ when $n = 2, 5$, and by π and πi when $n = 3$. (The metric on $\mathbb{C}$ here is given by $ds^2 = 2|dz|^2$). Apart from the $\mathbb{C}P^1$ case, these are the only examples of superconformal harmonic tori in $\mathbb{C}P^n$ known to us.

3 Toda equations and τ-primitive maps

In this section we discuss the generalization of Theorem 2.6 from the case of $SU(n+1)$ to that of an arbitrary compact simple Lie group G with maximal torus T. The approach is more Lie algebraic in flavour since we do not have the geometry of $\mathbb{C}P^n$ to help us in the general case.

We first discuss the general notion of a τ-primitive map from a Riemann surface S to G/T and indicate how such a map determines locally an essentially unique complex coordinate on S. We then discuss the correspondence between the local theory of such maps and solutions of the affine Toda equations for G.

The affine Toda equations and minimal surfaces

We begin by recalling some basic notions about Lie algebras and the structure of G/T. If g denotes the Lie algebra of G with t the Cartan subalgebra corresponding to T, we let $g^C = t^C \oplus \bigoplus_{\alpha \in \Delta} g^\alpha$ be the root space decomposition. We fix (once and for all) a Cartan-Weyl basis $\xi_\alpha \in g^\alpha, \alpha \in \Delta$, of the root spaces. In particular, we have the following relations:

$$\begin{aligned} \overline{\xi}_\alpha &= \xi_{-\alpha}, \\ [\xi_\alpha, \xi_{-\alpha}] &= \alpha^\#, \\ (\xi_\alpha, \xi_\beta) &= \delta_{\alpha,-\beta}, \end{aligned} \qquad (3.1)$$

where $\#: t^{*C} \to t^C$ is the isomorphism induced by the Killing form $(\,,\,)$ on g and defined by

$$\alpha(v) = (\alpha^\#, v), \qquad \alpha \in t^{*C}, \quad v \in t^C.$$

We also fix a set of simple roots $\alpha_1, \ldots, \alpha_\ell$, $\ell = \text{rank } G$. If $\theta = \sum_{p=1}^\ell m_p \alpha_p$, $m_p \in \mathbb{Z}^+$, denotes the highest root we define α_0 by $\alpha_0 = -\theta$. Then we define

$$\mathcal{M}_1 = \bigoplus_{p=0}^\ell g^{\alpha_p} \qquad (3.2)$$

as the sum of the simple root spaces and the root space corresponding to α_0.

There is an interesting alternative description of $\mathcal{M}_1$ using the canonical m-symmetric space structure [24] on G/T which we give in brief. The simple roots and the height of the highest root θ determine, up to conjugation, an inner automorphism $\tau: G \to G$ of order $m = \sum_{p=0}^\ell m_p$ [22], where by definition $m_0 = 1$. The differential of τ at the identity of G, which we also denote by τ, is an automorphism of g which acts on $\mathcal{M}_1$ via multiplication by $\zeta = e^{\frac{2\pi i}{m}}$. Explicitly we have $\tau = \text{Ad}\exp(2\pi i Z)$, with $Z = \frac{1}{m}\sum_{p=1}^\ell \eta_p$ where each $\eta_p \in it$ and $\alpha_q(\eta_p) = \delta_{qp}$. Note that the fixed point set G^τ of τ is the maximal torus T of G. In fact, τ is the automorphism of smallest order m which exhibits G/T as an m-symmetric space [24]. This gives us the $\mathbb{Z}_m$-grading

$$g^C = \bigoplus_{k \in \mathbb{Z}_m} \mathcal{M}_k, \qquad (3.3)$$

where $\mathcal{M}_k = \overline{\mathcal{M}}_{-k}$ is the ζ^k-eigenspace of τ on g^C. Each $\mathcal{M}_k$ is AdT-invariant and determines, in a natural way, a G-invariant subbundle $[\mathcal{M}_k]$ of the complexified tangent bundle $T(G/T)^C$ of G/T. Thus we have the identification

$$T(G/T)^C = \bigoplus_{k \in \mathbb{Z}_m \setminus \{0\}} [\mathcal{M}_k] \subset G \times_T g^C \qquad (3.4)$$

of $T(G/T)^C$ as a direct sum of G-invariant subbundles of $G \times_T g^C$, where T acts on G by right multiplication and on g^C via the adjoint representation.

Following [26], we say that an element $\xi \in \mathcal{M}_1$ is *cyclic* if

$$\xi = \sum_{p=0}^{\ell} a_p \xi_{\alpha_p} \qquad (3.5)$$

with each $a_p \in \mathbb{C}\setminus\{0\}$. Observe that the cyclic elements form an AdT-invariant subset of $\mathcal{M}_1$. An element of $[\mathcal{M}_1]$ will be said to be cyclic if it is obtained from a cyclic element of $\mathcal{M}_1$ via the action of G.

A map $\psi: S \to G/T$ is said to be τ-*primitive* if $d\psi(\partial/\partial z) \in [\mathcal{M}_1]$ and is cyclic at some point of S. A standard holomorphicity argument shows that, in this case, $d\psi(\partial/\partial z)$ is cyclic everywhere except on an isolated set of points.

It is shown in [2] that τ-primitive maps are equiharmonic, that is to say harmonic with respect to any G-invariant metric on G/T. Moreover, they project to equiharmonic maps under every homogeneous projection $\pi: G/T \to G/K$, where K is a subgroup of maximal rank.

Example. The group $SU(n+1)$ has rank n and a maximal torus T^n consists of the diagonal elements of $SU(n+1)$, so that $\mathfrak{t}$ is the space of purely imaginary diagonal matrices with zero trace. The complexification of $su(n+1)$ is $sl(n+1, \mathbb{C})$, and the roots are

$$\{\sigma_i - \sigma_j : i \neq j;\ i, j = 0, \ldots, n\}$$

where

$$\sigma_i(\text{diag}(y_0, \ldots, y_n)) = y_i, \qquad i = 0, \ldots, n.$$

We take

$$\{\sigma_i - \sigma_j \mid i > j;\ i, j = 0, \ldots, n\}$$

to be the positive roots, and $\alpha_p = \sigma_p - \sigma_{p-1}$, $p = 1, \ldots, n$, as the positive simple roots. Then $\theta = \sigma_n - \sigma_0 = \sum_{p=1}^{n} \alpha_p$ is the highest root, so we take $\alpha_0 = -\theta$. For each pair $(i, j), i, j \in \{0, \ldots, n\}, i \neq j$, let E_{ij} be the element of $sl(n+1, \mathbb{C})$ whose only non-zero entry is a 1 in the (i,j)-th place. Then E_{ij} spans the root space corresponding to $\sigma_i - \sigma_j$ and

$$\{E_{ij} : i \neq j;\ i, j = 0, \ldots, n\}$$

is a Cartan-Weyl basis. Hence, in the notation above, $\mathcal{M}_1$ is the vector subspace of $sl(n+1, \mathbb{C})$ spanned by $E_{1\,0}, \ldots, E_{n\,n-1}, E_{0\,n}$. Thus the elements of $\mathcal{M}_1$ are of the form $\sum_{i \in \mathbb{Z}_{n+1}} a_{i\,i-1} E_{i\,i-1}$, where $a_{1\,0}, \ldots, a_{n\,n-1}, a_{0\,n} \in \mathbb{C}$. Moreover, the cyclic elements are those in $\mathcal{M}_1$ for which each of $a_{1\,0}, \ldots, a_{n\,n-1}, a_{0\,n}$ is non-zero. This agrees with the situation described in Section 2, so that the superconformal maps into $\mathbb{C}P^n = SU(n+1)/S(U(1) \times U(n))$ are the harmonic maps corresponding to τ-primitive maps into $SU(n+1)$. In Sections 4 and 5 we discuss those harmonic maps into $S^{2m} = SO(2m+1)/SO(2m)$ and $S^6 = G_2/SU(3)$ which arise from τ-primitive maps into $SO(2m+1)/T$ and G_2/T respectively.

The affine Toda equations and minimal surfaces 71

Next we indicate how a τ-primitive map determines locally an essentially unique complex coordinate on S. The ring of $Ad\, G^{\mathbb{C}}$-invariant polynomials on $g^{\mathbb{C}}$ has homogeneous generators $P_1, \ldots, P_\ell$ with $\deg P_1 < \ldots < \deg P_\ell$. Moreover, by a result of [26], the cyclic elements in $\mathcal{M}_1$ are characterized as those on which P_ℓ does not vanish. The polynomials $P_1, \ldots, P_\ell$ extend to functions on $T(G/T)^{\mathbb{C}}$ and if $\psi: S \to G/T$ is τ-primitive then $P_\ell(d\psi(\partial/\partial z))$ may be shown to be holomorphic, and hence by a change of coordinate (away from those points where $d\psi(\partial/\partial z)$ is not cyclic) may be assumed to be any prescribed non-zero constant. Taking this to be $P_\ell(B)$, where $B = \sum_{p=0}^{\ell} \sqrt{m_p}\xi_{\alpha_p}$, this determines the coordinate z up to arbitrary translations and rotations by $\deg P_\ell$-th roots of unity. Such a coordinate will be called a *special complex coordinate*. In the case where $G = SU(n+1)$ the homogeneous polynomials $P_1, \ldots, P_n$ are the elementary symmetric functions of degree ≥ 2 of the eigenvalues and B is the permutation matrix of order $n+1$ referred to in Section 2.

By a (local) frame of a map $\psi: S \to G/T$ we mean a map $F: U \to G$ from an open subset $U \subset S$, with $\pi F = \psi$ where $\pi: G \to G/T$ is the homogeneous projection. Note that (cf. Lemma 2.4) ψ is τ-primitive if and only if

$$F^{-1}\frac{\partial F}{\partial z} = A_0 + A_1 \in \mathcal{M}_0 \oplus \mathcal{M}_1, \tag{3.6}$$

where $A_0 \in \mathcal{M}_0$, $A_1 \in \mathcal{M}_1$ and A_1 is cyclic at some point.

A local frame F will be called a *Toda frame* if there is a complex coordinate $z: U \to \mathbb{C}$ and a smooth map $\Omega: U \to it$ such that

$$F^{-1}\frac{\partial F}{\partial z} = \frac{\partial \Omega}{\partial z} + e^{\Omega}Be^{-\Omega} \in \mathcal{M}_0 \oplus \mathcal{M}_1, \tag{3.7}$$

where $B = \sum_{p=0}^{\ell} \sqrt{m_p}\xi_{\alpha_p} \in \mathcal{M}_1$.

The following theorem is proved in [6].

Theorem 3.1. *Let $\psi: S \to G/T$ be τ-primitive and let $x_0 \in S$ be a point such that $d\psi_{x_0}(\partial/\partial z)$ is cyclic. Then there is a local Toda frame F of ψ around x_0. Moreover, the complex coordinate z for the Toda frame is a special complex coordinate as described above, while the Toda frame F is unique up to multiplication by an element of the centre $Z(G)$ of G.*

Note that the corresponding result for $G = SU(n+1)$ (the first part of Theorem 2.6) in Section 2 was proved by using the harmonic sequence to explicitly produce the required special complex coordinate z. This is also done in the next section when the case of $SO(2n+1)$ is discussed.

We now consider the generalization of the final part of Theorem 2.6.

Theorem 3.2. *Let z be a complex coordinate on a simply connected open subset U of S and let $\Omega: U \to it$ be smooth. Then (3.7) has a G-valued solution F if and only if Ω satisfies the affine Toda equations (1.1) for G. In this case, $\psi = \pi F: U \to G/T$ is τ-primitive and F is a Toda framing of ψ.*

From (3.7), since F is G-valued, we have

$$F^{-1}\frac{\partial F}{\partial \bar{z}} = \bar{A}_0 + \bar{A}_1, \tag{3.8}$$

so that the integrability conditions for the existence of a G-valued solution of (3.7) are the Maurer-Cartan equations

$$\frac{\partial A_0}{\partial \bar{z}} - \frac{\partial \bar{A}_0}{\partial z} = [A_1, \bar{A}_1], \tag{3.9}$$

$$\frac{\partial A_1}{\partial \bar{z}} = [A_1, \bar{A}_0]. \tag{3.10}$$

It now follows from (3.1) that the Toda equations (1.1) are the integrability conditions for the existence of a G-valued solution of (3.7). The final statement of the theorem is clear from (3.7). □

As in the $SU(n+1)$ case, the above may be used to establish a global result for τ-primitive maps from a 2-torus. In fact, let T^2 be a 2-torus with universal cover $\tilde{\pi}: \mathbb{C} \to T^2$. If $\psi: T^2 \to G/T$ is a τ-primitive map then there is a linear complex coordinate z on $\mathbb{C}$ and a Toda frame $F: \mathbb{C} \to G$ which covers $\psi\tilde{\pi}$ for which (3.7) holds. Moreover, F factors through a finite cover $\tilde{T}^2$ of T^2 while the corresponding solution Ω of the affine Toda field equations (1.1) factors through T^2. This leads to the following result [6].

Theorem 3.3. *Every τ-primitive map $\psi: T^2 \to G/T$ is of finite type.*

4 The $SO(2m+1)$-case

In the previous section we discussed the relation between τ-primitive maps into G/T and solutions of the corresponding affine Toda equations for an arbitrary compact simple Lie group G. In this section we consider the case $G = SO(2m+1)$ and show that τ-primitive maps in this case correspond to a certain class, again called *superconformal*, of harmonic maps into S^{2m}. As in the $SU(n+1)$ case, the special complex coordinate systems arising from a τ-primitive map may be derived by using harmonic sequences. However, more care is needed in this case since the harmonic maps into S^{2m} are not necessarily linearly full, and the harmonic sequences involved are not orthogonally periodic in the linearly full case.

A maximal torus of $SO(2m+1)$ is given by

$$T^m = \{\text{diag}\,(1, R(\theta_1), \ldots, R(\theta_m)) \,|\, \theta_1, \ldots, \theta_m \in \mathbb{R}\}$$

where

The affine Toda equations and minimal surfaces

$$R(\theta) = \begin{pmatrix} \cos\theta & -\sin\theta \\ \sin\theta & \cos\theta \end{pmatrix}. \tag{4.1}$$

The Lie algebra t is thus given by

$$t = \{\text{diag}\,(0, \Theta(\omega_1), \ldots, \Theta(\omega_m))\,|\,\omega_1, \ldots, \omega_m \in \mathbb{R}\}$$

where

$$\Theta(\omega) = \begin{pmatrix} 0 & -\omega \\ \omega & 0 \end{pmatrix}.$$

Writing $\Omega: U \to it$ as

$$\Omega = -i\text{diag}\,(0, \Theta(\omega_1), \ldots, \Theta(\omega_{m-1}), \Theta(\eta)),$$

(the reason for using η here rather than ω_m will become clear later), the Toda equations (1.1) become

$$\frac{\partial^2 \omega_1}{\partial z \partial \bar{z}} + e^{2\omega_1} - e^{2(\omega_2 - \omega_1)} = 0$$

$$\frac{\partial^2 \omega_2}{\partial z \partial \bar{z}} + e^{2(\omega_2 - \omega_1)} - e^{2(\omega_3 - \omega_2)} = 0$$

$$\vdots \tag{4.2}$$

$$\frac{\partial^2 \omega_{m-2}}{\partial z \partial \bar{z}} + e^{2(\omega_{m-2} - \omega_{m-3})} - e^{2(\omega_{m-1} - \omega_{m-2})} = 0$$

$$\frac{\partial^2 \omega_{m-1}}{\partial z \partial \bar{z}} + e^{2(\omega_{m-1} - \omega_{m-2})} - \frac{1}{2} e^{2(\eta - \omega_{m-1})} - \frac{1}{2} e^{2(\eta + \omega_{m-1})} = 0$$

$$\frac{\partial^2 \eta}{\partial z \partial \bar{z}} + \frac{1}{2} e^{2(\eta - \omega_{m-1})} - \frac{1}{2} e^{2(\eta + \omega_{m-1})} = 0.$$

We first describe the subspace $\mathcal{M}_1$ of $so(2m+1)^\mathbb{C}$ which is needed to characterize τ-primitive maps into $SO(2m+1)/T^m$. For each $p = 1, \ldots, m$, let $\sigma_p: t \to i\mathbb{R}$ be defined by $\sigma_p(\text{diag}(0, \Theta(\omega_1), \ldots, \Theta(\omega_m))) = i\omega_p$. Then as positive simple roots we take $\sigma_1, \sigma_2 - \sigma_1, \ldots, \sigma_m - \sigma_{m-1}$. The highest root is then given by

$$\sigma_m + \sigma_{m-1}$$

$$= 2(\sigma_1 + (\sigma_2 - \sigma_1) + (\sigma_3 - \sigma_2) + \cdots + (\sigma_{m-1} - \sigma_{m-2})) + \sigma_m - \sigma_{m-1}.$$

In order to describe the subspace $\mathcal{M}_1$ of $so(2m+1)^{\mathbb{C}}$ we first define $u_0, \ldots, u_m$ to be the unit vectors in $\mathbb{C}^{2m+1} = \mathbb{R}^{2m+1} \otimes \mathbb{C}$ given by

$$u_0 = e_0, \quad u_p = \frac{1}{\sqrt{2}}(e_{2p-1} - ie_{2p}), \; p = 1, \ldots, m.$$

Then $u_0, u_1, \ldots, u_m, \bar{u}_1, \ldots, \bar{u}_m$ is a basis of $\mathbb{C}^{2m+1}$. Let $V_0, \ldots, V_m \in so(2m+1)^{\mathbb{C}}$ be defined as follows:

$$\begin{aligned}
V_p u_{p-1} &= u_p, & V_p u_\ell &= 0, & \ell &\neq p-1, \; p = 1, \ldots, m, \\
V_p \bar{u}_p &= -\bar{u}_{p-1}, & V_p \bar{u}_\ell &= 0, & \ell &\neq p, \; p = 1, \ldots, m,
\end{aligned}$$

$$\begin{aligned}
V_0 u_\ell &= 0, & \ell &= 0, \ldots, m-2, \\
V_0 \bar{u}_\ell &= 0, & \ell &= 1, \ldots, m, \\
V_0 u_{m-1} &= \bar{u}_m, & V_0 u_m &= -\bar{u}_{m-1}.
\end{aligned}$$

Then V_1 spans the eigenspace $\mathcal{A}_1$ corresponding to σ_1, V_p spans the eigenspace $\mathcal{A}_p$ corresponding to $\sigma_p - \sigma_{p-1}$ for $p = 2, \ldots, m$, and V_0 spans the eigenspace $\mathcal{A}_0$ corresponding to $-(\sigma_m + \sigma_{m-1})$. Then from (3.2)

$$\mathcal{M}_1 = \bigoplus_{p=0}^{m} \mathcal{A}_p.$$

We now describe a class of harmonic maps from S to S^{2m} which, (see Theorem 4.3 below), correspond to τ-primitive maps into $SO(2m+1)/T^m$. Suppose $f: S \to S^n$ is a smooth map from a connected Riemann surface S, and let $\pi: S^n \to \mathbb{R}P^n$ denote the standard Riemannian double covering and $\iota: \mathbb{R}P^n \hookrightarrow \mathbb{C}P^n$ the standard isometric totally geodesic inclusion. Then $\phi = \iota \pi f: S \to \mathbb{C}P^n$ is harmonic if and only if f is harmonic. (Also ϕ is linearly full if and only if f is linearly full). As in Section 2, we may construct the harmonic sequence of ϕ but now refer to this as the harmonic sequence of f. In this case, $f_0 = f$ is a global holomorphic section of L_0 and $L_{-p} = \bar{L}_p$ (see [9]). A harmonic map $f: S \to S^n$ will be called k-orthogonal if $\phi = \iota \pi f: S \to \mathbb{C}P^n$ is k-orthogonal in the sense of Section 2. The following lemma is proved in ([9]).

Lemma 4.1. *If, for some positive integer $k < \frac{1}{2}n + 1$, $f: S \to S^n$ is $(2k-1)$-orthogonal then f is $2k$-orthogonal. If n is even and if f is $(n+1)$-orthogonal then f is pseudo-holomorphic.*

Suppose then that $f: S \to S^{2m}$ is a harmonic map. If f is $(2m+1)$-orthogonal then by Lemma 4.1 f is pseudo-holomorphic. We define f to be *superconformal* if it is $2m$-orthogonal but not pseudo-holomorphic.

There are two types of superconformal harmonic maps into S^{2m}, namely those which are

(a) linearly full in S^{2m};

(b) linearly full in a totally geodesic S^{2m-1} in S^{2m}.

In case (b) the harmonic sequence is orthogonally periodic since $L_{p+2m} = L_p$ for all p, and any $2m$ consecutive line bundles $L_p, \ldots, L_{p+2m-1}$ are mutually orthogonal. Moreover, $L_m = \overline{L}_m$ and, for $m = 2$, the corresponding map $\widehat{f}: S \to S^{2m-1}$ is called the *polar* of f in [28]. (We remark that this usage of "polar" is different from that used in Definition 3.6 of [19]).

Notice that any conformal harmonic (i.e. minimal) map of a connected Riemann surface into S^3 or S^4 is superconformal or pseudo-holomorphic.

Lemma 4.2. *If $f: S \to S^{2m}$ is a superconformal harmonic map then $L_1, \ldots, L_{m-1}$ are mutually orthogonal lines whose span is an isotropic subspace of $\mathbb{C}^{2m+1}$ orthogonal to L_0. Moreover, L_m is not an isotropic line.*

The first part follows from the fact that $L_{-(m-1)}, \ldots, L_{m-1}$ are mutually orthogonal and $L_{-p} = \overline{L}_p$ for all p. For the second part, we note that if L_m were isotropic then $L_{-m}, \ldots, L_m$ would be mutually orthogonal, so that f would be $(2m+1)$-orthogonal which would contradict Lemma 4.1. □

Now let $\mathbf{B}_m$ denote the flag manifold whose elements are ordered sequences

$$X = (X_0, \ldots, X_m)$$

of mutually orthogonal complex lines in $\mathbb{C}^{2m+1}$ where X_0 is the complexification of a real line in $\mathbb{R}^{2m+1}$ and $X_1, \ldots, X_m$ span a maximal isotropic subspace of $\mathbb{C}^{2m+1}$. Note that each element $X = (X_0, \ldots, X_m) \in \mathbf{B}_m$ determines a unique unit vector $u_0 = \rho(X) \in \mathbb{R}^{2m+1}$ which spans X_0. For if $\frac{1}{\sqrt{2}}(u_{2p-1} - iu_{2p})$ is a unit basis vector for $X_p, p = 1, \ldots, m$, then u_0 is the unique unit vector in $\mathbb{R}^{2m+1}$ for which $u_0, \ldots, u_{2m}$ is a positively oriented orthonormal basis of $\mathbb{R}^{2m+1}$. This defines a projection $\rho: \mathbf{B}_m \to S^{2m}$. We also note that, similarly, $X \in \mathbf{B}_m$ is uniquely determined by $\rho(X)$ and $X_1, \ldots, X_{m-1}$.

The group $SO(2m+1)$ acts transitively on $\mathbf{B}_m$ and the stabilizer of each point is a maximal torus. Indeed, for $p = 0, \ldots, 2m$, let e_p be the $(p+1)$-st standard basis vector of $\mathbb{C}^{2m+1}$ (having a 1 in the $(p+1)$-st place and zeros elsewhere) and let $C_0 = \mathbb{C}e_0, C_p = \mathbb{C}(e_{2p-1} - ie_{2p})$ for $p = 1, \ldots, m$. Then T^m is the stabiliser of $C = (C_0, \ldots, C_m)$. Thus we may identify $\mathbf{B}_m$ with $SO(2m+1)/T^m$ and ρ with the projection of $SO(2m+1)/T^m$ onto S^{2m} induced by sending an element of $SO(2m+1)$ to its first column.

Let $f: S \to S^{2m}$ be a superconformal harmonic map. Then, by Lemma 4.2, there is a lift $\widetilde{f}: S \to SO(2m+1)/T^m$ defined by

$$\widetilde{f} = (X_0, \ldots, X_m), \tag{4.3}$$

where $X_p = L_p, p = 0, \ldots, m-1$ and X_m is determined by the condition that $\rho \tilde{f} = f$. Note that L_m is contained in the orthogonal complement $X_m \oplus \overline{X}_m$ of $L_{-(m-1)} \oplus \ldots \oplus L_{m-1}$. This is needed in the proof of the following theorem [6], which follows quite easily from (2.1) and the general results of [2].

Theorem 4.3. *Let $f: S \to S^{2m}$ be a superconformal harmonic map. Then the lift $\tilde{f}: S \to \mathbf{SO}(2m+1)/T^m$ given by (4.3) is τ-primitive. Conversely if $\psi: S \to \mathbf{SO}(2m+1)/T^m$ is τ-primitive, then $f = \rho \psi: S \to S^{2m}$ is a superconformal harmonic map with lift ψ.*

We recall from above that superconformal harmonic maps $f: S \to S^{2m}$ are either linearly full in S^{2m} or are linearly full in a totally geodesic $(2m-1)$-sphere $S^{2m} \cap (N)^\perp$, where N is a unit vector in $\mathbb{R}^{2m+1}$. These latter maps are characterized by the fact that $X_m \oplus \overline{X}_m$ contains the constant real line spanned by N (for then X_0 is orthogonal to N). The real vector in $X_m \oplus \overline{X}_m$ orthogonal to N determines the polar of f.

As in the $\mathbb{C}P^n$ case, the above theorem has as a consequence the following important result concerning harmonic tori. This is a generalization of a result in [20].

Corollary 4.4. *Every superconformal harmonic 2-torus $\phi: T^2 \to S^{2m}$ is of finite type.*

As in Section 2, we remark that there seems to be a dearth of examples. Apart from the examples of [23],[20] in S^4, the only other superconformal harmonic torus we know of is the flat example in S^5 written down in [10]. This corresponds to the zero solution of (4.2), as does the Clifford torus in S^3.

As in the $\mathbb{C}P^n$ case, we may use the geometry of the harmonic sequence of a superconformal harmonic map $f: S \to S^{2m}$ to produce directly the Toda framing and the associated complex coordinate for the corresponding τ-primitive map. This has also been done in [18]. Firstly, given a local complex coordinate z in some simply connected open subset of S we have the sequence of $\mathbb{C}^{2m+1}$-valued functions discussed in Section 2 with $f_0 = f$. Then from (2.1) and the fact that f is superconformal, it follows that (f_m, f_m) is a non-zero holomorphic function and hence we may suppose that z is chosen such that $(f_m, f_m) = 1$. (The coordinate z is thus uniquely determined up to arbitrary translations and multiplication by a $2m$-th root of unity). Now define a real positively oriented orthonormal frame $F_0, \ldots, F_{2m}$ on U and a real valued function η on U as follows:

$$f_0 = F_0, \quad \frac{f_p}{|f_p|} = \frac{1}{\sqrt{2}}(F_{2p-1} - iF_{2p}), \quad p = 1, \ldots, m-1,$$

$$f_m = \cosh \eta \, F_{2m-1} - i \sinh \eta \, F_{2m}.$$

We note that $F_0, \ldots, F_{2m}$ and η are uniquely determined by f_0 (up to ambiguities involving $2m$-th roots of unity in the choice of the coordinate z mentioned above). A straightforward calculation then shows that if $F = (F_0, \ldots, F_{2m}) \in SO(2m+1)$,

The affine Toda equations and minimal surfaces 77

$$F^{-1}\frac{\partial F}{\partial z} = \frac{\partial \Omega}{\partial z} + e^{\Omega} B e^{-\Omega}$$

where $i\Omega = \text{diag}(0, \Theta(\omega_1), \ldots, \Theta(\omega_{m-1}), \Theta(\eta))$ with $\Theta(\omega) = \begin{pmatrix} 0 & -\omega \\ \omega & 0 \end{pmatrix}$, $e^{\omega_p} = |f_p|$, $p = 1, \ldots, m-1$, and $B = \sum_{p=0}^{m} V_p$. We note that the case where $f: S \to S^{2m}$ is linearly full into a totally geodesic S^{2m-1} is characterized by $\eta \equiv 0$.

The functions $\omega_1, \ldots, \omega_{m-1}, \eta$ satisfying the affine Toda equations (4.2) for the group $SO(2m+1)$ have geometrical interpretations. In terms of the harmonic sequence $\{\phi_p\}$ of the corresponding superconformal map $\phi = \iota \pi f$ into $\mathbb{C}P^{2m}$, we have $\omega_0 = 0$ and $\omega_{-p} = -\omega_p$. The induced metric of f is $ds^2 = 2e^{2\omega_1}|dz|^2$ and we now give geometrical interpretations of the functions $\omega_2, \ldots, \omega_{m-1}, \eta$ in terms of the ellipses of curvature of the higher fundamental forms of f. Details of the following construction may be found in [8] where the pseudo-holomorphic case is considered. However, the relevant parts of that discussion carry over to the case of a superconformal harmonic map $f: S \to S^{2m}$.

The fibres of each line bundle $X_p, p = 1, \ldots, m$, determined by f (or $\tilde{f}$) as in (4.3), are isotropic lines in $\mathbb{C}^{2m+1}$. Hence each X_p may be identified with an oriented real 2-plane subbundle of $S \times \mathbb{R}^{2m+1}$. In this way we may identify f^*TS^{2m} with $X_1 \oplus \ldots \oplus X_m$. Indeed $X_1 \oplus \ldots \oplus X_p$ is the p-th osculating bundle $T(p)$ of f with $T(1) = TS$ [8]. The higher fundamental forms of f are defined inductively as follows: let $\mathcal{F}_1$ be the f^*TS^{2m}-valued form defined by

$$\mathcal{F}_1(Y) = df(Y). \tag{4.4}$$

Let $T(p-1)^\perp$ denote the orthogonal complement of $T(p-1)$ in f^*TS^{2m}. Then, for $2 \le p \le m$, $\mathcal{F}_p$ is the $T(p-1)^\perp$-valued p-form on S defined by

$$\mathcal{F}_p(Y_1, \ldots, Y_p) = [(\bar{\nabla}_{Y_p} \mathcal{F}_{p-1})(Y_1, \ldots, Y_{p-1})]^{T(p-1)^\perp}, \tag{4.5}$$

where $Y_1, \ldots, Y_p$ are vector fields on S and $\bar{\nabla}$ is covariant differentiation determined by the connections on $T(p-1)^\perp$ and S^{2m}.

It is not hard to show that $\mathcal{F}_p$ is a symmetric p-form, and considering the complex extension of $\mathcal{F}_p$ to $(TS^\mathbb{C})^{\otimes p}$ we have,

$$\mathcal{F}_p\left(\frac{\partial}{\partial z}^{\otimes p}\right) = f_p, \qquad p = 1, \ldots, m,$$

where f_p is the section of L_p in the harmonic sequence determined by taking $f_0 = f$.

Also, it follows from the fact that f is conformal harmonic (and hence minimal) that

$$\mathcal{F}_p\left(\frac{\partial}{\partial \bar{z}}^{\otimes r} \otimes \frac{\partial}{\partial z}^{\otimes (p-r)}\right) = 0, \qquad r = 1, \ldots, p-1, \qquad p = 2, \ldots, m.$$

Then, for $p = 1, \ldots, m$,

$$\mathcal{F}_p((\cos\theta\frac{\partial}{\partial x} + \sin\theta\frac{\partial}{\partial y})^{\otimes p}) = \mathcal{F}_p((e^{i\theta}\frac{\partial}{\partial z} + e^{-i\theta}\frac{\partial}{\partial \bar{z}})^{\otimes p}) \qquad (4.6)$$

$$= e^{pi\theta}\mathcal{F}_p\left(\frac{\partial}{\partial z}\right)^{\otimes p} + e^{-pi\theta}\mathcal{F}_p\left(\frac{\partial}{\partial \bar{z}}\right)^{\otimes p}. \qquad (4.7)$$

Thus, writing $f_p = a_p - ib_p$, we have

$$\mathcal{F}_p((\cos\theta\frac{\partial}{\partial x} + \sin\theta\frac{\partial}{\partial y})^{\otimes p}) = 2(a_p \cos p\theta + b_p \sin p\theta),$$

so that the image $\mathcal{F}_p(C(T_xS))$ under $\mathcal{F}_p$ of the unit circle $C(T_xS)$ in T_xS at each point $x \in S$ is an ellipse. Indeed, for $p = 1,\ldots,m-1$, since $(f_p, f_p) = 0$, $\mathcal{F}_p(C(T_xS))$ is a circle of radius

$$\left(\frac{1}{2}\right)^{\frac{p-1}{2}} \frac{|f_p|}{|f_1|^p} = \left(\frac{1}{2}\right)^{\frac{p-1}{2}} e^{\omega_p - p\omega_1}. \qquad (4.8)$$

For $p = m$, Lemma 4.2 shows that $(f_m, f_m) \neq 0$ so $\mathcal{F}_m(C(T_xS))$ is never a circle. However, as mentioned earlier, we may choose a local coordinate z so that $(f_m, f_m) = 1$, in which case we may write $f_m = \cosh\eta F_{2m-1} - i\sinh\eta F_{2m}$. Thus $\mathcal{F}_m(C(T_xS))$ is an ellipse with major and minor axes of lengths

$$\left(\frac{1}{2}\right)^{\frac{m}{2}-1} \frac{\cosh\eta}{e^{m\omega_1}}, \qquad \left(\frac{1}{2}\right)^{\frac{m}{2}-1} \frac{|\sinh\eta|}{e^{m\omega_1}}, \qquad (4.9)$$

respectively, so that $|\tanh\eta|$ is the eccentricity of the ellipse. The case $\eta = 0$, where the ellipse is degenerate, corresponds to the case where f has orthogonally periodic harmonic sequence and is linearly full in a totally geodesic S^{2m-1} in S^{2m}.

5 The G_2-case

We recall that the vector cross-product on $\mathbb{R}^7$ (see [7] for details relevant to the present context and for further references) derived from Cayley multiplication on $\mathbb{R}^8$ may be used to define a nearly Kähler structure on the unit sphere S^6 in $\mathbb{R}^7$ as follows. For each $x \in S^6$, let $J_x: T_xS^6 \to T_xS^6$ be defined by

$$J_xv = x \times v, \qquad x \in S^6, \; v \in T_xS^6.$$

This, together with the standard metric on S^6, defines a nearly Kähler structure whose automorphism group is the exceptional Lie group G_2, which is the subgroup of $SO(7)$ preserving the vector cross product structure on $\mathbb{R}^7$. We take $T^2 = T^3 \cap G_2$ as the maximal torus of G_2, where T^3 is the maximal torus of $SO(7)$ considered in Section 4.

The Cartan subalgebra t^2 of g_2 is the subspace of the Cartan subalgebra t^3 of $so(7)$ defined by $\omega_1 + \omega_2 = \omega_3$ and analysis of the root space decompositions shows that

$$\mathcal{M}_1(g_2) = \mathcal{M}_1(so(7)) \cap g_2. \tag{5.1}$$

Alternatively, one may deduce this from the general theory of Section 3 using the fact that G_2/T^2 is a 6-symmetric submanifold of the 6-symmetric space $SO(7)/T^3$, where in both cases the symmetry of order 6 is given by $\tau = Ad\exp(2\pi i Z)$ where $Z = \text{diag}(1, R(\frac{2\pi}{6}), R(\frac{2\pi}{3}), R(\pi))$ and $R(\theta)$ is given by (4.1). For, in both cases $\mathcal{M}_1$ is the $e^{2\pi i/6}$-eigenspace of the automorphism of the complexified Lie algebra given by τ. The following proposition is now clear.

Proposition 5.1. *Let $\psi: S \to G_2/T^2$ and let $\iota: G_2/T^2 \to SO(7)/T^3$ be the inclusion map. Then ψ is τ-primitive if and only if $\iota\psi$ is τ-primitive.*

We now describe those superconformal harmonic maps into S^6 whose lifts lie in G_2/T^2.

An *almost complex curve* in S^6 is a non-constant smooth map $f: S \to S^6$ whose differential is complex linear. Using a local complex coordinate z, this condition may be written as

$$f \times f_1 = i f_1. \tag{5.2}$$

In fact, (see [7]), each almost complex curve is either pseudo-holomorphic or superconformal and we have the following proposition.

Proposition 5.2. *Let $f: S \to S^6$ be a superconformal harmonic map with lift $\tilde{f}: S \to SO(7)/T^3$. Then f is almost complex if and only if $\tilde{f}(S) \subseteq G_2/T^2$.*

Denoting the restriction of the projection $\rho: SO(7)/T^3 \to S^6$ to G_2/T^2 also by ρ, the following theorem is now immediate.

Theorem 5.3. *Let $f: S \to S^6$ be an almost complex curve which is not pseudo-holomorphic. Then the lift $\tilde{f}: S \to G_2/T^2$ given by (4.3) is τ-primitive. Conversely, if $\psi: S \to G_2/T^2$ is τ-primitive then $f = \rho\psi: S \to S^6$ is almost complex with corresponding lift ψ.*

Pseudo-holomorphic almost complex curves have been discussed in [11].
Theorem 5.3 has the following consequence.

Corollary 5.4. *Every almost complex 2-torus in S^6 is either pseudo-holomorphic or of finite type.*

There is only one such example (up to G_2-congruence) that we know of, namely the flat almost complex torus in S^5 given in [7].

Finally we consider Toda framings and corresponding special complex coordinates for almost complex curves. As in Section 4, we first choose a local complex coordinate z in an open subset U of S such that $(f_3, f_3) = 1$. Next we define a real positively oriented frame $F_1, \ldots, F_7$ on U and a real valued function η on U as follows:

$$f_0 = F_1, \quad \frac{f_p}{|f_p|} = \frac{1}{\sqrt{2}}(F_{2p} - iF_{2p+1}), \quad p = 1, 2,$$

$$f_3 = -\cosh\eta\, F_7 - i\sinh\eta\, F_6.$$

If $|f_p| = e^{\omega_p}, p = 1, 2$, then $\omega_1 + \omega_2 = \eta$, and writing $F = (F_1, \ldots, F_7)$ we have [7] $F \in G_2$ and

$$F^{-1}\frac{\partial F}{\partial z} = \frac{\partial \Omega}{\partial z} + e^\Omega B e^{-\Omega},$$

where

$$\Omega = i\begin{pmatrix} 0 & \omega_1 & & & & & \\ -\omega_1 & & & & & & \\ & & & \omega_2 & & & \\ & & -\omega_2 & & & & \\ & & & & & \eta & \\ & & & & -\eta & & \end{pmatrix} \in it^2$$

and

$$B = \begin{pmatrix} & \frac{i}{\sqrt{2}} & \frac{1}{\sqrt{2}} & & & & \\ -\frac{i}{\sqrt{2}} & & & \frac{1}{2} & -\frac{i}{2} & & \\ -\frac{1}{\sqrt{2}} & & & \frac{i}{2} & \frac{1}{2} & & \\ & -\frac{1}{2} & -\frac{i}{2} & & & \frac{1}{2} & \frac{i}{2} \\ & \frac{i}{2} & -\frac{1}{2} & & & \frac{i}{2} & -\frac{1}{2} \\ & & & -\frac{1}{2} & -\frac{i}{2} & & \\ & & & -\frac{i}{2} & \frac{1}{2} & & \end{pmatrix} \in \mathcal{M}_1(g_2).$$

Bibliography

[1] M. Adler and P. van Moerbeke, *Completely integrable systems, Euclidean Lie algebras, and curves*, Adv. Math. **38** (1980), 267-317.

[2] M. Black, *Harmonic maps into homogeneous spaces*, Longman, Harlow, 1991.

[3] A.I. Bobenko, *All constant mean curvature tori in $\mathbb{R}^3$, S^3, H^3 in terms of theta-functions*, Math. Ann. **290** (1991), 209-245.

[4] A.I. Bobenko, *Surfaces in terms of 2 by 2 matrices. Old and new integrable cases*, this volume.

[5] J. Bolton, G.R. Jensen, M. Rigoli and L.M. Woodward, *On conformal minimal immersions of S^2 into $\mathbb{C}P^n$*, Math. Ann. **279** (1988), 599-620.

[6] J. Bolton, F. Pedit and L.M. Woodward, *Minimal surfaces and the Toda field model*, preprint, Universities of Durham and Massachussetts, Amherst.

[7] J. Bolton, L.Vrancken and L.M. Woodward, *On almost complex curves in the nearly Kähler 6-sphere*, Quarterly J. Math. (Oxford), to appear.

[8] J. Bolton and L.M. Woodward, *On immersions of surfaces into space forms*, Soochow J. Math. **14** (1988), 11-31.

[9] J. Bolton and L.M. Woodward, *Congruence theorems for harmonic maps from a Riemann surface into $\mathbb{C}P^n$ and S^n*, J. London Math. Soc. **45(2)** (1992), 363-376.

[10] J. Bolton and L.M. Woodward, *Minimal surfaces in $\mathbb{C}P^n$ with constant curvature and Kähler angle*, Math. Proc. Camb. Phil. Soc. **112** (1992), 287-296.

[11] R.L. Bryant, *Submanifolds and special structures on the octonians*, J. Diff. Geom. **17** (1982), 185-232.

[12] F.E. Burstall. *Harmonic tori in spheres and complex projective spaces*, in preparation.

[13] F.E. Burstall, D. Ferus, F. Pedit and U. Pinkall, *Harmonic tori in symmetric spaces and commuting Hamiltonian systems on loop algebras*, Ann. of Math. **138** (1993), 173-212.

[14] F.E. Burstall and F. Pedit, *Harmonic maps via Adler-Kostant-Symes theory*, this volume.

[15] F.E. Burstall and J.C. Wood, *The construction of harmonic maps into complex Grassmannians,*, J. Diff. Geom. **23** (1986), 255-297.

[16] E. Calabi, *Minimal immersions of surfaces into Euclidean spheres*, J. Diff. Geom.**1** (1967), 111-125.

[17] A.M. Din and W.J. Zakrzewski, *General classical solutions in the $\mathbb{C}P^{n-1}$ model*, Nuclear Phys. B **174** (1980), 397-406.

[18] A. Doliwa and A. Sym, *The non-linear σ-model in spheres*, preprint, Warsaw University, 1992.

[19] J. Eells and J.C. Wood, *Harmonic maps from surfaces to complex projective spaces*, Adv. in Math. **49** (1983), 217-263.

[20] D. Ferus, F. Pedit, U. Pinkall and I. Sterling, *Minimal tori in S^4*, J. Reine Angew. Math. **429** (1992), 1-47.

[21] A.P. Fordy, *Integrable equations associated with simple Lie algebras and symmetric spaces*, in: Soliton Theory: A Survey of Results, ed. A.P. Fordy, Manchester Univ. Press 1990, 315-337.

[22] S. Helgason, *Differential Geometry, Lie Groups, and Symmetric Spaces*, Academic Press, New York, San Francisco, London 1978.

[23] N.J. Hitchin, *Harmonic maps from a 2-torus to the 3-sphere*, J. Diff. Geom. **31** (1990), 627-710.

[24] A. Jimenez, *Addendum to "Existence of Hermitian n-symmetric spaces and non-commutative naturally reductive spaces"*, Math. Z. **197** (1988), 455-456.

[25] K. Kenmotsu, *On minimal immersions of $\mathbb{R}^2$ into $P^n(\mathbb{C})$*, J. Math. Soc. Japan **37** (1985), 665-682.

[26] B. Kostant, *The principal three dimensional subgroups and the Betti numbers of a complex simple Lie group*, Amer. J. Math. **81** (1959), 973-1032.

[27] B. Kostant, *The solution to a generalized Toda lattice and representation theory*, Adv. Math. **34** (1979), 195-338.

[28] H.B. Lawson, *Complete minimal surfaces in S^3*, Ann. of Math. **92** (1970), 335-374.

[29] A.N. Leznov and M.V. Saveliev. *Group-Theoretical Methods for Integration of Nonlinear Dynamical Systems*, Birkhäuser, Basel 1992.

[30] I. McIntosh, *Infinte dimensional Lie groups and the two-dimensional Toda lattice*, this volume.

[31] M. Melko and I. Sterling, *Integrable systems, harmonic maps and the classical theory of surfaces*, this volume.

[32] U. Pinkall and I. Sterling, *On the classification of constant mean curvature tori*, Ann. of Math. **130** (1989), 407-451.

[33] M. Toda, *Vibrations of a chain with nonlinear interaction*, J. Phys. Soc. Japan **22** (1967), 431-436.

[34] M. Toda, *Wave propagation in anharmonic lattices*, J. Phys. Soc. Japan **23** (1967), 501-506.

[35] M. Toda, *Theory of Nonlinear Lattices*, Springer, Berlin, Heidelberg, New York 1981.

[36] J.G. Wolfson, *On minimal surfaces in a Kähler manifold of constant holomorphic sectional curvature*, Trans. Amer. Math. Soc. **290** (1985), 627-646.

[37] J.C. Wood, *Harmonic maps into symmetric spaces and integrable systems*, this volume.

Surfaces in terms of 2 by 2 matrices. Old and new integrable cases

A.I. Bobenko

1 Introduction

Many of the equations which now are called integrable have been known in differential geometry for a long time. Probably the first was the famous sine-Gordon equation, which was derived to describe surfaces with constant negative Gaussian curvature. At that time many features of integrability of the sine-Gordon and other integrable equations were discovered [1], namely those which have clear geometrical interpretation (for example, the Bäcklund transform).

The theory of solitons appeared much later, in the 1960's. Though it was oriented basically to problems of mathematical physics, it deals in many cases with the same equations. Moreover, the starting point of this theory — the representation (which is called the Lax representation or Zakharov-Shabat representation) of the nonlinear equation in a form of compatibility condition

$$U_y(\lambda) - V_x(\lambda) + [U(\lambda), V(\lambda)] = 0$$

of two linear equations

$$\Psi_x = U(\lambda)\Psi, \qquad \Psi_y = V(\lambda)\Psi$$

also has a transparent geometrical origin. In differential geometry it is the Gauss-Codazzi equation represented as a compatibility condition of linear equations for the moving frame (the Gauss-Weingarten equations). The spectral parameter λ in this representation describes deformations of surfaces preserving their properties:

spectral parameter $\lambda \iff$ deformation parameter, (1.1)

---
[1] For the history of that period see the contribution of M.Melko and I.Sterling in this volume.

that is, integrable surfaces come in families! For the integrable equations of mathematical physics, which were found first, λ was interpreted as a spectral parameter in the corresponding linear problem. Therefore it was quite natural to investigate how the solution of this linear problem Ψ depends on λ. This free treatment of λ, independent of its geometrical interpretation, resulted in the construction of powerful analytical methods of solution for the nonlinear integrable equations. Some of these methods do not have a transparent geometrical interpretation and therefore are new to geometry. First of all here one should mention the finite gap integration method. Whereas the dressing procedure in the soliton theory and the Bäcklund transform essentially coincide and produce the same multisoliton solutions, only a few of the simplest solutions constructed by the finite gap integration method were known before. Although from the very beginning of the soliton theory a close relation with the differential geometry was clear, the first new essential results in classical differential geometry of surfaces were obtained in the late 1980's (see the survey [2] and [12]), when the methods of the finite gap theory were applied.

In the present paper we consider only the classical case of surfaces in a 3-dimensional Euclidean space. Our goal here is to reformulate the classical theory of surfaces in a form familiar to the soliton theory, which makes possible an application of the analytical methods of this theory to integrable cases. In Sect. 2 the moving frame for a general surface is described in terms of quaternions $\Psi \in \mathbf{H}_*$. Such a description is more convenient for the analytical treatment since it analytically characterizes the spin structure of the immersion and deals with 2×2 matrices. This analytic characterization of the spin structure of the minimal surfaces is given in Sect. 3 in terms of the Weierstrass representation. All the necessary results about the spinors on the Riemann surfaces are presented in the Appendix.

All the rest of the paper is devoted to a description of integrable cases and their deformation families (1.1). Some of these cases are well known, some are not well known and some are possibly new. The list below presents these integrable cases:

1) Minimal surfaces: $H = 0$;

2) Constant mean curvature surfaces: $H = \text{const}$;

3) Constant positive Gaussian curvature surfaces: $K = \text{const} > 0$;

4) Constant negative Gaussian curvature surfaces: $K = \text{const} < 0$;

5) Bonnet surfaces: surfaces, possessing nontrivial families of isometries preserving principal curvatures;

6) Surfaces with harmonic inverse mean curvature: $\partial_z \partial_{\bar{z}}(1/H) = 0$, z is a conformal variable of the first fundamental form;

7) Bianchi surfaces: $\partial_x \partial_y(1/\sqrt{-K}) = 0$, x, y are asymptotic coordinates;

8) Bianchi surfaces of positive curvature: $\partial_z \partial_{\bar{z}}(1/\sqrt{K}) = 0$, z is a conformal coordinate of the second fundamental form.

Only the cases 6 and 8 of this list can pretend to be new, and the case 8 is just another real form of the Bianchi surfaces (case 7). The surfaces with $\partial_z\partial_{\bar{z}}(1/H) = 0$ seems to be new, but upon seeing a wonderful 3-page long paper by Tzitzéica [16] on affine spheres, shown me by Sergey Tsarev, where the Bullough-Dodd-Jiber-Shabat equation $f_{uv} = e^f - e^{-2f}$ together with its representation (1.1) ("système complètement intégrable") are presented, one would have to have a lot of courage to claim that a new integrable case in differential geometry of surfaces is found.

Formulas for the moving frame of integrable surfaces can be integrated by an expression, first suggested by A. Sym for the constant negative curvature surfaces [15] (see also Sect. 8). Namely, the immersion function in many cases is given by

$$\Psi^{-1}\frac{\partial}{\partial\lambda}\Psi.$$

We prove that with slight modifications this immersion formula is valid for all the cases above except 1 and 5.

I wish to thank Rob Kusner, Ulrich Pinkall, Konrad Voss and Peter Zograf for valuable discussions.

The author was supported by the Alexander von Humboldt Stiftung and the Sonderforschungsbereich 288 during the preparation of this work.

2 Surfaces in a 3-dimensional Euclidean space

2.1 Differential equations of surfaces

Let $\mathcal{F}$ be a smooth surface in a 3-dimensional Euclidean space. The Euclidean metric induces a metric Ω on this surface, which in turn generates the complex structure of a Riemann surface. The surface is covered by domains $\mathcal{D}_i$ with $\cup_i \mathcal{D}_i = \mathcal{F}$, and in each of these there is defined a local coordinate $z_i : \mathcal{D}_i \to U_i \subset \mathbf{C}$. If the intersection $\mathcal{D}_i \cap \mathcal{D}_j \neq \emptyset$ is non-empty, the glueing functions $z_i \circ z_j^{-1}$ are holomorphic. Under such a parametrization, which is called *conformal*, the surface is given by a the vector-valued function:

$$F = (F_1, F_2, F_3) : \mathcal{R} \to \mathbf{R}^3,$$

and the metric is conformal: $\Omega = e^{u_i} dz_i d\bar{z}_i$. In the sequel we suppose that $\mathcal{F}$ is sufficiently smooth. We also omit the superscript i of the local coordinate z in case when it is not confusing.

The conformal parametrisation gives the following normalization of the function $F(z,\bar{z})$:

$$< F_z, F_z > = < F_{\bar{z}}, F_{\bar{z}} > = 0, \quad < F_z, F_{\bar{z}} > = \frac{1}{2}e^u, \qquad (2.1)$$

where the brackets mean the scalar product

$$< a, b > = a_1 b_1 + a_2 b_2 + a_3 b_3,$$

and F_z and $F_{\bar{z}}$ are the partial derivatives $\frac{\partial F}{\partial z}$ and $\frac{\partial F}{\partial \bar{z}}$ where

$$\frac{\partial}{\partial z} = \frac{1}{2}\left(\frac{\partial}{\partial x} - i\frac{\partial}{\partial y}\right), \quad \frac{\partial}{\partial \bar{z}} = \frac{1}{2}\left(\frac{\partial}{\partial x} + i\frac{\partial}{\partial y}\right).$$

The vectors $F_z, F_{\bar{z}}$ as well as the normal N,

$$< F_z, N > = < F_{\bar{z}}, N > = 0, \quad < N, N > = 1, \tag{2.2}$$

define a moving frame on the surface, which due to (2.1, 2.2) satisfies the following Gauss-Weingarten (GW) equations:

$$\sigma_z = \mathcal{U}\sigma, \quad \sigma_{\bar{z}} = \mathcal{V}\sigma, \quad \sigma = (F_z, F_{\bar{z}}, N)^T, \tag{2.3}$$

$$\mathcal{U} = \begin{pmatrix} u_z & 0 & Q \\ 0 & 0 & \frac{1}{2}He^u \\ -H & -2e^{-u}Q & 0 \end{pmatrix}, \quad \mathcal{V} = \begin{pmatrix} 0 & 0 & \frac{1}{2}He^u \\ 0 & u_{\bar{z}} & \bar{Q} \\ -2e^{-u}\bar{Q} & -H & 0 \end{pmatrix} \tag{2.4}$$

where

$$Q = < F_{zz}, N >, \quad < F_{z\bar{z}}, N > = \frac{1}{2}He^u. \tag{2.5}$$

The first and the second quadratic forms

$$< dF, dF > \;=\; < I\begin{pmatrix} dx \\ dy \end{pmatrix}, \begin{pmatrix} dx \\ dy \end{pmatrix} >, \quad z = x + iy,$$

$$- < dF, dN > \;=\; < II\begin{pmatrix} dx \\ dy \end{pmatrix}, \begin{pmatrix} dx \\ dy \end{pmatrix} >$$

are given by the matrices

$$I = e^u \begin{pmatrix} 1 & 0 \\ 0 & 1 \end{pmatrix}, \quad II = \begin{pmatrix} Q + \bar{Q} + He^u & i(Q - \bar{Q}) \\ i(Q - \bar{Q}) & -(Q + \bar{Q}) + He^u \end{pmatrix}.$$

The principal curvatures k_1 and k_2 are the eigenvalues of the matrix $II \cdot I^{-1}$, which gives the following expressions for the mean and the Gaussian curvatures:

$$H = \tfrac{1}{2}(k_1 + k_2) = \tfrac{1}{2}\,\text{tr}\,(II \cdot I^{-1}),$$
$$K = k_1 k_2 = \det(II \cdot I^{-1}) = H^2 - 4Q\bar{Q}e^{-2u}.$$

The Gauss-Codazzi (GC) equations, which are the compatibility conditions of equations (2.3, 2.4),

$$\mathcal{U}_{\bar{z}} - \mathcal{V}_z + [\mathcal{U}, \mathcal{V}] = 0,$$

have the following form (cf. [17]):

$$u_{z\bar{z}} + \frac{1}{2}H^2 e^u - 2Q\bar{Q}e^{-u} = 0, \tag{2.6}$$

$$Q_{\bar{z}} = \frac{1}{2}H_z e^u, \tag{2.7}$$

$$\bar{Q}_z = \frac{1}{2}H_{\bar{z}} e^u. \tag{2.8}$$

2.2 Quaternionic description

We construct and investigate surfaces in $\mathbf{R}^3$ by analytical methods. For this purpose it is more convenient to use 2×2 matrices instead of 3×3 matrices (2.4), therefore first we rewrite the equations (2.3) for the moving frame in terms of quaternions. This also allows us to control the spin structure of the immersion (see the next section) and makes the presentation familiar to the specialists in the theory of integrable equations.

Let us denote the algebra of quaternions by $\mathbf{H}$, the multiplicative quaternion group by $\mathbf{H}_* = \mathbf{H} \setminus \{0\}$ and their standard basis by $\{\mathbf{1}, \mathbf{i}, \mathbf{j}, \mathbf{k}\}$

$$\mathbf{ij} = \mathbf{k}, \quad \mathbf{jk} = \mathbf{i}, \quad \mathbf{ki} = \mathbf{j}. \tag{2.9}$$

The Pauli matrices σ_α are related with this basis as follows:

$$\sigma_1 = \begin{pmatrix} 0 & 1 \\ 1 & 0 \end{pmatrix} = i\,\mathbf{i}, \quad \sigma_2 = \begin{pmatrix} 0 & -i \\ i & 0 \end{pmatrix} = i\,\mathbf{j},$$

$$\sigma_3 = \begin{pmatrix} 1 & 0 \\ 0 & -1 \end{pmatrix} = i\,\mathbf{k}, \quad \mathbf{1} = \begin{pmatrix} 1 & 0 \\ 0 & 1 \end{pmatrix} \tag{2.10}$$

with the multiplication in (2.9) being just the matrix multiplication. We identify a 3-dimensional Euclidean space with the space of imaginary quaternions Im $\mathbf{H}$

$$X = -i \sum_{\alpha=1}^{3} X_\alpha \sigma_\alpha \in \text{Im } \mathbf{H} \longleftrightarrow X = (X_1, X_2, X_3) \in \mathbf{R}^3. \tag{2.11}$$

The scalar product of vectors in terms of matrices is then

$$< X, Y > = -\frac{1}{2} \text{tr} XY. \tag{2.12}$$

We also denote by F and N the matrices obtained in this way from the vectors F and N.

Let us take $\Phi \in \mathbf{H}_*$

$$\Phi = \begin{pmatrix} a & b \\ -\bar{b} & \bar{a} \end{pmatrix}, \quad |a|^2 + |b|^2 \neq 0, \tag{2.13}$$

which transforms the basis $\mathbf{i}, \mathbf{j}, \mathbf{k}$ into the basis F_x, F_y, N:

$$F_x = e^{u/2} \Phi^{-1} \mathbf{i} \Phi, \quad F_y = e^{u/2} \Phi^{-1} \mathbf{j} \Phi, \quad N = \Phi^{-1} \mathbf{k} \Phi. \tag{2.14}$$

Then

$$F_z = -i e^{u/2} \Phi^{-1} \begin{pmatrix} 0 & 0 \\ 1 & 0 \end{pmatrix} \Phi, \quad F_{\bar{z}} = -i e^{u/2} \Phi^{-1} \begin{pmatrix} 0 & 1 \\ 0 & 0 \end{pmatrix} \Phi, \tag{2.15}$$

and all the conditions (2.1) are automatically satisfied.

The quaternion Φ satisfies linear differential equations. To derive them we introduce matrices

$$U = \Phi_z \Phi^{-1}, \quad V = \Phi_{\bar{z}} \Phi^{-1}. \tag{2.16}$$

The compatibility condition $F_{z\bar{z}} = F_{\bar{z}z}$ for (2.15) implies

$$-ie^{u/2}\frac{u_{\bar{z}}}{2}\Phi^{-1}\begin{pmatrix} 0 & 0 \\ 1 & 0 \end{pmatrix}\Phi - ie^{u/2}\Phi^{-1}\left[\begin{pmatrix} 0 & 0 \\ 1 & 0 \end{pmatrix}, V\right]\Phi =$$
$$= -ie^{u/2}\frac{u_z}{2}\Phi^{-1}\begin{pmatrix} 0 & 1 \\ 0 & 0 \end{pmatrix}\Phi - ie^{u/2}\Phi^{-1}\left[\begin{pmatrix} 0 & 1 \\ 0 & 0 \end{pmatrix}, U\right]\Phi,$$

or, eqivalently,

$$u_{\bar{z}} = 2(V_{22} - V_{11}), \quad u_z = 2(U_{11} - U_{22}), \quad U_{21} = -V_{12},$$

where U_{kl} and V_{kl} are the matrix elements of U and V. In the same way the equalities (2.15) imply

$$F_{z\bar{z}} = \frac{1}{2}He^u N \quad \to \quad U_{21} = -V_{12} = \frac{1}{2}He^{u/2}$$
$$F_{zz} = u_z F_z + QN \quad \to \quad U_{12} = -Qe^{-u/2}$$
$$F_{\bar{z}\bar{z}} = u_{\bar{z}} F_{\bar{z}} + \bar{Q}N \quad \to \quad V_{21} = \bar{Q}e^{-u/2}$$

Now only the coeffitients U_{22} and V_{11} are still not determined. To fix them, we recall that Φ was defined by (2.14) up to a multiplication by a scalar factor. We normalize this factor by the condition

$$\det \Phi = e^{u/2}, \tag{2.17}$$

the reason for which is clarified in the next section. For the traces of U and V this implies

$$U_{11} + U_{22} = u_z/2, \quad V_{11} + V_{22} = u_{\bar{z}}/2.$$

Finally we get the following

Theorem 1 *Using the isomorphism (2.11), the moving frame $F_z, F_{\bar{z}}, N$ of the conformally parametrised surface (z is a conformal coordinate) is described by formulas (2.14),(2.15), where $\Phi \in H_*$ satisfies the equations (2.16) with U, V of the form*

$$U = \begin{pmatrix} \dfrac{u_z}{2} & -Qe^{-u/2} \\ \dfrac{1}{2}He^{u/2} & 0 \end{pmatrix}, \quad V = \begin{pmatrix} 0 & -\dfrac{1}{2}He^{u/2} \\ \bar{Q}e^{-u/2} & \dfrac{u_{\bar{z}}}{2} \end{pmatrix}. \tag{2.18}$$

Corollary 1 Φ *satisfies the Dirac equation*

$$e^{-u/2}\begin{pmatrix} 0 & \partial_z \\ -\partial_{\bar{z}} & 0 \end{pmatrix}\Phi = \frac{1}{2}H\Phi. \tag{2.19}$$

2.3 Spin structure

Let us now determine the dependence of Φ on the holomorphic coordinate z.

Lemma 1

$$\Phi_{sp} = \begin{pmatrix} \sqrt{dz} & 0 \\ 0 & \sqrt{d\bar{z}} \end{pmatrix} \Phi$$

is invariant under analytical changes of z.

Proof. Since $e^u dz d\bar{z}$ is invariant under analytical changes of $z \to w(z)$, u is transformed as follows

$$e^{u/2}(z) = e^{u/2}(w)\sqrt{w'\bar{w}'}, \quad w' = dw/dz. \tag{2.20}$$

The most general transformation of Φ compatible with the formulas for the moving frame (2.14, 2.15) is

$$\Phi(z) = \begin{pmatrix} c & 0 \\ 0 & \bar{c} \end{pmatrix} \Phi(w),$$

where

$$\frac{c}{\bar{c}} = \sqrt{\frac{w'}{\bar{w}'}}. \tag{2.21}$$

To get the last formula we have to take into account the transformation law of the metric (2.20). The normalization condition (2.17) implies for c

$$c\bar{c} = \sqrt{w'\bar{w}'},$$

which, combined with (2.21), completes the proof

$$c = \sqrt{w'}.$$

The local variation of $\Phi(z, \bar{z})$ is described by system (2.16, 2.18). These equations are compatible, therefore there is no local monodromy of Φ. We see that

$$\Phi_{sp} = \begin{pmatrix} \sqrt{dz} & 0 \\ 0 & \sqrt{d\bar{z}} \end{pmatrix} \begin{pmatrix} a & b \\ -\bar{b} & \bar{a} \end{pmatrix}$$

consists of two spinors $a\sqrt{dz}$ and $b\sqrt{dz}$. All the necessary information about spinors is presented in the Appendix. The spinors $a\sqrt{dz}$, $b\sqrt{dz}$ may change sign around cycles γ of $\mathcal{R}$.

Lemma 2 *The spin structures of the spinors $a\sqrt{dz}$ and $b\sqrt{dz}$ are the same. The flip numbers of these spinors coincide: $p^a(\gamma) = p^b(\gamma)$ for any contour $\gamma \subset \mathcal{R}$.*

Proof. This is straightforward. Since $\mathcal{F}$ is oriented, $F_z, F_{\bar{z}}, N$ are uniquely defined on $\mathcal{R}$, and formulas (2.14, 2.15) show that the only possible transformation of Φ_{sp} on passing around γ is

$$\Phi_{sp} \to (-1)^p \Phi_{sp}.$$

This spin structure, and generally the numbers $p(\gamma) \in \mathbf{Z}_2$, have a simple geometrical interpretation. Let us consider a closed contour $\Gamma \subset \mathcal{F}$. Γ together with the normal field N of the surface at this contour defines a closed orientable strip in space. Let $N_\Gamma \in \mathbf{Z}$ be the number of twists of this strip, or equivalently, the winding number of the contours Γ and $(\Gamma + \epsilon N)_\Gamma$, where ϵ is small. Isotopies of the band preserve N_Γ, whereas regular homotopies (self-intersections are allowed) preserve N_Γ (*mod* 2), which we from now on denote by

$$P(\Gamma) \in \mathbf{Z}_2$$

and call it the *parity of twists*.

Let us consider a little bit more general class of surfaces: surfaces with translational periods. For these surfaces the frame $F_z, F_{\bar{z}}, N$ is uniquely defined on $\mathcal{R}$, whereas the immersion function F can have periods around cycles on $\mathcal{R}$. All the previous considerations of spin structures are valid also for surfaces with periods, since only the formulas for the frames were used. The numbers $P(\Gamma) \in \mathbf{Z}_2$ can also be defined in this case. To do this we introduce a notion of a translation-holonomy strip.

Definition 1 *A translation-holonomy strip is a smooth curve in $\mathbf{R}^3$ equipped with a smooth normal field N, such that the orthonormal frames, consisting of the normal vector N and the tangential vector, coincide at the ends of the curve.*

If $F(s), N(s)$,

$$F : [0,1] \longrightarrow R^3, \quad N : [0,1] \longrightarrow S^2$$

is some parametrisation of the translation-holonomy strip, then

$$\frac{F_s}{|F_s|}(0) = \frac{F_s}{|F_s|}(1), \quad N(0) = N(1).$$

We consider smooth homotopies of translation-holonomy strips and define the parity of twists of a translation-holonomy strip S to be equal to a parity of twists of a closed strip S_0 smoothly homotopic to S

$$P(S) \stackrel{def}{=} P(S_0).$$

This is well-defined, i.e. $P(S)$ is independent of the choice of closed strip S_0 homotopic to S, which also shows that $P(S)$ is an invariant of the homotopy class.

Remark. The parity of twists of the straight strip $\{F(s) = \vec{f}s, \, N(s) = \vec{n}, \, (\vec{f}, \vec{n}) = 0; \vec{f}, \vec{n} = \text{const}\}$ is equal to 1!

Theorem 2 *If $F : \mathcal{R} \longrightarrow \mathbf{R}^3$ is a conformal parametrisation of a surface $\mathcal{F}$ with translational periods and $\gamma \subset \mathcal{R}$ is a closed contour, then the spinor flip number $p(\gamma)$ is equal to the parity of twists $P(\Gamma)$:*

$$p(\gamma) = P(\Gamma),$$

where $\Gamma \subset \mathcal{F}$ is the image of γ.

Proof. Let $\gamma(s)$, $s \in [0,1]$, $\gamma(0) = \gamma(1)$ be a parametrisation of the contour and $z = e^{2\pi i s}$ be a complex annular coordinate $|z(\gamma)| = 1$ on γ. Formulas (2.14) imply for the frame F_s, N along γ:

$$F_s = 2\pi e^{u/2} \Phi^{-1}(s) \begin{pmatrix} 0 & -e^{-2\pi i s} \\ e^{2\pi i s} & 0 \end{pmatrix} \Phi(s),$$

$$N = -i\Phi^{-1}(s)\sigma_3 \Phi(s).$$

By general rotation of the translation-holonomy strip in $\mathbf{R}^3$ we normalize $\Phi(0) = I$. Since the parity of twists $P(\Gamma)$ is preserved by the smooth homotopies of translation-holonomy strips, we can replace $e^{u/2}$ by 1 when calculating $P(\Gamma)$.

The curves $\Phi(s)$ have different topology for different flip numbers $p(\gamma)$:

a) $p(\gamma) = 0 \Rightarrow \Phi(1) = I$,

b) $p(\gamma) = 1 \Rightarrow \Phi(1) = -I$.

The variety H_* is simply connected, therefore by smooth homotopies of the translation-holonomy strips the curves $\Phi(s)$ can be transformed in the cases a),b) above respectively to

a) $\Phi(s) = I$,

b) $\Phi(s) = \begin{pmatrix} e^{-\pi i s} & 0 \\ 0 & e^{\pi i s} \end{pmatrix}.$

For the immersion of the translation-holonomy strip it yields

a) $F = -i \begin{pmatrix} 0 & e^{-2\pi i s} \\ e^{2\pi i s} & 0 \end{pmatrix}, \quad N = \mathbf{k} \Rightarrow P(\Gamma) = 0,$

b) $F = 2\pi s \mathbf{j}, \qquad\qquad\qquad N = \mathbf{k} \Rightarrow P(\Gamma) = 1,$

which proves $p(\gamma) = P(\Gamma)$.

Definition 2 *The spin structure of Φ_{sp} is called the spin structure of the immersion.*

Usually, if a surface is given by its immersion function it is difficult to answer the important geometric question of whether it is an embedding or not. Sometimes existence of the self-intersection can be proved by purely topological arguments analyzing the spin structure of the immersion. If $\mathcal{R}$ is a Riemann surface with G handles and K punctures or holes, then this spin structure is characterized by the numbers

$$[\alpha, \beta, \delta] \in \mathbb{Z}_2^{2G+K}$$

(see Appendix). The numbers α_i, β_i describe flips along the handles, whereas the δ_k describe flips around the holes or the punctures. As it is mentioned in the Appendix the parity of the spin structure $<\alpha, \beta> \in \mathbb{Z}_2$ is invariant with respect to the choice of the basis in $H_1(\mathcal{R}, \mathbb{Z})$. As it was proved in [13], it completely classifies regular homotopies of compact orientable immersions.

Theorem 3 *[13] The parity of the spin structure completely classifies compact orientable immersions with respect to regular homotopies, i.e. there is a smooth homotopy F_t, $t \in [0,1]$ of two immersions, F_0 and F_1, of a surface of genus G which at each moment remains an immersion if and only if the parities of the spin structures corresponding to F_0 and F_1 coincide. For embeddings, $<\alpha, \beta> = 0$.*

Corollary 2 *If F is an embedding with G handles and K holes or punctures, then*

$$<\alpha, \beta> = 0, \quad \delta_k = 0, \quad k = 1, \ldots, K.$$

To prove this statement we consider a sphere big enough to contain all the handles inside. Replacing the outside parts of the surface by the corresponding pieces of the sphere (smooth glueing) and applying Theorem 3 to this compact surface we get $<\alpha, \beta> = 0$. The embeddedness of the ends imply the vanishing of the $\delta's$.

3 Minimal surfaces

In the case of minimal surfaces ($H = 0$) the system (2.16, 2.18) can be solved. The elements $a(z), b(z)$ of Φ in (2.13) are holomorphic. The metric and the Hopf differential are expressed in the terms of the spinors $a(z), b(z)$ as follows:

$$e^{u/2} = |a|^2 + |b|^2, \quad Q = a_z b - b_z a.$$

Formulas (2.14, 2.15) for the frame yield

$$F_z = -i \begin{pmatrix} -ab & -b^2 \\ a^2 & ab \end{pmatrix}, \quad F_{\bar{z}} = -i \begin{pmatrix} -\bar{a}\bar{b} & \bar{a}^2 \\ -\bar{b}^2 & \bar{a}\bar{b} \end{pmatrix},$$

$$N = -i \frac{1}{|a|^2 + |b|^2} \begin{pmatrix} |a|^2 - |b|^2 & 2\bar{a}b \\ 2a\bar{b} & |b|^2 - |a|^2 \end{pmatrix}.$$

Finally, for the coordinates of the immersion and the Gauss map we obtain the Weierstrass representation:

$$\begin{aligned}
F_1 &= \mathrm{Re} \int^z (g^2 - 1)\eta, & N_1 &= \frac{2\,\mathrm{Re}\,g}{|g|^2 + 1}, \\
F_2 &= \mathrm{Im} \int^z (g^2 + 1)\eta, & N_2 &= \frac{2\,\mathrm{Im}\,g}{|g|^2 + 1}, \\
F_3 &= -2\mathrm{Re} \int^z g\eta, & N_3 &= \frac{|g|^2 - 1}{|g|^2 + 1},
\end{aligned} \qquad (3.1)$$

where $g = a/b$ is an analytic function and $\eta = b^2 dz$ is a holomorphic differential on $\mathcal{R}$.

Proposition 1 *The spin structure of the minimal immersion is given by the spinor $\sqrt{\eta}$, where η is the holomorphic differential in the Weierstrass representation (3.1). For any closed contour $\gamma \subset \mathcal{R}$ the flip number $p(\gamma)$ of $\sqrt{\eta}$ is equal to the parity of twists $P(\Gamma)$ of the corresponding normal strip $\Gamma = F(\gamma)$.*

4 Dual surfaces

Let us consider the special case when the Hopf differential is real

$$Q \in \mathbf{R}. \tag{4.1}$$

The second fundamental form is diagonal and the preimages of the curvature lines are the lines $x = $ const and $y = $ const on the parameter domain. A conformal parametrisation with this property is called *isothermal*.

Definition 3 *Surfaces which admit isothermal coordinates are called isothermal.*

In terms of arbitrary conformal coordinate Property (4.1) can be reformulated as follows:

$$Q(z, \bar{z}) = \frac{1}{2} q(z, \bar{z}) f(z), \tag{4.2}$$

where $f(z)$ is holomorphic and $q(z, \bar{z})$ is real.

Definition 4 *Let $F(z, \bar{z})$ be a conformal immersion of an isothermal surface $\mathcal{F}$ with Hopf differential of the form (4.2). Then a surface $\mathcal{F}^*$, defined via the immersion function $F^* : \mathcal{R} \to \mathbf{R}^3$ with the following formulas for the moving frame*

$$F_z^* = e^{-u} f F_{\bar{z}}, \quad F_{\bar{z}}^* = e^{-u} \bar{f} F_z, \quad N^* = N, \tag{4.3}$$

is called a dual surface

Proposition 2 *The immersion $F^* : \mathcal{R} \to \mathbf{R}^3$ defined above is a conformal parametrisation of an isothermal surface. The metric e^{u^*}, the mean curvature H^* and the Hopf differential Q^* of this surface are given by the formulas:*

$$e^{u^*} = e^{-u} f \bar{f}, \quad H^* = q, \quad Q^* = \frac{1}{2} H f. \tag{4.4}$$

Proof. The definition (4.3) of F^* is self-consistent since the equality $F_{z\bar{z}}^* = F_{\bar{z}z}^*$ is equivalent to $(e^{-u} f F_{\bar{z}})_{\bar{z}} = \frac{1}{2} e^{-u} f \bar{f} q N = (e^{-u} \bar{f} F_z)_z$. Here we use (4.2) and the Gauss-Weingarten equations for $F_{zz}, F_{\bar{z}\bar{z}}$. The conformality of F^* is evident. The expressions (4.4) are obtained by straightforward calculation, for example

$$Q^* = - < F_z^*, N_z^* > = -e^{-u} f < F_{\bar{z}}, (-HF_z - 2e^{-u} QF_{\bar{z}}) > = \frac{1}{2} H f,$$

which shows that $\mathcal{F}^*$ is also isothermal.

Remark. $\mathcal{F}^{**} = \mathcal{F}$ up to a scaling in $\mathbf{R}^3$.

5 Constant mean curvature (CMC) surfaces.

5.1 Formula for immersion

If the mean curvature of $\mathcal{F}$ is constant, then the Gauss-Codazzi equations

$$u_{z\bar{z}} + \frac{1}{2}H^2 e^u - 2Q\bar{Q}e^{-u} = 0, \qquad Q_{\bar{z}} = 0, \tag{5.1}$$

are invariant with respect to the transformation

$$Q \to Q^t = \lambda Q, \quad |\lambda| = 1, \tag{5.2}$$

where $\lambda = e^{2it}$ is a complex number of unit modulus which is the same for all points of $\mathcal{F}$. Integrating the equations for the moving frame with the coefficient Q replaced by $Q^t = \lambda Q$ we obtain a one-parameter family $\mathcal{F}^t$ of surfaces. The transformation (5.2) does not effect the metric and the mean curvature, therefore all the surfaces $\mathcal{F}^t$ are isometric and have the same constant mean curvature. Treating t as a deformation parameter we obtain a classical theorem of Bonnet.

Theorem 4 (Bonnet) *Every constant mean curvature surface has a one-parameter family of isometric deformations preserving both principal curvatures. The deformation is described by the transformations (5.2).*

The invariance of the principal curvatures follows from the fact that K is an isometric invariant $K = -2u_{z\bar{z}}e^{-u}$.

The quaternion Φ solving the system (2.16, 2.18) describes the moving frame $F_z, F_{\bar{z}}, N$ (2.14, 2.15) on the surface. In [2] it was shown that knowing the family $\Phi(z, \bar{z}, \lambda)$ for all $\lambda = e^{2it}$ allows us to integrate the formulas for the moving frame explicitly replacing the integration with respect to $z, \bar{z}$ by a differentiation with respect to t.

Theorem 5 *Let $\Phi(z, \bar{z}, \lambda = e^{2it})$ be a solution of the system*

$$\Phi_z = U(\lambda)\Phi, \qquad \Phi_{\bar{z}} = V(\lambda)\Phi,$$

$$U(\lambda) = \begin{pmatrix} \dfrac{u_z}{2} & -\lambda Q e^{-u/2} \\ \dfrac{1}{2}He^{u/2} & 0 \end{pmatrix},$$

$$V(\lambda) = \begin{pmatrix} 0 & -\dfrac{1}{2}He^{u/2} \\ \dfrac{1}{\lambda}\bar{Q}e^{-u/2} & \dfrac{u_{\bar{z}}}{2} \end{pmatrix}$$

(5.3)

belonging to the quaternion group $\Phi(z, \bar{z}, \lambda = e^{2it}) \in H_*$ with the norm $\det \Phi = e^{u/2}$. Then F and N, defined by the formulas

$$F = -\frac{1}{H}(\Phi^{-1}\frac{\partial}{\partial t}\Phi - i\Phi^{-1}\sigma_3\Phi), \qquad N = -i\Phi^{-1}\sigma_3\Phi, \tag{5.4}$$

describe a CMC surface with metric e^u the mean curvature H and Hopf differential $Q^t = e^{2it}Q$.

Conversely, let F be a conformal parametrisation of a CMC surface with metric e^u, mean curvature H and Hopf differential Q^t. Then F is given by Formula (5.4) where Φ is a solution of (5.3) as above.

Proof. First we note that both F and N take values in the imaginary quaternions ImH and therefore can be identified (2.11) with vectors in $\mathbf{R}^3$. The system (5.3) coincides with the quaternionic representation (2.18) for the equations for the moving frame with the Hopf differential λQ. Differentiating (5.4), we get

$$F_z = -\frac{1}{H}(\Phi^{-1}\frac{\partial U(\lambda)}{\partial t}\Phi - i\Phi^{-1}[\sigma_3, U(\lambda)]\Phi) = -ie^{u/2}\Phi^{-1}\begin{pmatrix} 0 & 0 \\ 1 & 0 \end{pmatrix}\Phi,$$

$$F_{\bar{z}} = -ie^{u/2}\Phi^{-1}\begin{pmatrix} 0 & 1 \\ 0 & 0 \end{pmatrix}\Phi,$$

which coincides with (2.15).

For a given e^u, H and Q^t the surface is determined up to an Euclidean motion of $\mathbf{R}^3$. The solution $\Phi(z,\bar{z},\lambda=e^{2it}) \in H_*, \det\Phi = e^{u/2}$ is defined up to multiplication on the right by a factor $R(\lambda) \in SU(2)$. This right multiplication $\Phi \to \Phi R(\lambda)$ describes all Euclidean motions of the surface

$$F \to R^{-1}FR + R^{-1}\frac{\partial}{\partial t}R, \qquad R \in SU(2), \quad R^{-1}\frac{\partial}{\partial t}R \in su(2). \tag{5.5}$$

The immersion function $F(z,\bar{z},\lambda = e^{2it_0})$ with a fixed $\lambda = e^{2it_0}$ (one CMC surface from the family) determines $\Phi(z,\bar{z},\lambda)$ uniquely up to the transformation

$$\Phi(z,\bar{z},\lambda) \to \Phi(z,\bar{z},\lambda)R(\lambda), \qquad R(\lambda) = \pm(I + O(t-t_0)^2), \quad t \sim t_0. \tag{5.6}$$

System (5.3) represents itself the Lax representation for the nonlinear equations (5.1). The Lax representation is a starting point of the integration procedure of the soliton theory, which allows us to construct explicit solutions of the corresponding nonlinear integrable equations. The main tool of this procedure is the study of the analytic properties of $\Phi(\lambda)$ with respect to the spectral parameter λ. Moreover, as a by-product of the integration procedure, $\Phi(\lambda)$ is also constructed. This explains why formula (5.4) for the CMC immersion, which seems not to have been known classically, is very useful for analytic treatment of the surfaces. It allows us to eliminate the double integration of the GW equations and supplies us with the final formula for the immersion. For the CMC tori case this helps to control the periodicity of the immersion and, finally, to describe all the CMC tori explicitly [2].

We also mention here a well known fact, which can be easily checked cf. [17].

Proposition 3 *The Gauss map $N : \mathcal{R} \to S^2$ of the CMC surface is harmonic, i.e.*

$$N_{z\bar{z}} = qN, \qquad q : \mathcal{R} \to \mathbf{R}.$$

Remark. In the neighborhood of a non-umbilic point $Q \neq 0$ by a conformal change of coordinate $z \to \tilde{z}(z)$ one can always normalize $Q = H/2$. In this parametrisation the Gauss equation and the system (5.3) become the elliptic sinh-Gordon equation

$$u_{z\bar{z}} + H \sinh u = 0$$

and its Lax representation.

5.2 Monodromy of Φ and balanced diagrams

Let $\Phi(z, \bar{z}, \lambda)$ be a solution of the system (5.3) as in Theorem 5, $\Phi(z, \bar{z}, \lambda = e^{2it}) \in H_*$, $\det \Phi = e^{u/2}$. It is defined on the universal covering of $\mathcal{R}$. Under passage around a closed contour γ on $\mathcal{R}$ this solution gets a monodromy

$$\Phi(z, \bar{z}, \lambda) \xrightarrow{\gamma} {}^\gamma\Phi(z, \bar{z}, \lambda) = \Phi(z, \bar{z}, \lambda) \, {}^\gamma M(\lambda).$$

Since the norm of Φ is preserved, M is unitary

$${}^\gamma M(\lambda = e^{2it}) \in SU(2).$$

Monodromy of the solution depends not on a particular cycle γ but on its homotopy class $[\gamma] \in \pi_1(\mathcal{R})$.

Lemma 3 *Let $F : \mathcal{R} \to \mathbf{R}^3$ be a CMC immersion defined by Formula (5.4) with $\Phi(z, \bar{z}, \lambda = e^{2it_0})$ and suppose that the image of the contour $\gamma \subset \mathcal{R}$ is a closed contour $\Gamma = F(\gamma)$ in $\mathbf{R}^3$. Then*

$${}^\gamma M(\lambda = e^{2it}) = \pm(I + A[\gamma](t - t_0)^2 + B[\gamma](t - t_0)^3 + O(t - t_0)^4), t \sim t_0, \quad (5.7)$$

where the sign is determined by the spin structure of the immersion (see Section 2.3)

Proof. Formula (5.4) yields the following transformation law for the immersion function under the passage around γ

$$F \longrightarrow {}^\gamma F = M^{-1} F M + M \frac{\partial}{\partial t} M_{|t=t_0}, \quad (5.8)$$

which implies ${}^\gamma M(t_0) = \pm I$, $\partial \, {}^\gamma M / \partial t_{|t=t_0} = 0$.

Since $M(\lambda = e^{2it}) \in SU(2)$, both A and B lie in the Lie algebra

$$A, B \in su(2)$$

and can be identified (2.11) with the vectors in $\mathbf{R}^3$.

Proposition 4 *Let $F : \mathcal{R} \to \mathbf{R}^3$ be a CMC immersion. Then to any homology class $[\Gamma] \in H_1(\mathcal{F}, \mathbf{Z})$ there can be associated two vectors $A[\Gamma], B[\Gamma]$. The maps $A : H_1(\mathcal{F}, \mathbf{Z}) \to \mathbf{R}^3$ and $B : H_1(\mathcal{F}, \mathbf{Z}) \to \mathbf{R}^3$ are homomorphisms. The group $E(3)$ of Euclidean motions*

$$F \longrightarrow F^R = R^{-1}FR + r, \quad R \in SU(2), \quad r \in su(2) \tag{5.9}$$

acts on A, B as follows $(A, B \in su(2))$:

$$A \longrightarrow A^R = R^{-1}AR, \quad R \longrightarrow B^R = R^{-1}BR + [A^R, r]. \tag{5.10}$$

Proof. Proving Theorem 5 we have shown that the immersion function $F(z, \bar{z}, \lambda = e^{2it_0})$ determines the solution $\Phi(z, \bar{z}, \lambda)$ uniquely up to the transformation (5.6). This transformation does not affect the terms $O((t-t_0)^2)$ and $O((t-t_0)^3)$ of the monodromy (5.7), therefore the vectors $A[\Gamma], B[\Gamma]$ characterize the immersion function F and not a special Φ. The Euclidean motion (5.9) transforms Φ as follows:

$$\Phi \longrightarrow \Phi^R = \Phi R(I + r(t - t_0)),$$

which implies

$$^\gamma M \longrightarrow {}^\gamma M^R = (I - r(t - t_0) + O((t-t_0)^2))R^{-1}\,{}^\gamma MR(I + r(t - t_0))$$

for the monodromy and (5.10) for A and B.

The multiplication law for the monodromy ${}^{\gamma_1 + \gamma_2}M = {}^{\gamma_1}M \cdot {}^{\gamma_2}M$ yields

$$A[\Gamma_1 + \Gamma_2] = A[\Gamma_1] + A[\Gamma_2], \quad B[\Gamma_1 + \Gamma_2] = B[\Gamma_1] + B[\Gamma_2], \quad \Gamma_i = F(\gamma_i).$$

This shows that A and B are the homomorphisms $H_1(\mathcal{F}, \mathbf{Z}) \to \mathbf{R}^3$.

The maps $A, B : H_1(\mathcal{F}, \mathbf{Z}) \to \mathbf{R}^3$ were constructed in [9] by the variational principle, which provides for these maps a more transparent geometrical and physical interpretations ($A[\Gamma]$ and $B[\Gamma]$ were called force and torque respectively). Namely, $A[\Gamma]$ and $B[\Gamma]$ transform exactly as linear and angular momentum for a moving body. If the translational force $A[\Gamma]$ is non-zero, then a natural "balancing" line in $\mathbf{R}^3$ can be associated to $[\Gamma]$ - the line along which the center of mass travels. For surfaces of revolution the balancing line is exactly the axis. For many cases it is proved that balancing line and homology class representative are "close" to each other. Moreover, sometimes one can proceed further and build a balanced diagram of the surface — a graph in $\mathbf{R}^3$ consisting of the segments of the balancing lines. Using this approach one can characterize the structure of CMC embeddings [10], in particular the localization of the surface with respect to its balancing diagram.

Remark. One more interpretation of the vector $A[\Gamma]$ can be given. For $t \sim t_0, t \neq t_0$ the immersion is not periodic. Formula (5.4) implies

$$^\gamma F = F + 2(t - t_0)A[\Gamma] + O((t-t_0)^2),$$

which allows us to suggest the following interpretation for $A[\Gamma]$:

$$A[\Gamma] = \frac{1}{2}\frac{\partial}{\partial t}({}^\gamma F - F)|_{t=t_0}.$$

Remark. There is one more map, which can be defined on homologies and is standard in the theory of solitons. The right multiplication $\Phi \to \Phi R(\lambda)$ transforms $^\gamma M$ as follows:

$$\gamma M \to R^{-1}(\lambda)^\gamma M R(\lambda).$$

The eigenvalues of $^\gamma M(\lambda)$, which we denote by $^{[\gamma]}m^{\pm 1}(t), |m| = 1$, are preserved by this transformation. To each element $[\gamma] \in H_1(\mathcal{R}, \mathbf{Z})$ there corresponds a function $^{[\gamma]}m : S^1 \to S^1$, which depends only on geometry of the surface and not on its disposition.

5.3 Three parallel surfaces

It is a classical result that surfaces parallel to a CMC surface and lying in the normal direction at distances $1/2H$ and $1/H$ are of constant Gaussian and of constant mean curvature respectively. To see how this fact comes in our description of surfaces we note that the formula (5.4) for the immersion is a sum of two vectors wherethe second one is a normal vector. It is thus natural to consider the surface described by the first term.

Proposition 5 *Let F a CMC surface, described by F, N, Φ as in Theorem 5. Then*

$$F^* = -\frac{1}{H}\Phi^{-1}\frac{\partial}{\partial t}\Phi = F + \frac{1}{H}N, \quad N^* = i\Phi^{-1}\sigma_3\Phi = -N \tag{5.11}$$

is a conformal parametrisation of another surface $\mathcal{F}^*$ and its Gauss map. The metric e^{u^*}, the mean curvature H^* and the Hopf differential Q^{t*} of this surface are given by

$$e^{u^*} = 4\frac{Q\bar{Q}}{H^2}e^{-u}, \quad H^* = H, \quad Q^{t*} = Q^t. \tag{5.12}$$

The surface $\mathcal{F}^*$ is dual to $\mathcal{F}$.

Proof. Differentiating (5.11) we get

$$\begin{aligned} F_z^* &= -\frac{1}{H}\Phi^{-1}\frac{\partial U}{\partial t}\Phi = \frac{2iQ}{H}e^{2it}e^{-u/2}\Phi^{-1}\begin{pmatrix} 0 & 1 \\ 0 & 0 \end{pmatrix}\Phi, \\ F_{\bar{z}}^* &= \frac{2i\bar{Q}}{H}e^{-2it}e^{-u/2}\Phi^{-1}\begin{pmatrix} 0 & 0 \\ 1 & 0 \end{pmatrix}\Phi, \end{aligned} \tag{5.13}$$

which shows the conformality of the parametrisation and orthogonality of $F_z^*, F_{\bar{z}}^*$ to N^*. Calculating scalar products via traces (2.12) we easily get (5.12)

$$\begin{aligned} e^{u^*} &= 2<F_z^*, F_{\bar{z}}^*> = \frac{4Q\bar{Q}}{H^2}e^{-u}\mathrm{tr}\left(\begin{pmatrix} 0 & 1 \\ 0 & 0 \end{pmatrix}\begin{pmatrix} 0 & 0 \\ 1 & 0 \end{pmatrix}\right) = \frac{4Q\bar{Q}}{H^2}e^{-u}, \\ H^* &= -2e^{-u^*}<F_z^*, N_{\bar{z}}^*> = \\ &= -\frac{2Qe^{-u^*}}{H}e^{2it}e^{-u/2}\mathrm{tr}\left(\begin{pmatrix} 0 & 1 \\ 0 & 0 \end{pmatrix}, [\sigma_3, V]\right) = H, \\ Q^{t*} &= <F_z^*, N_z^*> = -\frac{Q}{H}e^{2it}e^{-u/2}\mathrm{tr}\left(\begin{pmatrix} 0 & 1 \\ 0 & 0 \end{pmatrix}, [\sigma_3, U]\right) = Q^t. \end{aligned}$$

Equating Q^t and (4.2)

$$q = H, \quad f = 2Q^t H^{-1}$$

shows that (5.12) coincides with (4.4) and that (5.13) and (4.3) differ only by a sign. This proves the duality of $\mathcal{F}$ and $\mathcal{F}^*$.

The immersion function of the original CMC surface can also be written in a form of a logarithmic derivative with respect to t.

$$F = -\frac{1}{H}\Psi_{[2]}^{-1}\frac{\partial}{\partial t}\Psi_{[2]},$$

where $\Psi_{[2]}$ is a gauge transformed quaternion

$$\Psi_{[2]} = \begin{pmatrix} 1/\sqrt{\lambda} & 0 \\ 0 & \sqrt{\lambda} \end{pmatrix}\Phi, \quad \sqrt{\lambda} = e^{it}.$$

The coefficients in the linear equations for $\Psi_{[2]}$ are the same as for Φ, but λ is inserted in a different way

$$U_{[2]} = \begin{pmatrix} \frac{u_z}{2} & -Qe^{-u/2} \\ \frac{1}{2}\lambda H e^{u/2} & 0 \end{pmatrix}, V_{[2]} = \begin{pmatrix} 0 & -\frac{1}{2\lambda}He^{u/2} \\ \bar{Q}e^{-u/2} & \frac{u_{\bar{z}}}{2} \end{pmatrix}. \quad (5.14)$$

It is also natural to consider the intermediate case

$$\Psi_{[3]} = \begin{pmatrix} 1/\sqrt[4]{\lambda} & 0 \\ 0 & \sqrt[4]{\lambda} \end{pmatrix}\Phi, \quad \sqrt[4]{\lambda} = e^{it/2},$$

where the parameter λ enters symmetrically into the $U - V$ pair

$$U_{[3]} = \begin{pmatrix} \frac{u_z}{2} & -\sqrt{\lambda}Qe^{-u/2} \\ \frac{1}{2}\sqrt{\lambda}He^{u/2} & 0 \end{pmatrix},$$

$$V_{[3]} = \begin{pmatrix} 0 & -\frac{1}{2}\frac{1}{\sqrt{\lambda}}He^{u/2} \\ \frac{1}{\sqrt{\lambda}}\bar{Q}e^{-u/2} & \frac{u_{\bar{z}}}{2} \end{pmatrix}. \quad (5.15)$$

Renaming F, Φ, U, V by

$$F = F_{[1]}, \quad \Phi = \Psi_{[1]}, \quad U = U_{[1]}, \quad V = V_{[1]}$$

we can formulate the following already partially proved

Theorem 6 *The formulas*

$$F_{[i]} = -\frac{1}{H}\Psi_{[i]}^{-1}\frac{\partial}{\partial t}\Psi_{[i]}, \quad i = 1, 2, 3 \quad (5.16)$$

describe 3 parallel surfaces $\mathcal{F}_1, \mathcal{F}_2, \mathcal{F}_3$,

$$F_{[2]} = F_{[1]} + \frac{1}{H}N, \qquad F_{[3]} = F_{[1]} + \frac{1}{2H}N, \tag{5.17}$$

where

$$N = \Psi_{[i]}^{-1}\mathbf{k}\Psi_{[i]} \tag{5.18}$$

is their Gauss map. Surfaces $\mathcal{F}_1$ and $\mathcal{F}_2$ are of constant mean curvature H and dual $\mathcal{F}_1^* = \mathcal{F}_2$. The surface $\mathcal{F}_3$ is of constant Gaussian curvature $K = 4H^2$. Variation of t preserves both principal curvatures of $\mathcal{F}_1, \mathcal{F}_2, \mathcal{F}_3$; for $\mathcal{F}_1, \mathcal{F}_2$ it is an isometry, whereas for $\mathcal{F}_3$ the second fundamental form is preserved.

Proof. All statements about $\mathcal{F}_1$ and $\mathcal{F}_2$ are already proved. The calculation of the fundamental forms of $\mathcal{F}_3$ is quite similar to the calculations in Proposition 5. For the first fundamental form we get

$$A = <F_{[3]z}, F_{[3]z}> = -\frac{Q}{2H}e^{2it}, \quad B = <F_{[3]z}, F_{[3]\bar{z}}> = \frac{1}{2}\frac{Q\bar{Q}}{H^2}e^{-u} + \frac{1}{8}e^u.$$

For the determinant of the first fundamental form

$$<dF_{[3]}, dF_{[3]}> = A(dz)^2 + \bar{A}(d\bar{z})^2 + 2Bdzd\bar{z},$$

$$I_{[3]} = \begin{pmatrix} 2B + A + \bar{A} & i(A - \bar{A}) \\ i(A - \bar{A}) & 2B - (A + \bar{A}) \end{pmatrix},$$

this implies

$$\det I_{[3]} = \left(\frac{Q\bar{Q}}{H^2}e^{-u} - \frac{1}{4}e^u\right)^2. \tag{5.19}$$

Since

$$<F_{[3]z}, N_z> = <F_{[3]\bar{z}}, N_{\bar{z}}> = 0,$$

the parametrisation of $\mathcal{F}_3$ is conformal with respect to the second fundamental form

$$-<dF_{[3]}, dN> = -2<F_{[3]z}, N_{\bar{z}}> dzd\bar{z}.$$

Calculating

$$<F_{[3]z}, N_{\bar{z}}> = \frac{Q\bar{Q}}{H}e^{-u} - \frac{1}{4}He^u,$$

we see that the second fundamental form $II_{[3]}$ does not depend on t and the Gaussian curvature equals

$$K_{[3]} = \det II_{[3]}/\det I_{[3]} = 4H^2.$$

For the mean curvature of $\mathcal{F}_3$ we have

$$H_{[3]} = \frac{1}{2}\mathrm{tr}(II_{[3]} \cdot I_{[3]}^{-1}) = \left(\frac{1}{4}He^u - \frac{Q\bar{Q}}{H}e^{-u}\right)\frac{4B}{\det I_{[3]}} = 2H\frac{\frac{1}{4}H^2e^u + Q\bar{Q}e^{-u}}{\frac{1}{4}H^2e^u - Q\bar{Q}e^{-u}}.$$

Both principal curvatures of $\mathcal{F}_3$ are independent of t.

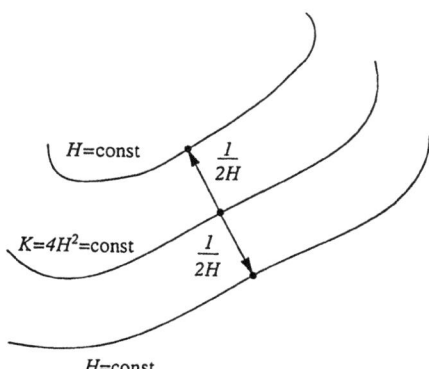

Fig.1 Three parallel surfaces

Remark. Transformations $\mathcal{F}_1 \to \mathcal{F}_2$ and $\mathcal{F}_1 \to \mathcal{F}_3$ can be singular. If $\mathcal{F}_1$ is smooth, then $\mathcal{F}_2 = \mathcal{F}_1^*$ is degenerate ($e^{u^*} = 0$) at images of umbilic points ($Q = 0$) of $\mathcal{F}_1$. The surface $\mathcal{F}_3$ is degenerate at images of flat points of $\mathcal{F}_1$; $K_{[1]} = 0$ implies $u_{z\bar{z}} = 0$ and, finally, $\det I_{[3]} = 0$ in (5.19).

6 Bonnet surfaces

In the present section we study the following

Problem. *To characterize non-trivial families of isometric surfaces having the same principal curvatures.* By a non-trivial family of surfaces we mean surfaces which do not differ by rigid motions. We suppose also that they do not contain umbilics and are sufficiently smooth. This problem was first studied by Bonnet, therefore we call these surfaces Bonnet surfaces. The most detailed results concerning these surfaces are presented in papers by E. Cartan [3] and S. Chern [4], where they were classified and, in particular, it was proved that they are Weingarten surfaces (the principal curvatures are algebraically related).

As we have shown already in Sect. 5.1, the CMC surfaces possess non-trivial isometries. In this section we exclude the CMC case and suppose that H is a non-trivial function on $\mathcal{F}$. To characterize other Bonnet surfaces let us note that since e^u and H are both preserved, only the Hopf differential Q can be varied by the deformations. Since the Gauss equation (2.6) guarantees that $Q\bar{Q}$ is also preserved, the deformation parameter is inserted, in the GW equations via the transformation

$$Q \to \lambda Q, \qquad \bar{Q} \to \frac{1}{\lambda}\bar{Q}. \tag{6.1}$$

In this way we get the $U - V$ system (5.3) (or $U_{[1]}, V_{[1]}$ in notations of Sect. 5.3), where now λ is allowed to depend on z and $\bar{z}$.

The transformation (6.1) does not effect the right hand sides of the Codazzi equations (2.7, 2.8), therefore the left hand sides should also be preserved:

$$(\lambda Q - Q)_{\bar{z}} = (\frac{\bar{Q}}{\lambda} - \bar{Q})_z = 0.$$

These equations can be easily solved:

$$Q = \frac{\bar{a} + \bar{b}}{b\bar{b} - a\bar{a}}, \qquad \lambda = \frac{1 + b/a}{1 + \bar{a}/\bar{b}}, \tag{6.2}$$

where both a and b are holomorphic. The transformation

$$a \to \frac{1}{1-k}(a + kb), \qquad \bar{a} \to \frac{1}{1-k}(\bar{a} + k\bar{b}),$$
$$b \to \frac{1}{1-k}(b + ka), \qquad \bar{b} \to \frac{1}{1-k}(\bar{b} + k\bar{a})$$

preserves Q and transforms λ

$$\lambda \to \lambda = \left(\frac{a+b}{\bar{a}+\bar{b}}\right)\left(\frac{\bar{b}+k\bar{a}}{a+kb}\right).$$

Now k is an independent parameter (k does not depend on z and $\bar{z}$). Deformations correspond to the case

$$|\Lambda| = 1 \iff |k| = 1.$$

One can see that Q in (6.2) is of the form (4.2). In terms of a new conformal variable $\tilde{z}$ (we avoid umbilic points):

$$\left(\frac{dz}{d\tilde{z}}\right)^2 = a + b$$

Q becomes real-valued

$$Q(\tilde{z}) = Q(z)\left(\frac{dz}{d\tilde{z}}\right)^2 = \frac{|a+b|^2}{|b|^2 - |a|^2}.$$

Further we omit the tilde and use the old notation z for this new isothermal coordinate. Finally Q and λ are parametrised by one holomorphic function

$$h(z) = \frac{1}{2}\frac{b-a}{b+a}$$

of an isothermal coordinate z:

$$\frac{1}{Q} = h + \bar{h}, \tag{6.3}$$

$$\lambda = \frac{1 - 2it\bar{h}}{1 + 2ith}, \qquad it = \frac{k-1}{k+1}, \tag{6.4}$$

where t (or equivalently k) is a deformation parameter. For unimodular $k = e^{2i\alpha}$ we have real valued $t = \tan\alpha$.

Proposition 6 *The Bonnet surfaces are isothermal and $1/Q$ is a harmonic function of an isothermal coordinate.*

The harmonicity of $1/Q$ is necessary but not sufficient. Substitution of (6.3) into the Codazzi equations (2.7, 2.8) yields

$$h_z H_z = \bar{h}_{\bar{z}} H_{\bar{z}}. \tag{6.5}$$

Since $h(z)$ is holomorphic, formula (6.5) shows that H depends on one variable s only:

$$H_s = H_w = H_{\bar{w}}, \tag{6.6}$$

where

$$s = w + \bar{w}, \qquad \frac{dw}{dz} = \frac{1}{h_z}. \tag{6.7}$$

For the metric this implies

$$e^u = 2\frac{Q_{\bar{z}}}{H_z} = -\frac{2(1/Q)_z}{(1/Q)^2 H_s \frac{dw}{dz}} = -\frac{2|h_z|^2}{(h+\bar{h})^2 H_s}. \tag{6.8}$$

Differentiating the metric twice:

$$u_{z\bar{z}} = 2\frac{|h_z|^2}{(h+\bar{h})^2} - \frac{1}{|h_z|^2}\left(\frac{H_{ss}}{H_s}\right)_s,$$

and substituting all this into the Gauss equation (2.6), we get

$$\left(\left(\frac{H_{ss}}{H_s}\right)_s - H_s\right) R^2 = 2 - \frac{H^2}{H_s}, \tag{6.9}$$

$$R = \frac{h + \bar{h}}{|h_z|^2} \tag{6.10}$$

The vanishing of $2H_s - H^2$ implies the vanishing of $H_s^2 - H_{sss} + H_{ss}^2/H_s$, therefore two cases are possible in principle:

i) $2H_s - H^2 = 0$,

ii) R depends on s only

$$R_w = R_{\bar w} \quad \Longleftrightarrow \quad h_z R_z = \bar h_{\bar z} R_{\bar z}. \tag{6.11}$$

The case (i) has no geometrical meaning, since formula (6.8) shows that H_s must be negative. Finally we end up with the following equation for $h(z)$, which is equivalent to (6.11)

$$\frac{\partial^2}{\partial z^2}\log(h+\bar h) = \frac{\partial^2}{\partial \bar z^2}\log(h+\bar h).$$

This equation shows that $(z = x + iy)$

$$\frac{\partial^2}{\partial x \partial y}\log(1/Q) = 0,$$

therefore $1/Q$ must be a product of two functions, depending on x and y respectively

$$\frac{1}{Q} = p(x)q(y).$$

The harmonicity condition allows us to find p and q explicitly, since the variables separate

$$\frac{p_{xx}}{p} = -\frac{q_{yy}}{q} = l, \tag{6.12}$$

where l does not depend on x and y. To normalize the solutions of (6.12) we have a reparametrization $z \to az + b + ic$ $(a, b, c \in \mathbf{R})$ and a general scaling of $\mathbf{R}^3$ at our disposal. Also, reconstructing $h(z)$ from p and q one can always add an imaginary constant to $h(z)$. This constant is not important for our calculations, since only the sum $h + \bar h$ and the derivative h_z are used, therefore we put this constant equal to zero. After suitable normalization one can easily prove the following

Lemma 4 *All non-constant normalized functions $h(z)$, generated by solutions of (6.12), are listed in the table below:*

h	w	s	$R(s)$
$-iz/2$	$2iz$	$-4y$	$-s$
e^z	$-e^{-z}$	$-(e^{-z}+e^{-\bar z})$	$-s$
iz^2	$\dfrac{1}{2i}\log z$	$\arg z$	$-\dfrac{1}{2}\sin 2s$
$2\sinh z$	$\dfrac{1}{2i}\log\dfrac{e^z-i}{e^z+i}$	$\dfrac{1}{2i}\log\left(\left(\dfrac{e^z-i}{e^z+i}\right)\left(\dfrac{e^{\bar z}-i}{e^{\bar z}+i}\right)\right)$	$-\dfrac{1}{2}\sin 2s$
$2\cosh z$	$\dfrac{1}{2}\log\dfrac{e^z-1}{e^z+1}$	$\dfrac{1}{2}\log\left(\left(\dfrac{e^z-1}{e^z+1}\right)\left(\dfrac{e^{\bar z}-1}{e^{\bar z}+1}\right)\right)$	$-\dfrac{1}{2}\sinh 2s$

where the corresponding coordinates w, s and the function $R(s)$ are also indicated.

Surfaces in terms of 2 by 2 matrices. Old and new integrable cases   105

Actually, not all the cases listed above are geometrically different. Only the cases corresponding to different $R(s)$ differ.

Lemma 5 *Surfaces corresponding to the same $R(s)$ and the same solution $H(s)$ of equation (6.9) belong to the same deformation family, i.e. the metrics and the mean curvatures of these surfaces coincide and the Hopf differentials differ by the transformation (6.1, 6.4).*

Proof. Rewriting characteristics of surfaces in terms of variables w we see, that the statement concerning the mean curvature is trivial since $s = w + \bar{w}$. The metric also depends on s only

$$e^{u(z,\bar{z})} dz d\bar{z} = -\frac{2|h_z|^2}{(h+\bar{h})^2 H_s} dz d\bar{z} = -\frac{2}{R^2 H_s} dw d\bar{w}. \qquad (6.13)$$

The calculation for the Hopf differentials is more complicated:

$$\lambda Q (dz)^2 = \frac{1}{R(s)} r(w, \bar{w}, t)(dw)^2, \quad r(w, \bar{w}, t) = \frac{h_z}{\bar{h}_{\bar{z}}} \frac{(1 - 2it\bar{h})}{(1 + 2ith)}.$$

Direct calculation for the cases 1-4 of the table above yields

$$r_1 = \frac{-\frac{i}{2}(1 + t_1 \bar{z})}{\frac{i}{2}(1 + t_1 z)} = -\frac{(1 + it_1 \bar{w}/2)}{(1 - it_1 w/2)}, \qquad (6.14)$$

$$r_2 = \frac{e^z(1 - 2it_2 e^{\bar{z}})}{e^{\bar{z}}(1 + 2it_2 e^z)} = \frac{(1 - \frac{i}{2t_2}\bar{w})}{(1 + \frac{i}{2t_2}w)},$$

$$r_3 = \frac{2iz(1 - 2t_3 \bar{z}^2)}{-2i\bar{z}(1 - 2t_3 z^2)} = -\frac{(e^{2i\bar{w}} - 2t_3 e^{-2i\bar{w}})}{(e^{-2iw} - 2t_3 e^{2iw})}, \qquad (6.15)$$

$$r_4 = \frac{2 \cosh z (1 - 4it_4 \sinh \bar{z})}{2 \cosh \bar{z} (1 + 4it_4 \sinh z)} = -\frac{\left(e^{2i\bar{u}} - \left(\frac{1 + 4t_4}{1 - 4t_4}\right) e^{-2i\bar{w}}\right)}{\left(e^{-2iw} - \left(\frac{1 + 4t_4}{1 - 4t_4}\right) e^{2iw}\right)},$$

where t_i denotes the deformation parameter t corresponding to the i-th case of the table. The identification

$$t_1 = -\frac{1}{t_2}, \quad 2t_3 = \frac{1 + 4t_4}{1 - 4t_4}$$

proves that the cases 1 and 2 as well as 3 and 4 are isomorphic.

For completeness let us write down the expression for the Hopf differential in case 5 in terms of the variable w:

$$\lambda Q_5 (dz)^2 = \frac{1}{R(s)} r_5(w, \bar{w}, t_5)(dw)^2,$$

$$r_5 = \frac{2\sinh z(1 - 4it_5 \cosh \bar{z})}{2\sinh \bar{z}(1 + 4it_5 \cosh z)} = -\frac{e^{2\bar{w}} - e^{-2\bar{w}}\dfrac{1 - 4it_5}{1 + 4it_5}}{e^{2w} - e^{-2w}\dfrac{1 + 4it_5}{1 - 4it_5}}. \tag{6.16}$$

So there are 3 cases to consider, which we denote by A, B and C following E. Cartan [3]:

$$\begin{aligned} \text{A}: \quad & R_A(s) = -\frac{1}{2}\sin 2s, \\ \text{B}: \quad & R_B(s) = -\frac{1}{2}\sinh 2s, \\ \text{C}: \quad & R_C(s) = -s. \end{aligned} \tag{6.17}$$

Till now the consideration was a local one and dealt with pieces of surfaces. It turns out that it is possible to combine all these pieces into global surfaces in such a way that the isometries preserving the principal curvatures are described by translations along the surface. For this purpose the coordinate w rather then the starting isothermal coordinate z is more convenient. Both the metric (6.13) and the mean curvature depend on the real part of w only, whereas a change of the imaginary part of w corresponds to the transformation (6.1, 6.4) of Q.

Theorem 7 *The non-trivial families of isometric surfaces having the same principal curvatures are the CMC surfaces and families A, B, C, which can be described as follows. The mean curvature $H(s) = H(w + \bar{w})$ is a solution of (6.9) with a negative derivative $H_s < 0$ (here and below the coefficient R should be replaced by R_A, R_B, R_C (6.17) for the cases A, B, C respectively), the metric equals*

$$e^{u(w,\bar{w})}dwd\bar{w} = -\frac{2}{R^2 H_s}dwd\bar{w}, \tag{6.18}$$

and the Hopf differentials for the families A, B, C are as follows:

$$\begin{aligned} Q_A(w,\bar{w})(dw)^2 &= -\frac{2\sin 2\bar{w}}{\sin 2(w+\bar{w})\sin 2w}(dw)^2, \\ Q_B(w,\bar{w})(dw)^2 &= \frac{2\sinh 2\bar{w}}{\sinh 2(w+\bar{w})\sinh 2w}(dw)^2, \\ Q_C(w,\bar{w})(dw)^2 &= -\frac{\bar{w}}{(w+\bar{w})w}(dw)^2. \end{aligned} \tag{6.19}$$

The isometries preserving the principal curvatures are given by the shift transformations of these surfaces

$$w \longrightarrow w + iT, \quad T \in \mathbf{R}. \tag{6.20}$$

Proof. Only formulas (6.19) for Q's and the transformation (6.20) need clarification. To get (6.19) we put $t_3 = 1/2$, $t_5 = 0$, $t_1 = \infty$ in (6.15, 6.16, 6.17). Clearly the parameters t_i can be fixed. The deformations described by them are given by the shift (6.20) as well. To identify t_i and T we substitute (6.20) in (6.19):

Surfaces in terms of 2 by 2 matrices. Old and new integrable cases 107

$$2t_3 = e^{-4T}, \quad \frac{1+4it_5}{1-4it_5} = e^{-4iT}, \quad t_1 = \frac{2}{T}.$$

The curvature lines on the w-domain for the cases A, B, C are presented in Fig.2.

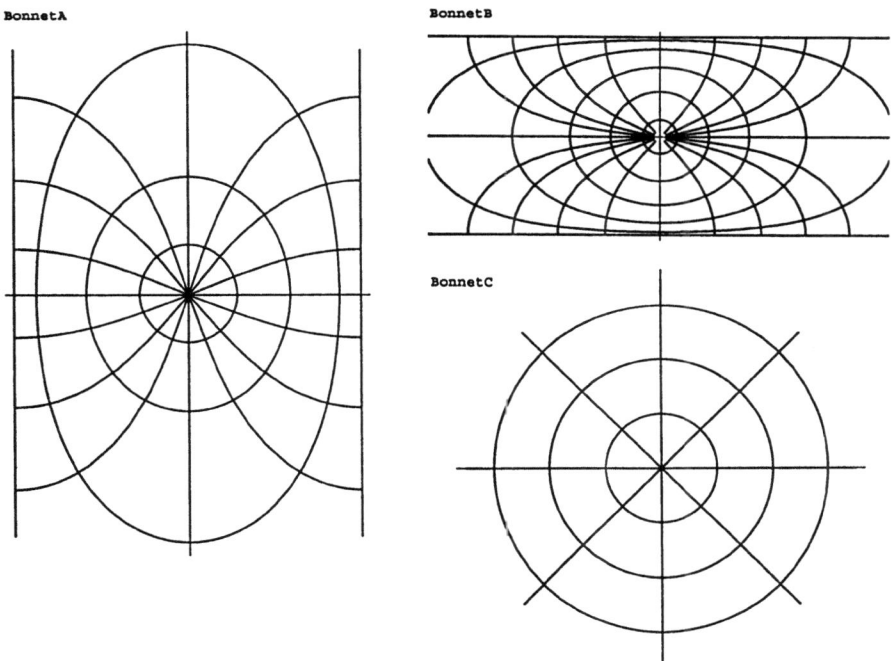

Fig. 2 Curvature lines in the w-domain for the cases A, B, C

A crucial point for the analytical description of the Bonnet surfaces is a solution of equation (6.9). The coefficient $R(s)$ in this equation is of the form

$$R(s) = -\frac{\sinh \alpha s}{\alpha} \tag{6.21}$$

for all the cases A ($\alpha = 2i$), B ($\alpha = 2$) and C ($\alpha \to 0$). Hazzidakis [6] was able to integrate equation (6.9) once[2] with $R(s)$ of the form (6.21):

$$\left[\left(\frac{H_{ss}}{H_s}\right) - 2\sigma_1\right]^2 + 4H\sigma_1 - 2\sigma\frac{H^2}{H_s} - 2H_s = C, \tag{6.22}$$

$$\sigma(s) = \left(\frac{\alpha}{\sinh \alpha s}\right)^2, \quad \sigma_1(s) = -\alpha \coth \alpha s,$$

where C =const. One should mention that equation (6.22) has extra solutions compared with (6.9).

---

[2] I am grateful to Prof. Voss, who pointed out to me this result of Hazzidakis

Equation (6.9) arises as a compatibility condition of the system (5.3), starting with which one can derive a Lax representation [3] for equation (6.9) similar to those, used for integration of the Painleve equations [5]. Following the usual terminology of the soliton theory equation (6.9) can be called an *integrable* ordinary differential equation. Nevertheless, a general integration of equation (6.9) seems not to be a simple problem. One can find a solution $H = r/s$, where r is an arbitrary constant, for the case C. The surface described by this solution is a cone.

Surfaces dual to the Bonnet surfaces can be defined, since the latter are isothermal. Propositions 2 and 6 show that the inverse mean curvature $1/H$ of these dual surfaces is harmonic. In the next section we show that the class of surfaces with harmonic $1/H$ is much bigger then the surfaces dual to the Bonnet families and this class can be put into the framework of soliton theory.

7 Surfaces with harmonic inverse mean curvature.

In Sect. 6 we generalized the Lax representation (5.3) (or $U_{[1]}, V_{[1]}$ in the notations of Sect. 5.3) to the case of λ depending on $z, \bar{z}$. Surfaces defined in this way retain the property of the CMC surfaces of possessing isometries preserving the principal curvatures and are described by solutions of an integrable ordinary differential equation. In the present section we suggest another generalization of the CMC surfaces, namely the property $H = \text{const}$ itself is generalized. Here we deal with the Lax representation $U_{[2]}, V_{[2]}$ (5.14) with λ depending on $z, \bar{z}$, which gives rise to integrable partial differential equations — the more usual situation for the soliton theory than the one of Sect. 6.

The compatibility condition

$$U_{[2]\bar{z}} - V_{[2]z} + [U_{[2]}, V_{[2]}] = 0 \qquad (7.1)$$

for the system (5.14) yields equation (2.6) and

$$Q_{\bar{z}} = \frac{1}{2}\left(\frac{H}{\lambda}\right)_z e^u, \qquad \bar{Q}_z = \frac{1}{2}(\lambda H)_{\bar{z}} e^u. \qquad (7.2)$$

We suppose that e^u, H and Q correspond to some surface and therefore the GC equations (2.6-2.8) are also satisfied. Subtraction of (7.2) from (2.7, 2.8) gives the equations

$$\left(H\left(\frac{1}{\lambda} - 1\right)\right)_z = 0, \qquad (H(\lambda - 1))_{\bar{z}} = 0,$$

which can be easily solved:

$$H = \frac{1}{h + \bar{h}}, \qquad \lambda = -\frac{\bar{h}}{h}, \qquad (7.3)$$

---
[3] One can reproduce the integration step (6.22) using this Lax representation

where $h(z)$ is holomorphic.
The transformation
$$h \to h + \frac{1}{2it}, \quad t \in \mathbf{R}$$
preserves H and transforms λ

$$\lambda \to \lambda = \frac{1 - 2i\bar{h}t}{1 + 2iht}, \tag{7.4}$$

where t is an independent of $z, \bar{z}$ parameter.

The form (7.3) of the mean curvature is equivalent to $1/H$ being a harmonic function:
$$\partial_z \partial_{\bar{z}} \left(\frac{1}{H}\right) = 0.$$

Here the complex structure is determined by the immersion. The parameter t in (7.4) can be considered as a deformation parameter.

Theorem 8 *Let $\mathcal{F}$ be a conformally parametrised $F : \mathcal{R} \to \mathbf{R}^3$ surface with metric e^u, Hopf differential Q and harmonic inverse mean curvature*

$$\frac{1}{H} = h(z) + \overline{h(z)}, \tag{7.5}$$

where $h(z)$ is holomorphic. Then the compatibility condition (7.1, 5.14) with λ of the form (7.4) is satisfied for all t. There is a one-parametric deformation family of surfaces $\mathcal{F}^t$, $t \in \mathbf{R}$ such that:

i) $\mathcal{F} = \mathcal{F}^{t=0}$;

ii) The metrics $e^{u^t} dz d\bar{z}$ of $\mathcal{F}^t$ are conformally equivalent

$$e^{u^t} dz d\bar{z} = \frac{e^u}{(1 + 2iht)^2(1 - 2i\bar{h}t)^2} dz d\bar{z};$$

iii) The inverse mean curvature $1/H^t$ of $\mathcal{F}^t$ is harmonic

$$\frac{1}{H^t} = h^t + \overline{h^t}, \tag{7.6}$$

where h^t is given by

$$h^t = \frac{h}{1 + 2iht}; \tag{7.7}$$

iv) The ratio of the principal curvatures k_1/k_2 is preserved by the deformation;

v) Let $\Psi(z, \bar{z}, t) \in H_*$ be a quaternion, normalized so that $\det \Psi$ is independent of t, which solves the system

$$\Psi_z = U_{[2]}\Psi, \quad \Psi_{\bar{z}} = V_{[2]}\Psi,$$

where λ is of the form (7.4). Then the immersion function

$$F = \Psi^{-1}\frac{\partial}{\partial t}\Psi \qquad (7.8)$$

describes a conformal parametrisation $F : \mathcal{R} \to \mathbf{R}^3$ of $\mathcal{F}^t$. Its Gauss map is given by

$$N = \Psi^{-1}\mathbf{k}\Psi, \quad \mathbf{k} = -i\sigma_3. \qquad (7.9)$$

Proof. The calculations are quite similar to those of Sect. 5. F and N given by formulas (7.8, 7.9) lie in $\text{Im}\mathbf{H}$. Differentiating F

$$F_z = \Psi^{-1}\frac{\partial U_{[2]}}{\partial t}\Psi, \qquad F_{\bar{z}} = \Psi^{-1}\frac{\partial V_{[2]}}{\partial t}\Psi,$$

we see, that the parametrisation is conformal and (2.1, 2.2) are satisfied. For the metric, the mean curvature and the Hopf differential of $\mathcal{F}^t$ determined by (7.8, 7.9) we get

$$\begin{aligned}
e^{u^t} &= 2<F_z, F_{\bar{z}}> = -\text{tr}\left(\frac{\partial U_{[2]}}{\partial t}\frac{\partial V_{[2]}}{\partial t}\right) = \\
&= \frac{1}{4}H^2 e^u \frac{\partial \lambda}{\partial t}\frac{\partial \lambda^{-1}}{\partial t}\text{tr}\left(\begin{pmatrix} 0 & 0 \\ 1 & 0 \end{pmatrix}\begin{pmatrix} 0 & 1 \\ 0 & 0 \end{pmatrix}\right) = \\
&= \frac{e^u}{(1-2i\bar{h}t)^2(1+2iht)^2}, \\
-\frac{1}{2}e^{u^t}H^t &= <F_z, N_{\bar{z}}> = \frac{i}{2}\text{tr}\left(\frac{\partial U_{[2]}}{\partial t}[\sigma_3, V_{[2]}]\right) = \\
&= -\frac{1}{2}e^u H \frac{1}{(1-2i\bar{h}t)(1+2iht)}, \\
Q^t &= -<F_z, N_z> = -\frac{i}{2}\text{tr}\left(\frac{\partial U_{[2]}}{\partial t}[\sigma_3, V_{[2]}]\right) = \frac{Q}{(1+2iht)^2}.
\end{aligned}$$

The transformation of the mean curvature

$$H^t = H(1+2iht)(1-2i\bar{h}t) \qquad (7.10)$$

can be rewritten in the form (7.6, 7.7). The Gaussian curvature $K = -2u_{z\bar{z}}e^{-u}$ transforms as follows:

$$K^t = K(1+2iht)^2(1-2i\bar{h}t)^2,$$

which, combined with (7.10), shows the conservation of $(H^t)^2/K^t$, or equivalently, proves (iv).

Surfaces in terms of 2 by 2 matrices. Old and new integrable cases

Remark. For a given H the holomorphic function $h(z)$ is determined uniquely up to an imaginary constant

$$h(z) \to h(z) + ic, \qquad c \in \mathbf{R}. \tag{7.11}$$

The family $\mathcal{F}^t$ is preserved by the transformation (7.10), which is equivalent to a redefinition of the parameter t only. Combining (7.7) with the transformations (7.11) for h and h^t we get a general Möbius transformation SL(2, $\mathbf{R}$) for ia

$$ih^t = \frac{\alpha(ih) + \beta}{\gamma(ih) + \delta}, \qquad \gamma = 2t, \qquad \alpha, \beta, \gamma, \delta \in \mathbf{R}, \qquad \alpha\delta - \beta\gamma = 1.$$

Remark. The Lax representation $U_{[2]}, V_{[2]}$ with λ of the form (7.4) allows us to apply the soliton theory to construct explicit examples of surfaces with the harmonic inverse mean curvature. In particular the Bäcklund transformation can be derived.

8 Surfaces with negative Gaussian curvature

8.1 Quaternionic description

The calculations of this section are similar to those of Sect. 2.2.

Let us consider a surface $\mathcal{F}$ with negative Gaussian curvature. For each regular point of $\mathcal{F}$ there are two directions in which the curvature vanishes. They are called the *asymptotic directions*. We use the asymptotic line parametrisation of $\mathcal{F}$

$$F : (x, y) \in \mathbf{R}^2 \to \mathbf{R}^3.$$

For this parametrisation the vectors F_x, F_y, F_{xx}, F_{yy} are orthogonal to the normal vector N

$$F_x, F_y, F_{xx}, F_{yy} \perp N.$$

The fundamental forms are as follows:

$$\begin{aligned} I &= \; <dF, dF> = A^2(dx)^2 + 2AB\cos\phi\, dxdy + B^2(dy)^2 \\ II &= \; -<dF, dN> = 2 <F_{xy}, N> dxdy, \end{aligned}$$

where ϕ is the angle between the asymptotic lines and

$$A = |F_x|, \quad B = |F_y|. \tag{8.1}$$

We consider weakly regular surfaces, i.e. suppose that A, B do not vanish. Let us suppose also that the Gaussian curvature is strictly negative

$$K = -\frac{1}{\rho^2}, \qquad \rho > 0. \tag{8.2}$$

For the second fundamental form this implies

$$-\frac{1}{\rho^2} = \frac{\det II}{\det I} = -\frac{<F_{xy},N>^2}{A^2 B^2 \sin^2 \phi}$$

and finally, choosing a suitable direction of the normal vector N, we get

$$II = 2AB\frac{\sin \phi}{\rho} dx\, dy.$$

Let $\Phi \in SU(2)$ be a unitary quaternion, which transforms the basis

$$A(i\cos\frac{\phi}{2} + j\sin\frac{\phi}{2}), \qquad B(i\cos\frac{\phi}{2} - j\sin\frac{\phi}{2}), \qquad \mathbf{k}$$

to the basis F_x, F_y, N:

$$\begin{aligned} F_x &= -iA\Phi^{-1}\begin{pmatrix} 0 & e^{-i\phi/2} \\ e^{i\phi/2} & 0 \end{pmatrix}\Phi, \\ F_y &= -iB\Phi^{-1}\begin{pmatrix} 0 & e^{i\phi/2} \\ e^{-i\phi/2} & 0 \end{pmatrix}\Phi, \\ N &= -i\Phi^{-1}\sigma_3\Phi. \end{aligned} \qquad (8.3)$$

Then the first fundamental form is as above.

To derive linear differential equations for Φ, as in Sect. 2.2 let us introduce the matrices

$$U = \Phi_x \Phi^{-1}, \qquad V = \Phi_y \Phi^{-1} \qquad (8.4)$$

lying in the imaginary quaternions $su(2)$. Orthogonality: $F_{xx} \perp N, F_{yy} \perp N$, i.e.

$$\begin{aligned} 0 &= <F_x, N_x> = \frac{A}{2}\mathrm{tr}\left(\begin{pmatrix} 0 & e^{-i\phi/2} \\ e^{i\phi/2} & 0 \end{pmatrix}[\sigma_3, U]\right), \\ 0 &= <F_y, N_y> = \frac{B}{2}\mathrm{tr}\left(\begin{pmatrix} 0 & e^{i\phi/2} \\ e^{-i\phi/2} & 0 \end{pmatrix}[\sigma_3, V]\right) \end{aligned}$$

shows that the off-diagonal parts of U and V are proportional to $\Phi F_x \Phi^{-1}$ and $\Phi F_y \Phi^{-1}$ respectively. One can calculate the coefficients of proportionality using the identities

$$<F_x, N_y> = <F_y, N_x> = -\frac{AB\sin\phi}{\rho}.$$

For U and V this implies

$$\begin{aligned} U &= -iu_3\sigma_3 - \frac{ia}{2}\begin{pmatrix} 0 & e^{-i\phi/2} \\ e^{i\phi/2} & 0 \end{pmatrix}, \\ V &= -iv_3\sigma_3 + \frac{ib}{2}\begin{pmatrix} 0 & e^{i\phi/2} \\ e^{-i\phi/2} & 0 \end{pmatrix}, \end{aligned}$$

where

$$a = \frac{A}{\rho}, \quad b = \frac{B}{\rho}, \tag{8.5}$$

and u_3, v_3 are some real coefficients. To calculate u_3, v_3 one should use the compatibility condition $F_{xy} = F_{yx}$ of (8.3):

$$-iA_y \begin{pmatrix} 0 & e^{-i\phi/2} \\ e^{i\phi/2} & 0 \end{pmatrix} + \frac{A}{2}\phi_y \begin{pmatrix} 0 & -e^{-i\phi/2} \\ e^{i\phi/2} & 0 \end{pmatrix} -$$

$$- iA \left[\begin{pmatrix} 0 & e^{-i\phi/2} \\ e^{i\phi/2} & 0 \end{pmatrix}, V \right] =$$

$$= -iB_x \begin{pmatrix} 0 & e^{i\phi/2} \\ e^{-i\phi/2} & 0 \end{pmatrix} + \frac{B}{2}\phi_x \begin{pmatrix} 0 & e^{i\phi/2} \\ -e^{-i\phi/2} & 0 \end{pmatrix} -$$

$$- iB \left[\begin{pmatrix} 0 & e^{i\phi/2} \\ e^{-i\phi/2} & 0 \end{pmatrix}, U \right],$$

which yields

$$u_3 = u - \frac{\phi_x}{4}, \quad u = \frac{B_x \cos\phi - A_y}{2B \sin\phi},$$

$$v_3 = v + \frac{\phi_y}{4}, \quad v = \frac{B_x - A_y \cos\phi}{2A \sin\phi}. \tag{8.6}$$

The diagonal part of the compatibility condition of (8.4)

$$U_y - V_x + [U, V] = 0 \tag{8.7}$$

yields the Gauss equation

$$\phi_{xy} + 2v_x - 2u_y - ab \sin\phi = 0, \tag{8.8}$$

whereas the off-diagonal part gives rise to other formulas for u and v:

$$u = \frac{a_y + b_x \cos\phi}{2b \sin\phi}, \quad v = -\frac{b_x + a_y \cos\phi}{2a \sin\phi}. \tag{8.9}$$

Comparing (8.6) and (8.9) we get the Codazzi equations

$$a_y + \frac{\rho_y}{2\rho}a - \frac{\rho_x}{2\rho}b\cos\phi = 0,$$

$$b_x + \frac{\rho_x}{2\rho}b - \frac{\rho_y}{2\rho}a\cos\phi = 0 \tag{8.10}$$

and one more representation for u, v

$$u = -\frac{\rho_y a}{4\rho b} \sin\phi, \quad v = \frac{\rho_x b}{4\rho a} \sin\phi. \tag{8.11}$$

The following theorem is proved:

Theorem 9 *Using the isomorphism (2.11), the moving frame F_x, F_y, N of the surface with negative Gaussian curvature $K = -1/\rho^2$ in the asymptotic line parametrisation is described by formulas (8.3), where $\Phi \in SU(2)$ satisfies equations (8.4) with U, V of the form*

$$U = -i(u - \frac{\phi_x}{4})\sigma_3 - \frac{ia}{2}\begin{pmatrix} 0 & e^{-i\phi/2} \\ e^{i\phi/2} & 0 \end{pmatrix},$$

$$V = -i(v + \frac{\phi_y}{4})\sigma_3 + \frac{ib}{2}\begin{pmatrix} 0 & e^{i\phi/2} \\ e^{-i\phi/2} & 0 \end{pmatrix}.$$

(8.12)

The coefficients u, v here can be written in one of the equivalent forms (8.6, 8.9, 8.11). Equations (8.8, 8.10) are the Gauss-Codazzi equations.

8.2 Surfaces with constant negative Gaussian curvature

For the constant curvature case $\rho = \text{const}$ the GC equations (8.8, 8.10) simplify a lot

$$\phi_{xy} - ab\sin\phi = 0, \qquad a_y = b_x = 0$$

(which implies also $u = v = 0$ in (8.2)). These equations are invariant with respect to the Lorentz transformation

$$a \to \lambda a, \qquad b \to b/\lambda, \qquad \lambda \in \mathbf{R},$$

(8.13)

which plays the role of the transformation (5.2) in the CMC case.

Treating

$$\lambda = e^t, \qquad t \in \mathbf{R}$$

as a deformation parameter we get the following

Theorem 10 *Every surface with constant negative Gaussian curvature has a one-parameter family of deformations preserving the second fundamental form*

$$II = 2\rho ab\sin\phi \, dxdy,$$

the Gaussian curvature and the angle ϕ between the asymptotic lines. The deformation is described by the transformation (8.13).

The quaternion Φ solving the system (8.4) with U, V of the form (8.12), where one should put $u = v = 0$, describes the moving frame of the surface $\mathcal{F}^t$ with $|F_x| = \rho a e^t, |F_y| = \rho b e^{-t}$. As it was shown by A. Sym [15], knowing the family $\Phi(x, y, \lambda)$ for all $\lambda = e^t$ allows us to integrate the formulas for the moving frame explicitly replacing the integration with respect to x, y by a differentiation with respect to t.

Surfaces in terms of 2 by 2 matrices. Old and new integrable cases 115

Theorem 11 *Let* $\Phi(x, y, \lambda = e^t) \in SU(2)$ *be a solution of the system (8.4) with*

$$U = \begin{pmatrix} \frac{i\phi_x}{4} & -\frac{ia}{2}\lambda e^{-i\phi/2} \\ -\frac{ia}{2}\lambda e^{i\phi/2} & -\frac{i\phi_x}{4} \end{pmatrix}, \quad V = \begin{pmatrix} -\frac{i\phi_y}{4} & \frac{ib}{2\lambda}e^{i\phi/2} \\ \frac{ib}{2\lambda}e^{-i\phi/2} & \frac{i\phi_y}{4} \end{pmatrix}. \quad (8.14)$$

Then F and N defined by the formulas

$$F = 2\rho \Phi^{-1}\frac{\partial}{\partial t}\Phi, \quad N = -i\Phi^{-1}\sigma_3 \Phi \quad (8.15)$$

describe a constant negative Gaussian curvature surface with the fundamental forms

$$\begin{aligned} I &= \rho^2(\lambda^2 a^2 dx^2 + 2ab\cos\phi\, dx dy + \lambda^{-2}b^2(dy)^2), \\ II &= 2\rho ab \sin\phi\, dx\, dy. \end{aligned} \quad (8.16)$$

A surface with constant negative Gaussian curvature in asymptotic line parametrisation with the fundamental forms (8.16) is described by formula (8.15), where Φ is as above.

Proof. Both F and N lie in ImH. The system (8.14) coincides with (8.12) for $\mathcal{F}^t$. Differentiating (8.15), we get

$$F_x = 2\rho\Phi^{-1}\frac{\partial U}{\partial t}\Phi, \quad F_y = 2\rho\Phi^{-1}\frac{\partial V}{\partial t}\Phi,$$

which coincides with (8.3) for $\mathcal{F}^t$. The proof of the second part of the theorem is identical to the proof of the corresponding part of Theorem 5.

For the weakly regular surfaces, i.e. the surfaces with $A \neq 0, B \neq 0$ for all x, y, the conformal change of coordinates $x \to \tilde{x}(x), y \to \tilde{y}(y)$ reparametrises the surface so, that the asymptotic lines are parametrised by arc-lengths (generally different for x and y directions)

$$A = |F_x| = \text{const}, \quad B = |F_y| = \text{const}.$$

In this parametrisation (which is called a *Chebyshev net* if $A = B$) the Gauss equation and the system (8.14) become the sine-Gordon equation with the standard Lax representation. Lastly we mention also a well known fact, which can easily be checked.

Proposition 7 *The Gauss map $N : \mathbf{R}^2 \to S^2$ of the surface with $K = -1/\rho^2 = \text{const} < 0$ is Lorentz-harmonic, i.e.*

$$N_{xy} = qN, \quad q : \mathbf{R}^2 \to \mathbf{R}.$$

It forms in S^2 the same kind of Chebyshev net as the immersion function does in $\mathbf{R}^3$

$$|N_x| = a, \quad |N_y| = b.$$

8.3 Bianchi surfaces

To generalize the Lax representation (8.14) to the case of λ depending on x, y let us first rewrite the GC equations of a surface in asymptotic line parametrisation in the form

$$\begin{aligned} \phi_{xy} + 2v_x - 2u_y - ab\sin\phi &= 0, \\ a_y \sin\phi - 2av\cos\phi - 2bu &= 0, \\ b_x \sin\phi + 2bu\cos\phi + 2av &= 0, \end{aligned} \quad (8.17)$$

as they come in the compatibility equation $U_y - V_x + [U, V] = 0$ with the $U - V$ pair (8.12).

If λ is inserted into the $U - V$ pair (8.12) in the same way as in Sect. 8.2

$$\begin{aligned} U^{Bi}(\lambda) &= -i(u - \frac{\phi_x}{4})\sigma_3 - \frac{ia\lambda}{2}\begin{pmatrix} 0 & e^{-i\phi/2} \\ e^{i\phi/2} & 0 \end{pmatrix}, \\ V^{Bi}(\lambda) &= -i(v + \frac{\phi_y}{4})\sigma_3 + \frac{ib}{2\lambda}\begin{pmatrix} 0 & e^{i\phi/2} \\ e^{-i\phi/2} & 0 \end{pmatrix}, \end{aligned} \quad (8.18)$$

the Gauss equation is preserved, whereas the two last equations in (8.17) transform as follows:

$$\lambda a_y \sin\phi + \lambda_y a \sin\phi - 2\lambda av\cos\phi - 2b\lambda^{-1}u = 0,$$
$$\lambda^{-1}b_x \sin\phi + (\lambda^{-1})_x b \sin\phi + 2(\lambda^{-1})bu\cos\phi + 2\lambda av = 0. \quad (8.19)$$

We suppose that ρ, ϕ, a, b correspond to the some surface, therefore the GC equations (8.17) are also satisfied. Subtracting (8.17) from (8.19) and using formulas (8.11) for u, v, we get the equations

$$\lambda_y = (\lambda - \frac{1}{\lambda})\frac{\rho_y}{2\rho}, \qquad (\frac{1}{\lambda})_x = (\frac{1}{\lambda} - \lambda)\frac{\rho_x}{2\rho},$$

which can be easily solved:

$$\rho = f(x) + g(y), \qquad \lambda = \sqrt{-\frac{g(y)}{f(x)}}, \quad (8.20)$$

where $f(x)$ and $g(y)$ are two arbitrary functions.

The transformation

$$f \to f - \frac{1}{2t}, \qquad g \to g + \frac{1}{2t}, \qquad t \in \mathbf{R}$$

preserves ρ and transforms λ

$$\lambda \longrightarrow \tilde{\lambda} = \sqrt{\frac{1 + 2tg}{1 - 2tf}}, \quad (8.21)$$

Surfaces in terms of 2 by 2 matrices. Old and new integrable cases   117

where t is an independent of x, y parameters.

The form (8.20) of the Gaussian curvature $K = -1/\rho^2$ is equivalent to $1/\sqrt{-K}$ being Lorentz-harmonic

$$\partial_y \partial_x \left(\frac{1}{\sqrt{-K}} \right) = 0. \tag{8.22}$$

Surfaces with a curvature of this form were investigated by Bianchi [1], therefore they are called Bianchi surfaces, although may be it would be fairer to call them *Peterson surfaces* after Peterson, who probably studied this problem first (see a comment of Stäckel in [14] [4]).

Theorem 12 *Let $\mathcal{F}$ be a surface with negative Gaussian curvature $K = -1/\rho^2$, $F(x, t)$ its asymptotic line parametrisation and*

$$I = <dF, dF> = \rho^2(a^2(dx)^2 + 2ab\cos\phi\, dx\, dy + b^2(dy)^2),$$
$$II = -<dF, dN> = 2\rho ab \sin\phi\, dx\, dy$$

its fundamental forms. If $\rho = 1/\sqrt{-K}$ is Lorentz-harmonic (8.22)

$$\rho(x, y) = f(x) + g(y),$$

then the compatibility condition $U_y^{Bi} - V_x^{Bi} + [U^{Bi}, V^{Bi}] = 0$ with λ of the form (8.21) is satisfied for all t. There is a one-parametric deformation family of surfaces $\mathcal{F}^t$, $t \in \mathbf{R}$ such that:

i) $\mathcal{F} = \mathcal{F}^{t=0}$;

ii) *the fundamental forms of $\mathcal{F}^t$ are as follows:*

$$I^t = <dF^t, dF^t> =$$
$$= (\rho^t)^2((a^t)^2(dx)^2 + 2a^t b^t \cos\phi\, dx\, dy + (b^t)^2(dy)^2),$$
$$II^t = -<dF^t, dN^t> = 2\rho^t a^t b^t \sin\phi\, dx\, dy,$$
$$a^t = \lambda a, \quad b^t = \frac{1}{\lambda} b, \quad \rho^t = \frac{\rho}{(1 + 2tg)(1 - 2tf)}.$$

In particular, the angle ϕ between the asymptotic lines is preserved, $\rho^t = 1/\sqrt{-K^t}$ is Lorentz-harmonic $\rho^t(x, y) = f^t(x) + g^t(y)$, where f^t and g^t are given by the formulas

$$f^t = \frac{f}{1 - 2tf}, \quad g^t = \frac{g}{1 + 2tg};$$

iii) *let $\Psi(x, y, t) \in SU(2)$ be a solution of the system*

$$\Psi_x = U^{Bi}\Psi, \quad \Psi_y = V^{Bi}\Psi,$$

---
[4] I am grateful to Prof. Voss for showing me this paper

where λ is of the form (8.21). Then the immersion function

$$F = 2\Psi^{-1}\frac{\partial}{\partial t}\Psi \qquad (8.23)$$

describes an asymptotic line parametrisation of $\mathcal{F}^t$. Its Gauss map is given by

$$N = \Psi^{-1}\mathbf{k}\Psi, \qquad \mathbf{k} = -i\sigma_3.$$

We omit the proof of this theorem, which is given by a direct calculation analogous to ones used to prove Theorems 8 and 11.

The Lax representation (8.18, 8.21) for the Bianchi surfaces was obtained in [11]. Using this representation the Bäcklund-Darboux transformation [11] and finite gap solutions [8] were constructed.

9 Surfaces with positive Gaussian curvature

At the end of the Sect. 5 the parametrisation conformal with respect to the second fundamental form appeared, when the surface $\mathcal{F}_3$ of constant positive Gaussian curvature was described. This parametrisation can be used to describe general surfaces with $K > 0$ since the second fundamental form of these surfaces is positive. All calculations in this case are quite parallel to the calculations of Sect. 8. Here we present the final results, which can be obtained by a simple replacement of the symbols of Sect. 8:

$$\rho \longrightarrow -i\sigma, \qquad x \longrightarrow z, \qquad y \longrightarrow \bar{z},$$
$$-ia \longrightarrow c, \qquad -ib \longrightarrow \bar{c}, \qquad i\phi \longrightarrow \psi.$$

The Gaussian curvature is positive

$$K = \frac{1}{\sigma^2} \qquad \sigma > 0,$$

and the fundamental forms are follows:

$$I \;=\; <dF, dF> \;=\; \sigma^2(c^2(dz)^2 + 2c\bar{c}\cosh\psi\, dz\, d\bar{z} + \bar{c}^2(d\bar{z})^2),$$

$$II \;=\; -<dF, dN> \;=\; 2\sigma c\bar{c}\sinh\psi\, dz\, d\bar{z}. \qquad (9.1)$$

The mean curvature in this parametrisation is

$$H = \frac{1}{\sigma}\coth\psi.$$

Surfaces in terms of 2 by 2 matrices. Old and new integrable cases 119

Theorem 13 *The moving frame $F_z, F_{\bar{z}}, N$ of the surface with positive Gaussian curvature $K = 1/\sigma^2$ in the parametrisation conformal with respect to the second fundamental form is described by the following formulas*

$$F_z = -i\sigma c \Phi^{-1} \begin{pmatrix} 0 & e^{-\psi/2} \\ e^{\psi/2} & 0 \end{pmatrix} \Phi,$$

$$F_{\bar{z}} = -i\sigma \bar{c} \Phi^{-1} \begin{pmatrix} 0 & e^{\psi/2} \\ e^{-\psi/2} & 0 \end{pmatrix} \Phi,$$

$$N = -i\Phi^{-1}\sigma_3 \Phi,$$

where $\Phi \in SU(2)$ satisfies the GW equations

$$\Phi_z = U\Phi, \qquad \Phi_{\bar{z}} = V\Phi \qquad (9.2)$$

with U, V of the form

$$U = (\frac{\psi_z}{4} + \mathrm{u})\sigma_3 + \frac{c}{2}\begin{pmatrix} 0 & e^{-i\psi/2} \\ e^{i\psi/2} & 0 \end{pmatrix},$$

$$V = (-\frac{\psi_{\bar{z}}}{4} + \mathrm{v})\sigma_3 - \frac{\bar{c}}{2}\begin{pmatrix} 0 & e^{\psi/2} \\ e^{-\psi/2} & 0 \end{pmatrix}.$$

The coefficients u, v here are given by

$$\mathrm{u} = \frac{\sigma_z c}{4\sigma \bar{c}}\sinh\psi, \qquad \mathrm{v} = -\frac{\sigma_{\bar{z}} \bar{c}}{4\sigma c}\sinh\psi.$$

The Gauss-Codazzi equations are as follows:

$$\psi_{z\bar{z}} + 2\mathrm{u}_{\bar{z}} - 2\mathrm{v}_z + c\bar{c}\sinh\psi = 0,$$
$$c_{\bar{z}} + \frac{\sigma_{\bar{z}}}{2\sigma}c - \frac{\sigma_z}{2\sigma}\bar{c}\cosh\psi = 0,$$
$$\bar{c}_z + \frac{\sigma_z}{2\sigma}\bar{c} - \frac{\sigma_{\bar{z}}}{2\sigma}c\cosh\psi = 0.$$

The following theorem is just a reformulation of the statements of Theorem 6 concerning the surface $\mathcal{F}_3$

Theorem 14 *Every surface with constant positive Gaussian curvature $K = 1/\sigma^2$ has a one-parameter deformation family preserving the Gaussian and the mean curvatures and the second fundamental form. This deformation is described by the transformation*

$$c \longrightarrow \lambda c, \qquad \bar{c} \longrightarrow \frac{1}{\lambda}\bar{c}, \qquad \lambda = e^{it}, \qquad t \in \mathbf{R}.$$

Let $\Phi(z, \bar{z}, \lambda = e^{it}) \in SU(2)$ be a solution of the system (9.2) with

$$U = \begin{pmatrix} \frac{\psi_z}{4} & \lambda\frac{c}{2}e^{-\psi/2} \\ \lambda\frac{c}{2}e^{\psi/2} & -\frac{\psi_z}{4} \end{pmatrix}, \qquad V = \begin{pmatrix} -\frac{\psi_{\bar{z}}}{4} & -\frac{\bar{c}}{2\lambda}e^{\psi/2} \\ -\frac{\bar{c}}{2\lambda}e^{-\psi/2} & \frac{\psi_{\bar{z}}}{4} \end{pmatrix}.$$

Then F and N, defined by the formulas

$$F = -2\sigma\Phi^{-1}\frac{\partial}{\partial t}\Phi, \qquad N = -i\Phi^{-1}\sigma_3\Phi,$$

describe a surface with constant positive Gaussian curvature and the fundamental forms

$$\begin{aligned} I &= \sigma^2(\lambda^2 c^2 (dz)^2 + 2c\bar{c}\cosh\psi\, dz\, d\bar{z} + \lambda^{-2}c^{-2}(d\bar{z})^2), \\ II &= 2\sigma c\bar{c}\sinh\psi\, dz\, d\bar{z}. \end{aligned}$$

The surfaces of positive curvature analogous to the Bianchi surfaces are described by the following

Theorem 15 *Let $\mathcal{F}$ be a surface with positive Gaussian curvature $K = 1/\sigma^2$ and (9.1) be its fundamental forms. If $\sigma = 1/\sqrt{K}$ is harmonic*

$$\sigma(z,\bar{z}) = h(z) + \overline{h(z)},$$

where $h(z)$ is holomorphic, then the compatibility condition

$$U_{\bar{z}}^{Bi} - V_z^{Bi} + [U^{Bi}, V^{Bi}] = 0$$

with the matrices

$$\begin{aligned} U^{Bi} &= (\frac{\psi_z}{4} + u)\sigma_3 + \frac{c\lambda}{2}\begin{pmatrix} 0 & e^{-\psi/2} \\ e^{\psi/2} & 0 \end{pmatrix}, \\ V^{Bi} &= (\frac{-\psi_{\bar{z}}}{4} + v)\sigma_3 - \frac{\bar{c}}{2\lambda}\begin{pmatrix} 0 & e^{\psi/2} \\ e^{-\psi/2} & 0 \end{pmatrix}, \end{aligned}$$

and λ of the form

$$\lambda = \sqrt{\frac{1 - 2it\bar{h}}{1 + 2ith}}$$

is satisfied for all t. There is a one-parameter deformation family of surfaces $\mathcal{F}^t, t \in \mathbf{R}$, such that:

i) $\mathcal{F} = \mathcal{F}^{t=0}$;

ii) the fundamental forms of $\mathcal{F}^t$ are (9.1), where one should replace $c, \bar{c}, \sigma$ by

$$c \to c^t = \lambda c, \quad \bar{c} \to \bar{c}^t = \lambda^{-1}\bar{c}, \quad \sigma \to \sigma^t = \frac{\sigma}{(1 + 2iht)(1 - 2i\bar{h}t)}.$$

In particular, $\sigma^t = 1/\sqrt{K^t}$ remains harmonic $\sigma^t(z,\bar{z}) = h^t(z) + \overline{h^t(z)}$, where $h^t(z)$ is given by

$$h^t = \frac{h}{1 + 2iht}.$$

The ratio of the principal curvatures is preserved by the deformation;

iii) let $\Psi(z,\bar{z},t) \in SU(2)$ be a solution of the system

$$\Psi_z = U^{\text{Bi}}(t)\Psi, \qquad \Psi_{\bar{z}} = U^{\text{Bi}}(t)\Psi.$$

Then the immersion function

$$F = 2\Psi^{-1}\frac{\partial}{\partial t}\Psi$$

describes the parametrisation of $\mathcal{F}^t$ conformal with respect to the second fundamental form. Its Gauss map is given by

$$N = \Psi^{-1}\mathbf{k}\Psi.$$

The formulas for the Bianchi surfaces of positive curvature are similar to the formulas of Sect. 7 for the surfaces with harmonic inverse mean curvature. It would be interesting to find a geometrical relation between these surfaces.

10 Appendix. Spinors and spin structures on Riemann surfaces

Let $\mathcal{R}$ be a Riemann surface (compact or not) and z a local coordinate on $\mathcal{R}$.

Definition 5 *Differentials $f(z,\bar{z})\sqrt{dz}$ and $f(z,\bar{z})\sqrt{d\bar{z}}$ of order $1/2$ are called spinors on $\mathcal{R}$ if $f(z,\bar{z})$ have no local monodromy (no branch points) on $\mathcal{R}$.*

These two types of spinors can be considered in exactly the same way. ¿From now on we concentrate on the type $f\sqrt{dz}$, which means that under conformal changes $z \to w(z)$ of the local parameter $f(z,\bar{z})$ transforms as follows:

$$f(z,\bar{z}) = f(w,\bar{w})\sqrt{\frac{dw}{dz}}. \tag{10.1}$$

Let us fix some spinor and investigate structures induced by it. Locally $f(z,\bar{z})$ is defined up to a sign (but it has no branch points !). We show that it supplies closed contours on $\mathcal{R}$ with nontrivial $\mathbf{Z}_2$ numbers, allowing us to define a spin structure — a quadratic form $H_1(\mathcal{R}, \mathbf{Z}_2) \to \mathbf{Z}_2$

Let $\gamma \in \mathcal{R}$ be an embedding of S^1 in $\mathcal{R}$, i.e. a smooth closed contour on $\mathcal{R}$ without self-intersections (we shall also call such contours *simple* contours). In a small neighborhood $A(\gamma)$ of γ, which is topologically an annulus, we introduce a complex coordinate z. Then it is possible to control the global behaviour of the spinor along γ. The function $f(z,\bar{z})$ is not necessarily single-valued on $A(\gamma)$: it can have monodromy +1 or -1 around γ. The transformation law (10.1) shows that if z

and w are two different coordinates on $A(\gamma)$ then the monodromies of $f(z,\bar z)$ and $f(w,\bar w)$ along γ are different if $\sqrt{dw/dz}$ changes sign on passing around γ. To avoid this ambiguity we ask for z to be a map from $A(\gamma)$ to an annular domain on a complex plane. Then the number $p(\gamma) \in \mathbf{Z}_2$ is defined by the monodromy of the spinor, written in terms of this annular coordinate

$$f(z,\bar z) \xrightarrow{\gamma} (-1)^{p(\gamma)} f(z,\bar z).$$

We say that spinor changes or does not change sign along γ if $p(\gamma)$ equals -1 or +1 respectively. It is not difficult to show that:

1) $p(\gamma)$ is independent of the annular map z,

2) $p(\gamma)$ depends only on γ, but not on $A(\gamma)$,

3) $p(\gamma)$ depends only on the isotopy class of γ, i.e. if γ_1 is isotopic to γ_2 (there is a smooth deformation of γ_1 to γ_2, which is an embedding at each stage), then $p(\gamma_1) = p(\gamma_2)$.

Let us consider some characteristic examples.

Example 1. Let γ be an embedding of S^1 in the complex z-plane. Then $\sqrt{dz}$ does not change sign along γ

$$\sqrt{dz} \xrightarrow{\gamma} \sqrt{dz}.$$

Example 2. Let $C = \mathbf{C}/\{z \to z + 2\pi i\}$ be a cylinder represented as the complex z-plane factored by a shift. To prove that $\sqrt{dz}$ changes a sign around the cycle γ of the cylinder

$$\sqrt{dz} \xrightarrow{z \to z+2\pi i} -\sqrt{dz}$$

we note that z is not the annular coordinate. In terms of coordinate $w = e^z$ the spinor $\sqrt{dz}$ looks as follows

$$\sqrt{dz} = \frac{\sqrt{dw}}{\sqrt{w}}.$$

On the w-plane γ is a loop $|w| = const$ around the origin, so w-coordinate maps γ to an annulus. As we have seen in the first example, $\sqrt{dw}$ does not change sign around γ, whereas $\sqrt{w}$ does.

Example 3. Let $T = \mathbf{C}/\{z \to z + a, \ z \to z + b\}$, Im $a/b \neq 0$, be a torus. Example 2 shows that $\sqrt{dz}$ flips under passage around the cycles of the torus

$$\begin{aligned}\sqrt{dz} &\xrightarrow{z \to z+a} -\sqrt{dz},\\ \sqrt{dz} &\xrightarrow{z \to z+b} -\sqrt{dz},\\ \sqrt{dz} &\xrightarrow{z \to z+a+b} -\sqrt{dz},\end{aligned} \quad (10.2)$$

since the annular coordinate w for any of the 3 contours above is constructed as in Example 2.

Surfaces in terms of 2 by 2 matrices. Old and new integrable cases

Flip numbers associated with immersed contours in $\mathcal{R}$ can be introduced in exactly the same way. Let γ be an immersion of S^1 in $\mathcal{R}$ and z a conformal map of γ to a simple contour in $\mathbf{C}$. Then the flip number $p(\gamma)$ is defined by the monodromy of the spinor, written in the z-coordinate as

$$f(z,\bar{z}) \xrightarrow{\gamma} (-1)^{p(\gamma)} f(z,\bar{z}). \tag{10.3}$$

For example, if γ_k is a closed contour in the z-plane having k points of self-intersection, then the flip number of $\sqrt{dz}$ is equal to $p(\gamma_k) = (-1)^k$.

This number $p(\gamma)$ is a characteristic not of the particular immersion γ but of the regular homotopy class $[\gamma]$ of γ (two immersions γ_1 and γ_2 are called *regularly homotopic* if there is a deformation of γ_1 and γ_2 which is an immersion at each stage).

Definition 6 *A flip number $p([\gamma]) \in \mathbf{Z}_2$ of a regular homotopy class $[\gamma]$ is the monodromy (10.3) of the spinor under passage around γ, parametrised by annular coordinate.*

Now we define another structure associated with the spinor - a map s from $\mathbf{Z}_2$-homologies of $\mathcal{R}$ to $\mathbf{Z}_2$. The group $H_1(\mathcal{R}, \mathbf{Z}_2)$ coincides with the cobordism group of embedded sets of simple contours in $\mathcal{R}$. The elements of this group are cobordant classes of embedded sets of simple contours in $\mathcal{R}$. Two embedded sets of simple contours are called *cobordant* if one can be transformed into another by isotopies of contours and "touching transformations"

$$)(\;\; = \;\; \asymp$$

when the contours touch.

Proposition 8 *There is a map $s : H_1(\mathcal{R}, \mathbf{Z}_2) \to \mathbf{Z}_2$ defined by the rule*

$$s(\alpha) = \sum p(\alpha_k) \tag{10.4}$$

where $\alpha = \sum \alpha_k$ is a representation of an element $\alpha \in H_1(\mathcal{R}, \mathbf{Z}_2)$ by a sum of simple non-intersecting contours α_k.

Proof. We have seen already the invariance of $p(\gamma)$ with respect to isotopies. To prove that s is well-defined by the above we must prove the invariance of s with respect to the touching transformation. To show this we use the following

Lemma 6 *Let $\mathcal{P}$ be a pair of pants, i.e. a Riemann surface, which is topologically a sphere with 3 discs removed. The boundary $\partial \mathcal{P} = \alpha + \beta + \gamma$ consists of 3 simple non-intersecting contours. Then*

$$p(\alpha) + p(\beta) + p(\gamma) = 0 \tag{10.5}$$

(all the equalities here and below are in $\mathbf{Z}_2$).

Proof. Let z be a conformal map of $\mathcal{P}$ to the complex plane. The image of it is a plane domain $P \subset \mathbf{C}$, which is topologically a disc with 2 holes. The coordinate z is an annular coordinate for all simple cycles on $\mathcal{P}$. Identity (10.5) is the identity for monodromy of $f(z,\bar{z})$ in the plane domain P.

It is easy to see that the touching transformation always transforms two simple contours into one simple contour, which is cobordant to their sum, or vice versa. These three contours form a pair of pants, the equality (10.5) for which completes the proof of the proposition.

Lemma 7 *Let T be a torus or a torus with a hole and a, b be a canonical basis of cycles ($a \circ b = 1$) of T. Then the form s calculated at the element $na+mb \in H_1(T, \mathbf{Z}_2)$ is given by*

$$s(na + mb) = ns(a) + ms(b) + nm. \tag{10.6}$$

Proof. Since we consider $H_1(T, \mathbf{Z}_2)$, the only equality to be proved is

$$s(a+b) = s(a) + s(b) + 1.$$

Let z be a flat coordinate of the torus. As we have seen in Example 3, $\sqrt{dz}$ changes sign along any simple contour on T. This means that $f(z,\bar{z})$ acquires the factors $(-1)^{s(a)+1}$ and $(-1)^{s(b)+1}$ along the cycles a and b respectively. The group of $f(z,\bar{z})$ is multiplicative, therefore the factor along $a+b$ is equal to $(-1)^{s(a)+s(b)}$. To calculate s at $a+b$ we should multiply the monodromy of $f(z,\bar{z})$ by (-1) since $\sqrt{dz}$ changes sign along $a+b$:

$$f(z,\bar{z})\sqrt{dz} \xrightarrow{a+b} (-1)^{s(a)+s(b)+1} f(z,\bar{z})\sqrt{dz},$$

which completes the proof.

Corollary 3 *For any two cycles γ_1, γ_2 on a torus or on a torus with a hole,*

$$s(\gamma_1 + \gamma_2) = s(\gamma_1) + s(\gamma_2) + \gamma_1 \circ \gamma_2, \tag{10.7}$$

where $\gamma_1 \circ \gamma_2$ is the intersection number.

Proof. Let $\gamma_1 = n_1 a + m_1 b$, $\gamma_2 = n_2 a + m_2 b$. The following calculation proves (10.7):

$$\begin{aligned} s(\gamma_1 + \gamma_2) &= (n_1+n_2)s(a) + (m_1+m_2)s(b) + (n_1+n_2)(m_1+m_2) = \\ &= (n_1 s(a) + m_1 s(b) + n_1 m_1) + (n_2 s(a) + m_2 s(b) + n_2 m_2) + \\ &+ (n_1 m_2 + n_2 m_1) = s(\gamma_1) + s(\gamma_2) + \gamma_1 \circ \gamma_2. \end{aligned}$$

Proposition 9 *$s : H_1(\mathcal{R}, \mathbf{Z}_2) \to \mathbf{Z}_2$ is a quadratic form, i.e. for any two elements $\alpha, \beta \in H_1(\mathcal{R}, \mathbf{Z}_2)$,*

$$s(\alpha + \beta) = s(\alpha) + s(\beta) + \alpha \circ \beta, \tag{10.8}$$

where $\alpha \circ \beta$ is the intersection number.

Proof. Let us fix some standard basis of cycles on a Riemann surface $\mathcal{R}$ of genus G with K holes or punctures and let

$$\alpha = \alpha_1 + \ldots + \alpha_G + \gamma,$$
$$\beta = \beta_1 + \ldots + \beta_G + \delta$$

be a decomposition of α and β in $H_1(\mathcal{R}, \mathbf{Z}_2)$, where α_i, β_i are the simple contours corresponding to the i-th handle of $\mathcal{R}$, whereas γ and δ correspond to homology of holes and punctures. These elements have the following intersection numbers:

$$\alpha_i \circ \beta_i = 1 \quad \alpha_i \circ \alpha_j = \beta_i \circ \beta_j = \alpha_i \circ \beta_j = 0, \quad i \neq j,$$
$$\alpha_i \circ \gamma = \alpha_i \circ \delta = \beta_i \circ \gamma = \beta_i \circ \delta = \gamma \circ \delta = 0.$$

Let us take a part of $\mathcal{R}$ corresponding to the i-th handle. This part can be chosen to be topologically a torus with a hole. Corollary 3, applied to such a part, gives

$$s(\alpha_i + \beta_i) = s(\alpha_i) + s(\beta_i) + \alpha_i \circ \beta_i.$$

Using this we easily get (10.8):

$$s(\alpha + \beta) = \sum_i s(\alpha_i + \beta_i) + s(\gamma + \delta) =$$
$$= \sum_i (s(\alpha_i) + s(\beta_i)) + \sum_i \alpha_i \circ \beta_i + s(\gamma) + s(\delta) =$$
$$= s(\sum_i \alpha_i + \gamma) + s(\sum_i \beta_i + \delta) + \alpha \circ \beta.$$

Definition 7 *The quadratic form $s : H_1(\mathcal{R}, \mathbf{Z}_2) \to \mathbf{Z}_2$ is called a spin structure of the spinor $f(z, \bar{z})\sqrt{dz}$*

Let $\mathcal{R}$ be a Riemann surface with G handles and K punctures and d a contour a homologous to zero surrounding all the punctures. Let us chose a basis a_n, b_n, d_k of $H_1(\mathcal{R}, \mathbf{Z})$ in such a way, that the cycles $d_k, k = 1, \ldots, K$ surround each its own puncture, the cycles $a_n, b_n, n = 1, \ldots, G$ form a canonical basis of the compact part of $\mathcal{R}$ and do not intersect d. Then to the spinor there correspond characteristics $[\alpha, \beta, \delta] \in \mathbf{Z}_2^{2G+K}$

$$\alpha = (\alpha_1, \ldots, \alpha_G), \quad \beta = (\beta_1, \ldots, \beta_G), \quad \delta = (\delta_1, \ldots, \delta_K),$$

where $\alpha_n = s(a_n), \ \beta_n = s(b_n), \ \delta_k = s(d_k)$.
Since d is homologous to zero the sum of $\delta's$ in $\mathbf{Z}_2$ always vanishes

$$\delta_1 + \ldots + \delta_K = 0.$$

All the other characteristics are independent, which shows that there are 2^{2G+K-1} if $K \neq 0$ and 2^{2K} if $K = 0$ different spin structures.

The characteristics $[\alpha, \beta, \delta]$ depend on the choice of the basis of the compact part of $\mathcal{R}$ and on the arrangement of the punctures.

Definition 8 *The scalar product*

$$< \alpha, \beta > = \sum_{i=1}^{G} \alpha_i \beta_i \in \mathbb{Z}_2$$

is called the parity of the spin structure.

Proposition 10 *The parity of the spin structure and the number of $\delta_k{'}s$, which are equal to zero are independent of the choice of basis in $H_1(\mathcal{R}, \mathbb{Z})$.*

The independence of the number of vanishing $\delta_k{'}s$ is evident. The invariance of the parity of the spin structure needs some calculation, which can be found, for example, in [7]. One should consider a symplectic transformation, which relates canonical bases a, b and a', b' and then prove $< \alpha, \beta > = < \alpha', \beta' >$ using (10.8).

Bibliography

[1] L. Bianchi, *Lezioni di geometria differentiale*, Spoerri, Pisa 1902.

[2] A. I. Bobenko, *Constant mean curvature surfaces and integrable equations*, Uspekhi Matem. Nauk **46**:4 (1991) 3-42; English transl.: Russian Math. Surveys **46**:4 (1991), 1-45.

[3] E. Cartan, *Sur les couples de surfaces applicables avec conservation des courbures principales*, Bull. Sc. Math. **66** (1942), 1-30; reprinted in: Oeuvres Completes, Partie III, vol. 2, 1591-1620.

[4] S. Chern, *Deformation of surfaces preserving principal curvatures*, in: Differential Geometry and Complex Analysis, ed. I. Chavel and H. Farkas, Springer, Berlin 1985, 155-163.

[5] A. Its and V. Novokshenov, *The isomonodromic deformation method in the theory of Painlevé equations*, Lecture Notes in Math. 1191, Springer, Berlin 1986.

[6] J. N. Hazzidakis, *Biegung mit Erhaltung der Hauptkrümmungsradien*, J. reine u. angew. Math. **117** (1897), 42-56.

[7] J. Igusa, *Theta-functions*, Grundlagen Math. Wiss. 194, Springer, Berlin 1972.

[8] D. A. Korotkin and V. A. Reznik, *Bianchi surfaces in $\mathbb{R}^3$ and deformation of hyperelliptic curves*, Matem. Zametki **52** (1992), 78-88, 158.

[9] N. Korevaar, R. Kusner and B. Solomon, *The structure of complete embedded surfaces with constant mean curvature*, J. Diff. Geom. **30** (1989), 465-503.

[10] N. Korevaar and R. Kusner, *The global structure of constant mean curvature surfaces*, Invent. Math. (1993) (to appear).

[11] D. Levi and A. Sym, *Integrable systems describing surfaces of non-constant curvature*, Phys. Lett. **149A**, 381-387.

[12] M. Melko and I. Sterling, this volume.

[13] U. Pinkall, *Regular homotopy classes of immersed surfaces*, Topology **24** (1985), 421-434.

[14] P. Stäckel, *Biegungen und conjugirte Systeme*, Math. Ann. **49** (1897), 255-310.

[15] A. Sym, *Soliton surfaces and their application* in: Soliton geometry from spectral problems, Lecture Notes in Physics 239, Springer, Berlin 1985, 154-231

[16] G. Tzitzéica, *Sur une nouvelle classe de surfaces*, C. R. Acad. Sci. Paris, **150** (1910), 955-956, 1227-1229.

[17] J.C. Wood, *Harmonic maps into symmetric spaces and integrable systems*, this volume.

Integrable systems, harmonic maps and the classical theory of surfaces

M. Melko and I. Sterling

1 Harmonic Maps and Surface Theory

Many geometers in the 19th and early 20th century studied surfaces in $\mathbf{R}^3$ with particular conditions on the curvature. Examples include minimal surfaces, surfaces of constant mean curvature and surfaces of constant Gauss curvature. The typical observation was that one could introduce a special coordinate chart (i.e. curvature lines or asymptotic lines) in order to reduce the compatability conditions of the given class of surfaces to some nonlinear P.D.E. In this way, the description of a given class of surfaces is reduced to the study of the solution space of the corresponding P.D.E. The following table illustrates this correspondence. Here, the compatability conditions are given in curvature line coordinates.

Geometry	Soliton P.D.E.
constant mean curvature $H = 1/2$	$\omega_{xx} + \omega_{yy} = -\sinh \omega$
constant Gauss curvature $K = 1$	
constant Gauss curvature $K = -1$	$\omega_{xx} - \omega_{yy} = \sin \omega$
minimal surfaces $H = 0$	$\omega_{xx} + \omega_{yy} = e^\omega$

Table 1

The classical geometers were able to give detailed descriptions of those solutions that satisfy additional conditions (e.g. solutions corresponding to surfaces of rotation), but there was no methodology at hand by which one could characterize the entire space, or at least a reasonably large subset thereof. A brief discussion of their results is relegated to the third section of this paper.

The intervening period of time has seen the development of the theory of integrable systems, or soliton theory. The idea of this theory is to find a bi-Hamiltonian system (of infinite dimension) that has the given P.D.E. (or set of P.D.E.'s) as a compatability condition. This system turns out to be completely integrable, moreover, one can apply the theory of algebraic curves to construct action-angle coordinates on the corresponding phase space.

This theory, which appears to be the correct approach to describing the space of solutions corresponding to a particular class of surfaces, is outlined in the following section.

Here, we shall concern ourselves with the problem of constructing the preliminary data that is needed for the development in Section 2. This is done via harmonic maps. A more detailed account for the case that ε is the Minkowski metric may be found in [21].

In what follows, we assume that $\mathbf{R}^2$ has a metric, which we denote by ε, and we denote the corresponding Laplacian on $(\mathbf{R}^2, \varepsilon)$ by Δ. We are mainly interested in the case that ε is either the standard Euclidean metric or the standard Minkowski metric. We denote the standard metric on S^2 by δ.

A map $\psi : (\mathbf{R}^2, \varepsilon) \to (S^2, \delta)$ is then defined to be *harmonic* if its *tension field* vanishes, that is

$$\tau(\psi) := \mathrm{trace}_\varepsilon (\nabla \, d\psi) = 0 \; ,$$

where ∇ denotes the connection on the vector bundle $T^*(\mathbf{R}^2) \otimes \psi^*(TS^2)$ induced by the metric $\varepsilon^* \otimes \delta$.

The above equation arises as the Euler-Lagrange equation for the variational problem of the energy integral

$$\mathrm{E}_\mathrm{U}(\psi) := \frac{1}{2} \int_\mathrm{U} \| \, d\psi \, \|^2_{\varepsilon \otimes \delta} \, dx \, dy \; .$$

Here, U denotes an open set in $\mathbf{R}^2$ (with compact closure), and $\| \; \|_{\varepsilon \otimes \delta}$ denotes the norm on $T^*(\mathbf{R}^2) \otimes \psi^*(TS^2)$ induced by $\varepsilon^* \otimes \delta$.

It should be pointed out that this variational problem is only formal in the case that ε is not positive definite. Also, the harmonicity of ψ is invariant under conformal transformations (because the domian is two-dimensional).

The following well-known fact is the key to our approach:

Proposition 1.1 *A smooth map* $\psi : (\mathbf{R}^2, \varepsilon) \to (S^2, \delta)$ *is harmonic if and only if*

$$\Delta \psi = \rho \psi \; , \qquad \rho : \mathbf{R}^2 \to \mathbf{R} \; . \tag{1.1}$$

The idea of the proof is simply to consider ψ as a map into $\mathbf{R}^3$, and then to compute the Euler-Lagrange equation with the constraint $\|\psi\|_\delta = 1$, see also [39].

It is natural to try to make use of the symmetric space structure of S^2 in order to put the *harmonic map equation* (1.1) in some tractable form. For that reason we introduce the following notation conventions: First, we consider $\mathrm{G} = S^3$ as being the group of unit quaternions. We denote the quaternions by $\mathbf{H}$ and write $\{1, \mathrm{i}, \mathrm{j}, \mathrm{k}\}$ for the usual basis. Furthermore, we denote the Lie algebra of G by $\mathcal{G} := T_1 S^3$, which

Integrable systems, harmonic maps and the classical theory of surfaces

we identify with the set of imaginary quaternions. Given these identifications, we may write the symmetric splitting of $\mathcal{G}$ corresponding to S^2 by $\mathcal{G} = \mathcal{K} \oplus \mathcal{P}$, where $\mathcal{K} := \mathbf{R}\,\mathrm{i}$, and $\mathcal{P} := \mathbf{R}\,\mathrm{j} \oplus \mathbf{R}\,\mathrm{k}$.

Since $\mathbf{R}^2$ is contractible, the pullback $\psi^*(S^3)$ is necessarily a trivial circle bundle over $\mathbf{R}^2$. This is equivalent to saying that a smooth map $\psi : \mathbf{R}^2 \to S^2$, always has a smooth lift $\Psi : \mathbf{R}^2 \to S^3$ such that the following diagram commutes:

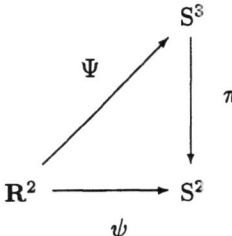

It follows that the maps ψ and Ψ are related by the identity

$$\psi = \Psi\,\mathrm{i}\,\Psi^{-1} \,. \tag{1.2}$$

We wish now to characterize equation (1.1) in terms of the derivative of the lift Ψ. Translating $d\Psi$ back to the identity from the left side gives us a $\mathcal{G}$-valued connection 1-form $A := \Psi^{-1} d\Psi$. The connection A may be decomposed into its $\mathcal{K}$ and $\mathcal{P}$ parts. Specifically,

$$A = A'\,dx + A''\,dy \,, \quad A' = A'_0 + A'_1 \,, \quad A'' = A''_0 + A''_1 \,, \tag{1.3}$$

where, A', A'' are $\mathcal{G}$-valued functions, A'_0, A''_0 are $\mathcal{K}$-valued functions, and A'_1, A''_1 are $\mathcal{P}$-valued functions on $\mathbf{R}^2$.

The $*$-operator on the exterior algebra $\mathcal{E}^*(\mathbf{R}^2)$ induced by the metric ε extends linearly to an endomorphism of the graded Lie algebra of $\mathcal{G}$-valued exterior forms $\mathcal{E}^*(\mathbf{R}^2, \mathcal{G}) \simeq \mathcal{E}^*(\mathbf{R}^2) \otimes \mathcal{G}$.

Definition 1.2 *The connection A is admissible if it satisfies the pair of equations*

$$dA + \frac{1}{2}[A, A] = 0 \,, \quad d*A_1 + [A_0, *A_1] = 0 \,, \tag{1.4}$$

where $A_1 := A'_1 dx + A''_1 dy$ *and* $A_2 := A'_2 dx + A''_2 dy$.

The first equation in (1.4) (the *integrability* or *flatness condition*) guarantees that the connection A integrates to a map Ψ. The second equation (the *harmonicity condition*) is equivalent to (1.1).

We denote the space of harmonic maps by $\mathcal{L}$, and the space of admissible connections by $\mathcal{C}$. Further, we define the gauge group to be $C^\infty(\mathbf{R}^2, S^1)$, where S^1 denotes the complex numbers of unit modulus. An element g in $C^\infty(\mathbf{R}^2, S^1)$ acts on a map Ψ, resp. a connection A by the rule

$$\Psi \mapsto \Psi g^{-1}, \qquad A \mapsto -dg\, g^{-1} + g\, A\, g^{-1}.$$

Since a map $\psi = \Psi i \Psi^{-1}$ is invariant under such gauge tranformations, one easily obtains the following:

Proposition 1.3 *There is a one-to-one correspondence between the space of harmonic maps $\mathcal{L}$ and the moduli space $\mathcal{M}$ of equivalence classes of admissible connections under the action of the gauge group.*

We have found an equivalent formulation of the harmonic map equation (1.1) in terms of the pair of equations (1.4). Now we combine the pair of equations (1.4) into the single equation (1.6) by extending the range of values of the connection A. Let $\mathcal{G}^\mathbf{C} := \mathcal{G} \otimes \mathbf{C} \simeq \mathcal{SL}(\mathbf{C}^2)$ denote the complexification of $\mathcal{G}$, and set $\mathcal{K}^\mathbf{C} := \mathcal{K} \otimes \mathbf{C}$, $\mathcal{P}^\mathbf{C} := \mathcal{P} \otimes \mathbf{C}$. Then the corresponding bracket induces a complex Lie algebra structure on the ring extension $\tilde{\mathcal{G}}^\mathbf{C} := \mathcal{G}^\mathbf{C}[\lambda, \lambda^{-1}]$, where λ takes values in $\mathbf{C}_* := \mathbf{C} \setminus \{0\}$. It is easily shown that the space

$$\tilde{\mathcal{G}}^\mathbf{C}_\sigma = \{ \sum_a \lambda^a \xi_a \in \tilde{\mathcal{G}}^\mathbf{C} \mid \xi_a \in \mathcal{K}^\mathbf{C} \text{ if } a \text{ even}, \xi_a \in \mathcal{P}^\mathbf{C} \text{ if } a \text{ odd} \}$$

induced by the symmetric splitting $\mathcal{G} = \mathcal{K} \oplus \mathcal{P}$ is a Lie subalgebra of $\tilde{\mathcal{G}}^\mathbf{C}$. Complex conjugation extends in the obvious way to $\mathcal{G}^\mathbf{C}$, and therefore induces a conjugation operator on the algebras $\tilde{\mathcal{G}}^\mathbf{C}$ and $\tilde{\mathcal{G}}^\mathbf{C}_\sigma$. We denote all of these conjugation operaors by $(\bar{\ })$.

We now extend A to a family of connections A^λ with values in $\tilde{\mathcal{G}}^\mathbf{C}_\sigma$ by the rule

$$(A^\lambda)' := A_0' + \lambda A_1', \quad (A^\lambda)'' := A_0'' + \lambda^{-1} A_1'', \quad \lambda \in \mathbf{C}_*. \tag{1.5}$$

For reasons to be explained later, λ will be referred to as the *spectral parameter*. The following fundamental observation was first due to Pohlmeyer [27]:

Proposition 1.4 *A is an element of the space of admissible connections $\mathcal{C}$ if and only if its associated loop A^λ satisfies*

$$dA^\lambda + \frac{1}{2}[A^\lambda, A^\lambda] = 0, \qquad \forall\, \lambda \in \mathbf{C}_*. \tag{1.6}$$

In order to recover the the geometry determined by the metric ε, we need to restrict ourselves to some real form of the complex Lie algebra $\tilde{\mathcal{G}}^\mathbf{C}_\sigma$. These real forms can be obtained as the fixed point sets of appropriate involutions on $\tilde{\mathcal{G}}^\mathbf{C}_\sigma$. If we denote the involution corresponding to the Euclidean metric by ι_E, and that corresponding to the Minkowski metric by ι_M, we have

$$\iota_E(\xi^\lambda) = \overline{\xi^{1/\bar\lambda}}, \qquad \iota_M(\xi^\lambda) = \overline{\xi^{\bar\lambda}}.$$

Suppose that $\xi^\lambda = \sum_a \lambda^a \xi_a$, with $\xi_a \in \mathcal{K}^{\mathbf{C}}$ if a is even, and $\xi_a \in \mathcal{P}^{\mathbf{C}}$ if a is odd. Then ξ^λ is invariant under ι_E, resp. ι_M, if $\lambda \in S^1$ resp. $\lambda \in \mathbf{R}_*$, and, in both cases, $\xi_a \in \mathcal{K}$ if a is even, and $\xi_a \in \mathcal{P}$ if a is odd. In the first case, the real form of $\tilde{\mathcal{G}}_\sigma^{\mathbf{C}}$ is an example of a twisted loop algebra. In the second case, one obtains another example of a Euclidean Lie algebra. Both of these algebras can be equiped with ad-invariant inner products.

If the associated family of connections defined by (1.5) is restricted to lie in the appropriate real form of $\tilde{\mathcal{G}}_\sigma^{\mathbf{C}}$, one easily obtains the following:

Corollary 1.5 *To every harmonic map $\psi : \mathbf{R}^2 \to S^2$, there is a naturally associated one-parameter family of harmonic maps ψ^λ. When ε is the Euclidean metric on $\mathbf{R}^2$, $\lambda \in S^1$, and when it is the Minkowski metric, $\lambda \in \mathbf{R}_*$.*

Now we wish to consider the problem of realizing a given harmonic map as the Gauss map of some surface in Euclidean $\mathbf{R}^3$.

With appropriate regularity conditions on the harmonic map ψ, one can put a corresponding admissible connection into a normal form. This process involves a conformal transformation of $(\mathbf{R}^2, \varepsilon)$ and a gauge transformation. Each normal connection represents an equivalence class of harmonic maps under the action of the group of conformal diffeomorphisms of $(\mathbf{R}^2, \varepsilon)$.

For the case that ε is the Minkowski metric, the appropriate regularity is that ψ is "weakly regular", i. e. one assumes that the differential $d\psi$ never vanishes on the characteristic directions. This implies that the corresponding surface has a global parametrization in asymptotic coordinates. It then follows that A can be put in the form

$$A' = -i\Omega_u - \frac{1}{2}e^{-i\Omega} k e^{i\Omega}, \qquad A'' = i\Omega_v + \frac{1}{2}e^{i\Omega} k e^{-i\Omega},$$

where $\Omega = \omega/4$, and $u = x + y$ and $v = x - y$ are asymptotic coordinates. The integrability condition in (1.4) is then easily computed to be the (real) sine-Gordon equation:

$$\omega_{uv} = \omega_{xx} - \omega_{yy} = \sin \omega.$$

Similarly, if ε is the Euclidean metric, we assume that the eigenvalues of $d\psi|_{(x,y)}$, considered as a linear map into $T_{(x,y)}S^2$ are never equal. Geometrically, this condition means that the corresponding surface has no umbilics, and hence global curvature line coordinates. The normal form for A is essentially the same, and the integrability condition becomes the elliptic sinh-Gordon equation:

$$\omega_{xx} + \omega_{yy} = \sinh \omega.$$

Note that the real sine-Gordon equation and the elliptic sinh-Gordon equation are both real forms of the complex sine-Gordon equation.

Proposition 1.6 *Let ψ^λ be the family of harmonic maps associated to a harmonic map ψ, and let Ψ^λ be its lift to S^3. If ε is the Euclidean metric on $\mathbf{R}^2$, then*

$$\varphi^\lambda = \frac{d}{d\tau}|_{\tau=t}\Psi^\lambda\,(\Psi^\lambda)^{-1}\,,\qquad \lambda = e^{i\tau}$$

is a surface of constant Gauss curvature $K = +1$ in $\mathbf{R}^3$ for every t.
Similarly, if ε is the Minkowski metric on $\mathbf{R}^2$, then

$$\varphi^\lambda = \frac{d}{d\tau}|_{\tau=t}\Psi^\lambda\,(\Psi^\lambda)^{-1}\,,\qquad \lambda = e^{\tau}$$

is a surface of constant negative Gauss curvature $K = -1$ in $\mathbf{R}^3$ for every t.

Furthermore, in both cases, if $A = \Psi^{-1}d\Psi$ is in normal form, then $\mathrm{Ad}_{\Psi^\lambda}$ is a canonical frame for each surface φ^λ.

The formulas for the parametrized surface φ^λ given above are originally due to Sym [33] (for $K = -1$) and Bobenko [5] (for $K = +1$ or $H = 1/2$). (Sym has derived similar expressions for a variety of other examples.) We remind the reader that proofs for the case that ε is the Minkowski metric can be found in [21].

As noted in Table 1 the soliton P.D.E. corresponding to surfaces of constant positive Gauss curvature $K = +1$ is also the elliptic sinh-Gordon equation. Geometrically, this equivalence follows from Bonnet's Theorem that to every surface with $K = +1$ there is a parallel surface with $H = 1/2$. A surface with a linear relationship between the Gauss curvature K and the mean curvature H is called a linear Weingarten surface. Table 1 gives a complete list, up to homothety and parallelism, of the nontrivial linear Weingarten surfaces.

So far we have restricted ourselves to linear Weingarten surfaces in three-dimensional Euclidean space. In closing, we would like to mention a few other examples of interesting classes of surfaces that correspond to well-known soliton-P.D.E.'s. These are listed in Table 2, below.

A conformal harmonic immersion of $\mathbf{R}^2$ into the Lorentzian 4-Sphere is a minimal immersion. By means of Lie's sphere geometry one can construct a correspondence between such minimal immersions and Willmore surfaces in $\mathbf{R}^3$. (For a survey of Willmore surfaces, see [26].) The corresponding compatabilty conditions for the surface can be expressed in terms of the Toda system given in Table 2. The Willmore torus (Figure 9) constructed by Ferus and Pedit [16] is one of the simplest nontrivial examples constructed via this method.

If one idealizes a smoke ring as a curve in $\mathbf{R}^3$, then the surface swept out by its evolution over time (Figure 10) is another example of a class of surfaces arising from a soliton equation. In this case, the compatability condition is the nonlinear Schrödinger equation, and the corresponding surface is called a Hasimoto surface (see [17] and also [24]).

Bianchi also studied classes of surfaces whose compatability equations are of a more general character than the Weingarten surfaces discussed above. These results appear in the original version of his book, but not in the German translation. These "Bianchi surfaces" have also been studied by Antoni Sym et. al. These surfaces are discussed in more detail in [7].

Examples of such "soliton geometries" also arise in affine surface theory [34].

Geometry	Soliton P.D.E.
Willmore surfaces	$\omega_{xx} + \omega_{yy} = 2\,e^{-2\omega}\cos\eta - 2\,e^{\omega}$
	$\eta_{xx} + \eta_{yy} = 2\,e^{-2\omega}\sin\eta$
Hasimoto surfaces	$\omega_{xx} = -i\omega_t + \omega|\omega|^2$
Bianchi surfaces	see Bobenko's contribution [7]
affine spheres	$\omega_{xy} = e^{\omega} - e^{-2\omega}$

Table 2

2 Hamiltonian systems, Spectral Theory and Flat Connections.

The purpose of this section is to indicate the correspondence between the flat connectons discussed in the previous section and the theory of Hamiltonian systems.

Let's begin with the assumption that $\tilde{\mathcal{G}}$ is some Lie algebra, possibly of infinite dimension, equiped with an ad-invariant inner product $\langle\ ,\ \rangle$. In the case when $\tilde{\mathcal{G}}$ is finite dimensional, it possesses a canonical Poisson structure. If $C^{\infty}(\tilde{\mathcal{G}})$ is used to denote the space of smooth functions on $\tilde{\mathcal{G}}$, and ∇ is used to denote the gradient on the Euclidean space $(\tilde{\mathcal{G}}, \langle\ ,\ \rangle)$, then this Poisson structure is the bilinear map $\{\ ,\ \}: C^{\infty}(\tilde{\mathcal{G}}) \times C^{\infty}(\tilde{\mathcal{G}}) \to C^{\infty}(\tilde{\mathcal{G}})$ given by

$$\{f, g\}(\chi) := \langle \chi, [\nabla f|_\chi, \nabla g|_\chi] \rangle, \tag{2.1}$$

where $[\ ,\ ]$ denotes the bracket on $\mathcal{G}$. It is known that this bracket is nondegenerate on the adjoint orbits of $\tilde{G}$ (the group corresponding to $\tilde{\mathcal{G}}$) in $\tilde{\mathcal{G}}$, making them into symplectic manifolds.

We now drop the assumption that $\tilde{\mathcal{G}}$ is finite dimensional, and suppose that $L \in \text{End}(\tilde{\mathcal{G}})$ is a linear operator which is self-adjoint with respect to $\langle\ ,\ \rangle$. Suppose, further, that L has a finite-dimensional invariant subspace, which we denote by $\mathbf{V}$, and that the restriction of $\xi \mapsto \text{ad}^{\tilde{\mathcal{G}}}_{L(\xi)}(\xi)$ to $\mathbf{V}$ defines a vector field on $\mathbf{V}$. Here, $\text{ad}^{\tilde{\mathcal{G}}}$ is used to denote the adjoint representation of $\tilde{\mathcal{G}}$. The pullback of the bracket (2.1) to the subspace $\mathbf{V}$ via the inclusion map immediately gives us a finite-dimensional Poisson manifold $(\mathbf{V}, \{\ ,\ \})$. If we further define the Hamiltonian h to be the quadratic function

$$h(\chi) = \frac{1}{2}\langle \chi, L(\chi) \rangle,$$

we have that $(\mathbf{V}, \{\ ,\ \}, h)$ is a Hamiltonian system. Since L is self-adjoint, $\nabla h|_\chi = L(\chi)$. Hence Hamilton's equations take the form

$$\dot{\chi} = -\text{ad}^{\tilde{\mathcal{G}}}_{L(\chi)}(\chi). \tag{2.2}$$

Note that the conditions on L are equivalent to saying that both the gradient and skew-gradient of h are tangent to **V**. It is easily shown that the solution curves to (2.2) have the form

$$\chi(t) = (\Psi(t))^{-1} \xi \Psi(t) , \qquad (2.3)$$

where $\Psi : \mathbf{R} \to \tilde{G}$ is a curve in the group $\tilde{G}$ of $\tilde{\mathcal{G}}$ with $\Psi(0) = 1$, and $\chi(0) = \xi \in \mathbf{V}$ (see [21]).

Note, in particular, that the form of (2.3) implies that the spectrum of $\rho(\chi)$ is independent of the time t, where $\rho : \tilde{\mathcal{G}} \to \text{End}(\mathbf{W})$ denotes a representation of $\tilde{\mathcal{G}}$ on some finite-dimensional vector space **W**. Furthermore, if we complexify **W** to obtain $\mathbf{W}^\mathbf{C} = \mathbf{W} \otimes \mathbf{C}$, we may think of (2.3) as defining a deformation of the decomposition of $\rho(\xi)$, considered as a linear map on $\mathbf{W}^\mathbf{C}$, into eigenspaces.

At this point, we assume that the $\tilde{\mathcal{G}}$ is some Euclidean algebra obtained as an extension of $\mathcal{G}$, the Lie algebra of imaginary quaternions, e.g. either of the real forms of $\tilde{\mathcal{G}}_\sigma^\mathbf{C}$ described in Section 1. Then we have that (2.3) depends on the spectral parameter, i.e.

$$\chi^\lambda(t) = (\Psi^\lambda(t))^{-1} \xi^\lambda \Psi^\lambda(t) . \qquad (2.4)$$

The evaluation map $\xi^\lambda \mapsto \rho(\xi^{\lambda=\lambda_0})$ gives us a representation of $\tilde{\mathcal{G}}$ into $\text{End}(\mathbf{W}^\mathbf{C})$. In analogy with the previous remark, equation (2.4) implies that the affine algebraic curve

$$\Gamma(\xi^\lambda) := \{(\mu, \lambda) \in \mathbf{C} \times \mathbf{C}_* \mid \det(\mu I - \rho(\chi^\lambda(t))) = 0\} \qquad (2.5)$$

where I is the identity on $\mathcal{G}$, is independent of the time t. This curve is called the *spectral curve* associated to the loop ξ^λ. In most of the cases discussed in Section 1, $\Gamma(\xi^\lambda)$ turns out to be hyperelliptic.

By observing the behaviour of the eigenspaces of $\text{ad}^\mathcal{G}(\chi^\lambda(t))$ in $\mathcal{G}^\mathbf{C}$ as a function of t, we may interpret (2.4) as a deforming complex line bundle in $\Gamma(\xi^\lambda) \times \mathcal{G}^\mathbf{C}$. This gives a representation of a solution to (2.3) as a line in the Jacobian of the corresponding spectral curve.

There remains the question of complete integrability of this system. Equation (2.2) is not, in general, completely integrable with the given Poisson structure.

The standard method of obtaining an integrable system is to substitute the given Lie bracket on $\tilde{\mathcal{G}}$ with an another bracket. This new bracket is given by

$$[\xi, \eta]_\mathrm{R} := [\mathrm{R}(\xi), \eta] + [\xi, \mathrm{R}(\eta)]$$

where $\mathrm{R} \in \text{End}(\tilde{\mathcal{G}})$ is required to satisfy the modified classical Yang-Baxter equation

$$\mathrm{R}([\mathrm{R}(\xi), \eta] + [\xi, \mathrm{R}(\eta)]) - [\mathrm{R}(\xi), \mathrm{R}(\eta)] - \alpha[\xi, \eta] = 0 , \quad \alpha \in \mathbf{R}$$

This equation is a sufficient condition for insuring that $[\ ,\ ]_\mathrm{R}$ satisfies the Jacobi identity. A linear endomorphism R that satisfies the Yang-Baxter equation is called an "r-matrix" (see also [39, 9]).

Integrable systems, harmonic maps and the classical theory of surfaces

We denote the canonical Poisson structure associated to $(\tilde{\mathcal{G}}, [\ ,\ ]_R)$ by $\{\ ,\ \}_R$. This new bracket has the property that (certain) conserved quantities of $(\mathbf{V}, \{\ ,\ \}_R, h)$ commute with respect to it. This allows us to apply Arnold's theorem on integrable systems to obtain a dynamical system on a torus (or cylinder). In fact, the branch points of $\Gamma(\xi^\lambda)$, for $\xi^\lambda \in \mathbf{V}$, together with linear coordinates on (a real subtorus of) its Jacobian $\text{Jac}(\xi^\lambda)$ essentially give action-angle coordinates on $\mathbf{V}$.

In this way $\mathbf{V}$ can be given the structure of a bi-Hamiltonian system (see [22] for the definition).

By way of an example, we consider the case of surfaces with constant Gauss curvature $K = -1$ (i.e. when the metric ε on $\mathbf{R}^2$ is the Minkowski metric).

Writing $\tilde{\mathcal{G}} = \lambda^{-1}\mathcal{G}[\lambda^{-1}] \oplus \mathcal{G} \oplus \lambda\mathcal{G}[\lambda]$, we define the needed R-matrix to be $R := \Pi_+ - \Pi_-$, where $\Pi_- : \tilde{\mathcal{G}} \to \lambda^{-1}\mathcal{G}[\lambda^{-1}]$ resp. $\Pi_+ : \tilde{\mathcal{G}} \to \lambda\mathcal{G}[\lambda]$ denotes the projection of a Laurent polynomial onto its tail of negative resp. positive terms. For each nonnegative integer n, we define linear operators L'_n, L''_n on $\tilde{\mathcal{G}}$ by the rules

$$L'_n := \frac{1}{2}(R+1) \circ M_{\lambda^{1-n}}, \qquad L''_n := \frac{1}{2}(R-1) \circ M_{\lambda^{n-1}}, \qquad (2.6)$$

where $M_{p(\lambda)}$ denotes multiplication by the polynomial $p(\lambda)$.

We further define a nested sequence of finite dimensional subspaces of $\tilde{\mathcal{G}}_\sigma$ by the rule

$$\tilde{\mathcal{G}}_\sigma^n = \{\sum_{-n}^{n} \lambda^a \xi_a \in \tilde{\mathcal{G}} \mid \xi_a \in \mathcal{K} \text{ if } a \text{ even}, \xi_a \in \mathcal{P} \text{ if } a \text{ odd}\}.$$

It is easily verified that the restriction of the above operators to $\tilde{\mathcal{G}}_\sigma^n$ is given by

$$L'_n(\xi^\lambda) = \frac{1}{2}\xi_{n-1} + \lambda\xi_n, \qquad L''_n(\xi^\lambda) = -\frac{1}{2}\xi_{1-n} - \lambda^{-1}\xi_{-n} \qquad (2.7)$$

where $\xi^\lambda = \sum_a \lambda^a \xi_a \in \tilde{\mathcal{G}}_\sigma^n$. In particular, they leave $\tilde{\mathcal{G}}_\sigma^n$ invariant for n odd.

Let us denote the real form of $\tilde{\mathcal{G}}_\sigma^{\mathbf{C}}$ corresponding to ε by $\xi^\lambda \in \tilde{\mathcal{G}}_\sigma^\varepsilon$. An element $\xi^\lambda \in \tilde{\mathcal{G}}_\sigma^\varepsilon$ will be called an *admissible loop in normal form* if $\xi^\lambda = \sum_{-n}^{n} \lambda^a \xi_a$ for n odd, $\xi_{-n} = -(1/2)\,e^{i\omega_0}\,k$ and $\xi_n = -(1/2)\,e^{-i\omega_0}\,k$, where $\omega_0 \in \mathbf{R}$.

Recall that we defined (u, v) to be characteristic coordinates on $(\mathbf{R}^2, \varepsilon)$. For notational convenience, let ad denote the adjoint representation of $\tilde{\mathcal{G}}_\sigma^\varepsilon$.

We are now ready to state the main result of this section:

Theorem 2.1 *Let $\xi^\lambda \in \tilde{\mathcal{G}}_\sigma^n$ be an admissible loop in normal form. Then there exists a unique smooth map $\chi^\lambda : \mathbf{R}^2 \to \tilde{\mathcal{G}}_\sigma^n$ satisfying the system of Lax equations*

$$\chi_u = -\text{ad}_{L'_n(\chi^\lambda)}(\chi^\lambda), \quad \chi_v = -\text{ad}_{L''_n(\chi^\lambda)}(\chi^\lambda), \quad \chi^\lambda(0,0) = \xi^\lambda. \qquad (2.8)$$

Furthermore, if $A^\lambda = (A^\lambda)'\, du + (A^\lambda)''\, dv$, where

$$(A^\lambda)' := L'_n(\chi^\lambda), \qquad (A^\lambda)'' := L''_n(\chi^\lambda), \qquad (2.9)$$

then A^λ is the family of connections associated to a connection in normal form.

Remark: The solution χ^λ to the above system of equations also has a geometrical interpretation. It is referred to as a *polynomial Killing field* because it is a Laurent polynomial in λ and an infinitesimal symmetry of the family of connections A^λ. Furthermore, one has that the kernel of $\text{ad}^\mathcal{G}(\chi^{\lambda=1})$ determines the direction of the axis for the corresponding surface $\varphi^{\lambda=1}$. This axis is apparent in the series of pictures at the end of this article.

3 History

Wente, in his 1984 paper [37] (Figure 4), solved the long standing problem of Hopf:

Is a constant mean curvature surface in $\mathbf{R}^3$ (soap bubble) that is complete and compact necessarily a round sphere?

Wente proved that there exist soap bubbles (Wente tori) with the same topological type as a torus. He had learned from Eisenhart's textbook [12] the relationship between soap bubbles and the sinh-Gordon equation discussed in Section 1. Had he read a few more pages in Eisenhart he would have seen that the classical geometers had already found the solutions he would spend years reconstructing via analytical methods [38]. These solutions play a central role in Wente's work.

In fact, in 1985, after Wente's construction, Abresch [1] rediscovered the methods of Enneper (and his school) [13] and [14], used over a century earlier. Abresch gave explicit formulas, in terms of elliptic integrals, of Wente tori. The simplifying ansatz of Enneper was the geometric condition that one family of curvature lines be planar. As mentioned in Section 1, to each constant Gauss curvature $K = +1$ surface, there is a parallel surface with constant mean curvature $H = 1/2$. So there are in some sense three cases: $K = -1$, $K = +1$, and $H = 1/2$. And Enneper had three students: Lenz, Bockwoldt and Voretzsch (resp!). And each wrote a doctoral dissertation [19], [8] and [35]. The dissertation of Voretzsch is the most impressive. The last six pages are tables of numbers giving the x, y, and z coordinates of a constant mean curvature surface. It is important to note that it was not known until Wente's paper in 1984 that some of these examples in fact close up to form compact soap bubbles. As early as 1982 Walter [36] attempted to construct compact examples using Enneper's method. In 1986 Abresch [2] extended Enneper's method. This second paper of Abresch played a pivotal role in the modern development of the theory.

Enneper's solutions, in the case $K = -1$, correspond to the solutions generated by (symmetric) genus two hyperelliptic spectral curves [21] (Figure 3). Solutions corresponding to genus one spectral curves arise geometrically by assuming the stronger condition that the surface is a surface of revolution. The rotationally invariant examples were discovered much earlier (in 1839) by Minding (Figure 1). The analogous constant mean curvature surfaces of revolution were found (in 1841) by Delaunay (Figure 2).

In 1883, Dobriner [10], wrote a long paper essentially repeating Enneper's 1868 work. The editor allowed (in the same issue) a scathing rebuttal by Enneper [15], criticizing Dobriner for being 15 years behind the times!
Nevertheless in 1886 Dobriner [11], solved a much harder problem posed by Enneper:

Characterize those constant Gauss curvature $K = -1$ surfaces with one family of spherical lines of curvature.

In this elegant paper, Dobriner classifies all such surfaces and derives an explicit representation of such surfaces by means of theta functions. Dobriner's solutions correspond to the solutions generated by (symmetric) genus three hyperelliptic spectral curves [21],[24] (Figures 7 and 8).

An excellent historical survey of surfaces of constant curvature is Reckziegel [28]. The earlier works of Bonnet, Dini and Joachimsthal are discussed. Also discussed are the examples of Kuen [18] (Figure 5) and Sievert [29] (Figure 6) which correspond to simple solitons and are given explicitly via trigonometric functions. One of Kuen's beautiful surfaces graces the cover of Gerd Fischer's book (see [28]).

The most detailed and comprehensive work from the classical perspective is that of Steuerwald [32]. This paper also gives a complete discussion of Bäcklund's transformation and it's relationship to Enneper surfaces.

Recall that the sine-Gordon equation, in characteristic coordinates, is given by

$$\partial_u \partial_v \omega = \sin \omega ,$$

where ω is a real-valued function on the plane. In [20], Lie observed that if ω is a solution to this equation, then $\tilde{\omega}(u,v) := \omega(\lambda u, \lambda^{-1} v)$ is also a solution for all nonzero real numbers λ. By virtue of the fundamental theory of surfaces, there is a $K = -1$ surface associated to each $\tilde{\omega}$. This collection of surfaces, constructed via Sym's formula in Section 1 is the *associated family* of the surface corresponding to ω. The parameter λ plays the role of the spectral parameter in the soliton methods described in this paper.

Some of the standard references for surface theory deal with the work of Enneper and Dobriner. These include Darboux and Eisenhart. In 1893 Adam (probably Darboux's student) also studied Enneper surfaces [3]. The Italian edition of Bianchi's textbook [4] is more inclusive than the German translation. For example, Bianchi's multisoliton like surfaces can be found there (Figure 11 and also [30], [31]).

A classification of constant mean curvature tori using soliton theory is given in Pinkall-Sterling [23]. These surfaces can be generated by theta functions using Bobenko's representation [6].

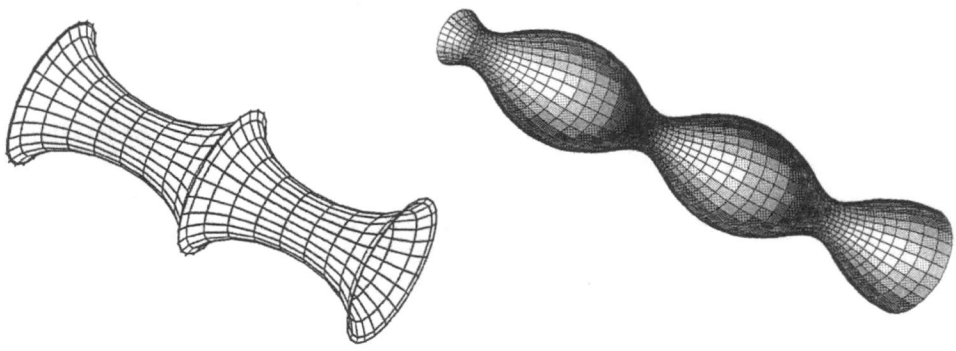

Fig. 1 Minding, K = -1   Fig. 2 Delaunay, H = $\frac{1}{2}$

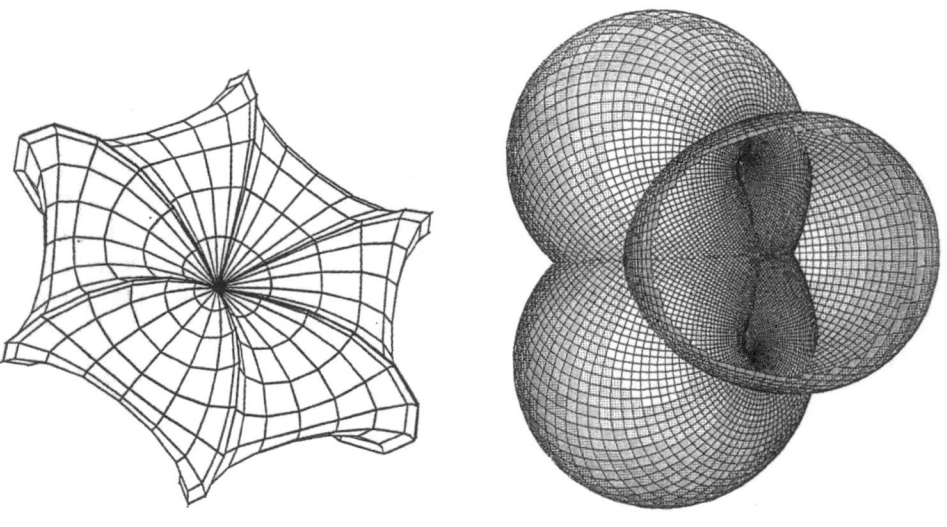

Fig. 3 Enneper, K = -1   Fig. 4 Wente, H = $\frac{1}{2}$

Integrable systems, harmonic maps and the classical theory of surfaces 141

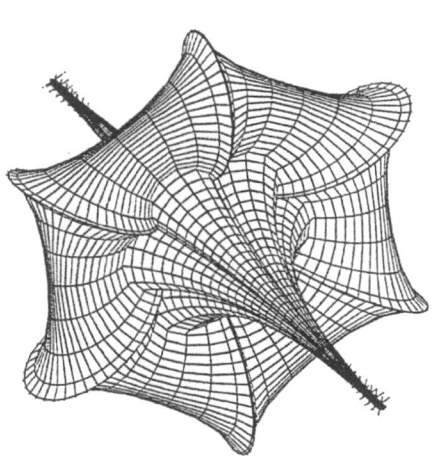

Fig. 5 Kuen, $K = -1$

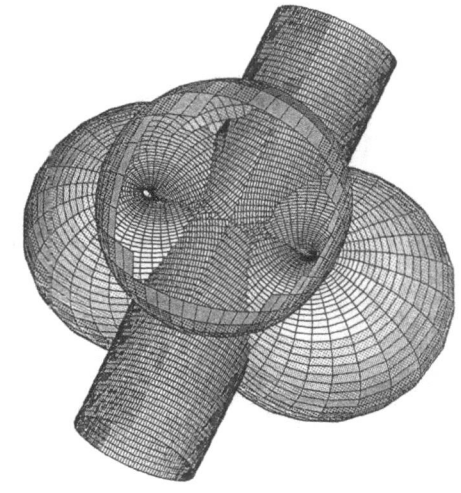

Fig. 6 Sievert, $H = \frac{1}{2}$

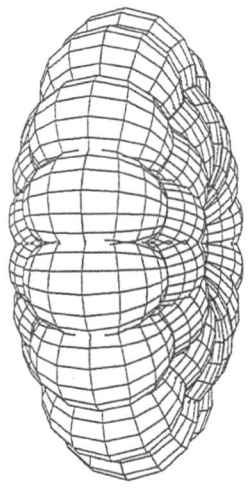

Fig. 7 Dobriner, $K = +1$

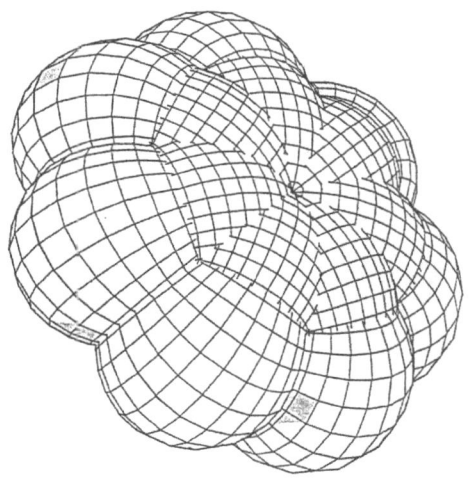

Fig. 8 Dobriner, $H = \frac{1}{2}$

Fig. 9 Ferus-Pedit, Willmore

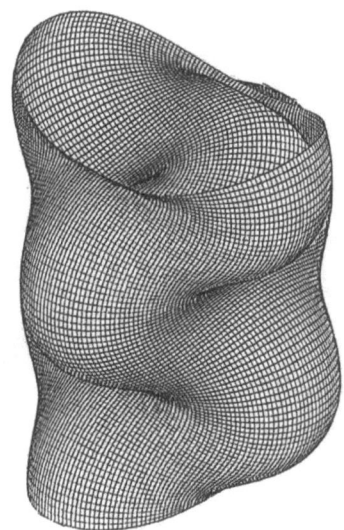

Fig. 10 Pinkall-Sterling, Hasimoto

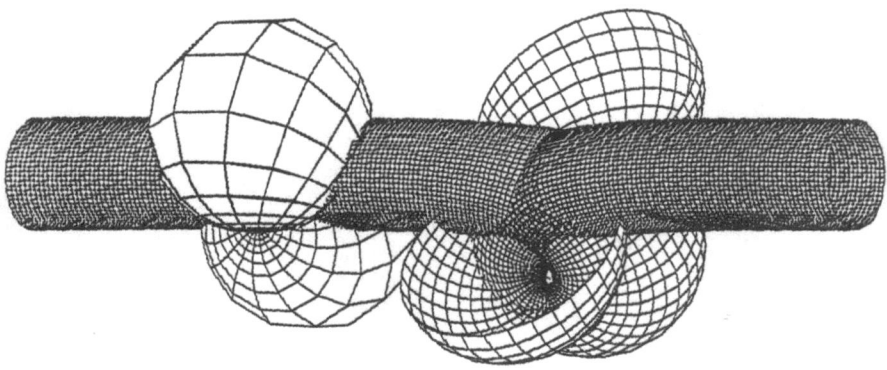

Fig. 11 Bianchi, Multibubbleton

Bibliography

[1] U. Abresch, *Constant mean curvature tori in terms of elliptic functions*, J. reine angew. Math. **374**(1987), 169-192.

[2] U. Abresch, *Old and new periodic solutions of the sinh-Gordon equation*, Seminar on new results in non-linear partial differential equations, Vieweg, Wiesbaden 1987.

[3] P. Adam, *Sur les surfaces isothermique à lignes de courbure planes dans un système ou dans les deux systèmes*, Ann. Sci. Éc. Norm. Sup. **10** (3) (1893), 319-358.

[4] L. Bianchi, *Lezioni di geometria differenziale*, Spoerri, Pisa (1902).

[5] A.I. Bobenko, *Constant mean curvature surfaces and integrable equations*, Russian Math. Surveys **46**:4(1991), 1-45.

[6] A.I. Bobenko, *All constant mean curvature tori in R^3, S^3, H^3 in terms of theta-functions*, Math. Ann. **290** (1991), 209-245.

[7] A.I. Bobenko, *Surfaces in terms of 2 by 2 matrices. Old and new integrable cases*, this volume.

[8] G. Bockwoldt, *Über die Enneperschen Flächen mit konstantem positivem Krümmungsmass*, Dissertation, Universität Göttingen (1878).

[9] F.E. Burstall and F. Pedit, *Harmonic maps via Adler-Kostant-Symes theory*, this volume.

[10] H. Dobriner, *Über die Flächen mit einem System sphärischer Krümmungslinien*, J. reine angew. Math. (Crelle's Journal) **94**(1883), 116-161.

[11] H. Dobriner, *Die Flächen Constanter Krümmung mit einem System Sphärischer Krümmungslinien dargestellt mit Hilfe von Theta Functionen Zweier Variabeln*, Acta Math. **9**(1886), 73-104.

[12] L. P. Eisenhart, *A treatise on the differential geometry of curves and surfaces*, Dover Publications, Inc., New York 1909.

[13] A. Enneper, *Analytisch-geometrische Untersuchungen*, Göttinger Nachr.(1868), 258-277 and 421-443.

[14] A. Enneper, *Untersuchungen über die Flächen mit planen und sphärischen Krümmungslinien*, Abh. Königl. Ges. Wissensch. Göttingen, **23**(1878) and **26**(1880).

[15] A. Enneper, *Über die Flächen mit einem System sphärischer Krümmungslinien*, J. reine angew. Math. (Crelle's Journal) **94**(1883), 329-341.

[16] D. Ferus, and F. Pedit, *S^1-equivariant minimal tori in S^4 and S^1-equivariant Willmore tori in S^3*, Math. Z. **204**(1990), 269-282.

[17] H. Hasimoto, *A soliton on a vortex filament*, J. Fluid Mech. **51**(1972), 477-485.

[18] T. Kuen, *Flächen von constantem negativen Krümmungsmass nach L. Bianchi*, Sitzungsber. der Bayer. Akad. (1884).

[19] E. Lenz, *Über Ennepersche Flächen konstanten negativen Krümmungsmasses*, Dissertation, Universität Göttingen (1879).

[20] S. Lie, *Zur Theorie der Flächen konstanter Krümmung*, Archiv for Mathematik og Naturvidenskab (I u. II)(1879) and (III)(1880), Kristiania.

[21] M. Melko, and I. Sterling, *Application of Soliton Theory to the Construction of Pseudospherical Surfaces in R^3*, Ann. of Glob. Anal. (to appear).

[22] P. Olver, *Applications of Lie Groups to Differental Equations*, Springer-Verlag, Berlin, 1986.

[23] U. Pinkall, and I. Sterling, *On the classification of constant mean curvature tori*, Ann. of Math. **130** (1989), 407-451.

[24] U. Pinkall, and I. Sterling, *Computational aspects of smoke ring deformations*, in preparation.

[25] U. Pinkall, and I. Sterling, *Computational aspects of soap bubble deformations*, Proc. Amer. Math. Soc. (to appear).

[26] U. Pinkall, and I. Sterling, *Willmore surfaces*, Math. Intell. **9**(1987), 38-43.

[27] K. Pohlmeyer, *Integrable Hamiltonian systems and interactions through quadratic constraints*, Comm. Math. Phys. **46**(1976), 207-221.

[28] H. Reckziegel, *Surfaces of Constant Curvature*, Mathematical Models (edited by Gerd Fischer), Vieweg, Wiesbaden 1986, 30-41.

[29] H. Sievert, *Über die Zentralflächen der Enneperschen Flächen konstanten Krümmungsmasses*, Dissertation, Universität Tübingen (1886).

[30] I. Sterling, and H. Wente, *Existence and Classification of Constant Mean Curvature Multibubbletons of Finite and Infinite Type*, Indiana J. Math. (to appear).

[31] I. Sterling, and H. Wente, *Renormalized Energy for Multisolitons and the Langer-Perline Invariant*, University of Toledo preprint (1993).

[32] R. Steuerwald, *Über Enneperische Flächen und Bäcklundsche Transformation*, Abh. d. Bayer. Akad. d. Wiss. (1936).

[33] A. Sym, *Soliton surfaces and their application (Soliton geometry from spectral problems)*, Lect. Notes in Phys. **239**(1985), 154-231.

[34] G. Tzitzeica, *Sur une Nouvelle Classe de Surfaces*, C. R. Acad. Sci. Paris, **150** (1910), 955-956.

[35] M. Voretzsch, *Untersuchung einer speziellen Fläche constanter mittlerer Krümmung bei welcher die eine der beiden Schaaren der Krümmungslinien von ebenen Curven gebildet wird*, Dissertation, Universität Göttingen (1883).

[36] R. Walter, *Zum H-Satz von H. Hopf*, preprint, Universität Dortmund,1982.

[37] H. Wente, *Counterexample to a conjecture of H. Hopf*, Pacific J. Math. **121**(1986), 193-243.

[38] H. Wente, personal conversation.

[39] J.C. Wood, *Harmonic maps into symmetric spaces and integrable systems*, this volume.

SIGMA AND CHIRAL MODELS

The Principal chiral model as an integrable system

M. Mañas

1 Introduction

The study of harmonic maps is an important subject of research not only in Differential Geometry, but also in Theoretical Physics and Mathematical Physics under the name of chiral fields [9]. These are maps with values in nonlinear manifolds such as Lie groups, Grassmannians, projective spaces, spheres, Stiefel manifolds, etc; therefore the equations defining these maps are nonlinear. The two–dimensional case can be solved exactly (with the exception of the Stiefel manifold case, up to now). In Mathematical Physics this was known for the non–linear σ–model since [21], but it was the Russian school in integrable systems who made an exhaustive study of the principal chiral model (chiral fields with values in a Lie group), see [11, 18, 10]. The Zakharov–Shabat dressing method was applied in [31, 32] to construct solutions of this model in a systematic way. The analysis performed in those papers concerns chiral fields from a Minkowski space–time $\mathbb{R}^{1,1}$, hence one is dealing with a hyperbolic evolution equation, and the solutions found there were of soliton type, see [18] for a detailed exposition of the dynamics of these solitons. The integrable character of the principal chiral model is also reflected in the existence of a doubly infinite family of local conservations laws, see [8, 19, 4, 5, 7]. Both aspects can be deduced from the zero–curvature formulation [21, 31, 32] of the model and the consequent Birkhoff factorization technique for constructing solutions. In [1] one can find a beautiful analysis of some particular soliton type solution of the $U(N)$ chiral model, in particular Morse theory is used to describe the dynamics of the one–soliton, and also there is a description of the factorization problem in terms of an infinite–dimensional Grassmannian.

In Differential Geometry the interest in centred on harmonic maps from an Euclidean domain to some manifold, say a Lie group. The seminal paper [26] contains a brilliant analysis of these maps when the Lie group is $U(N)$. The so called uniton solutions can have finite energy, the flag factors [3, 29], and therefore be defined as maps from the two-sphere S^2 into $U(N)$, in fact one can characterizes these solutions by the minimal uniton number, which is less than N. This is an unexpected aspect of the theory, and a rather different situation to that of solitons, in that case there being no upper bound in the soliton number. Let us mention that the approach to the factorization problem of [26] and that of [1] is quite similar. In [29] a method is presented to obtain all the harmonic maps from S^2 to $U(N)$, in the cases $N = 3, 4$ this overlap with the results of [20]. In [24] there is a Grassmannian model for this construction, and in [12] we find a clear exposition of the factorization problem, the natural action, and an infinite-dimensional Grassmannian description. Some of these results can be extended from the Riemann sphere to an arbitrary Riemann surface, this is connected with the so called Hitchin equations or two-dimensional self-duality equations [15]. We want also to remark that in [28] the finite energy 1-unitons were connected to holomorphic bundles over the compactification of certain minitwistor space.

From the above exposition is quite clear that two streams of research regarding chiral fields or harmonic maps from two-dimensional space into Lie groups exists. It appears that the work of the Russian school is not well known to geometers. The purpose of this contribution is to give an exposition of the integrable system aspects of chiral fields or harmonic maps. We are mainly interested in the zero-curvature representation, the associated factorization problem and its relation to the dressing method and with the conservation laws, that will be presented in Sections §3,4,5, and finally in particular solutions arising from this factorization problem §6–12. We review in §7,10 the papers of Zakharov and Mikhailov [31, 32] giving the constructions of soliton type solutions with values in $U(N), O(p, q), Sp(N)$, and in §8,11 we extend these results to Euclidean spaces. In Sections 6,9 we present the reality conditions and finally in Section 12 we give some simple examples.

The main conclusion one obtains is the following. The first part of [26] was essentially known in integrable systems theory since the works [31, 32]. Nevertheless, the second part of that work gives a relevant analysis of finite energy solutions that is completely new in integrable systems theory.

We must remark that there are other relevant aspects of the chiral model as an integrable system that for reasons of space we do not report here. We are thinking now in the Krichever-Cherednik algebro-geometric method for constructing solutions, the so called finite gap potentials, of the principal chiral model [5, 7]. Notice that this might be an interesting area of research — let us mention that in [7] there is an application of this scheme to the self-dual Yang-Mills equations, but till recently [16, 17] there was not a clear analysis of the solutions obtained in this way, for example in [17] the instanton solutions were reproduced as a degenerate case. Solitons are degenerate cases of the finite gap potentials, but can unitons be obtain within this scheme? Another aspect which we have not treated here is the anisotropic principal chiral field, which is also integrable [7, 10]. In this case the action or energy for the harmonic map is anisotropic in the Lie group, and now the factorization problem is not of a Birkhoff type and needs to be formulated in an elliptic curve. Can unitons

The Principal chiral model as an integrable system

be constructed in this anisotropic case? Finally the relation of the principal chiral model with other integrable systems is not touched, see [14, 27] for an analysis of the relation with self-duality equations in four dimensions.

2 The principal chiral model

Let $\mathfrak{g}$ be a complex Lie algebra and consider the functions $u(\xi,\eta), v(\xi,\eta) \in \mathfrak{g}$ depending on $(\xi,\eta)^t \in \mathbb{C}^2$. The principal chiral field equations are (cf. [30])

$$u_\eta + \frac{1}{2}[u,v] = 0,$$

$$v_\xi - \frac{1}{2}[u,v] = 0,$$

where we have use the notation $u_\eta = \partial u/\partial \eta$ and $v_\xi = \partial v/\partial \xi$. This system of equations is equivalent to

$$u_\eta - v_\xi + [u,v] = 0, \qquad (2.1)$$
$$u_\eta + v_\xi = 0. \qquad (2.2)$$

Eq.(2.1) allows one to find locally a chiral field $s(\xi,\eta)$ with values in the corresponding Lie group G such that

$$u = s_\xi \cdot s^{-1}, \quad v = s_\eta \cdot s^{-1},$$

then Eq.(2.2) implies for the chiral field s the following non-linear partial differential equation

$$(s_\xi \cdot s^{-1})_\eta + (s_\eta \cdot s^{-1})_\xi = 0. \qquad (2.3)$$

We say that u,v are the ξ,η-left currents for the chiral field s, respectively. Notice that given a chiral field s any function $s \cdot Q$, with $Q \in G$ a constant, is a chiral field as well. Observe that the left currents are the fundamental endomorphims, up to a normalization constant, appearing in [26, 29].

Suppose now that G is a compact semi–simple Lie group, and that we restrict our attention to $\xi, \eta \in \mathbb{R}$, then we can consider the action $\mathcal{S}$ of the map $s : \mathcal{R} \subset \mathbb{R}^{1,1} \to G$, being $\mathcal{R}$ an open subset, which is defined in terms of the left currents of s as

$$\mathcal{S}(s) = \kappa \int_\mathcal{R} d\xi d\eta \mathrm{B}(u,v),$$

where κ is a normalization constant and B is the Cartan–Killing bilinear form in $\mathfrak{g}$. The critical points of the action $\mathcal{S}$ are the solutions of (2.3). We are dealing with a Lorentzian relativistic system defined over a two–dimensional Minkowski space–time $\mathbb{R}^{1,1}$, and so the coordinates ξ and η can be considered as null–coordinates or light cone coordinates, the standard space–time coordinates in $\mathbb{R}^{1,1}$ are $x = 1/2(\xi + \eta)$ and $t = 1/2(\xi - \eta)$. In these coordinates the non–linear hyperbolic partial differential equation satisfied by the chiral field s is

$$(s_x \cdot s^{-1})_x - (s_t \cdot s^{-1})_t = 0,$$

being the action function

$$S(s) = \kappa \int_{\mathcal{R}} dxdt (\mathrm{B}(s_t \cdot s^{-1}, s_t \cdot s^{-1}) - \mathrm{B}(s_x \cdot s^{-1}, s_x \cdot s^{-1})),$$

where $\mathcal{R}$ can be taken as $\mathbb{R} \times [t_1, t_2]$.

Performing a Wick rotation $t \to it = y$, we can consider $t \in i\mathbb{R}$. Thus $\xi = x + iy, \eta = x - iy = \bar{\xi}$, where $(x, y) \in \mathbb{R}^2$. We are dealing with an Euclidean two-dimensional space–time. The non–linear elliptic partial differential equation for s is

$$(s_x \cdot s^{-1})_x + (s_y \cdot s^{-1})_y = 0, \qquad (2.4)$$

which is the Euler–Lagrange equation for the critical points of the following energy function, or Dirichlet functional,

$$\mathcal{E}(s) = \kappa \int_{\mathcal{R}} dxdy (\mathrm{B}(s_x \cdot s^{-1}, s_x \cdot s^{-1}) + \mathrm{B}(s_y \cdot s^{-1}, s_y \cdot s^{-1})),$$

where $\mathcal{R} \subset \mathbb{R}^2$ is an open subset. Solutions to Eq.(2.4) are called harmonic maps from $\mathcal{R}$ into the group G.

If we consider the groups $G = SU(N), SO(N)$ we can impose the constraint $s = \mathrm{id} - 2\pi$, where $\pi^2 = \pi$, which is equivalent to s being idempotent: $s^2 = \mathrm{id}$. The operator $\pi(\xi, \eta)$ is a projector into the subspace $V(\xi, \eta) := \mathrm{Im}\, \pi(\xi, \eta) \subset \mathbb{C}^N, \mathbb{R}^N$. The map s is continuous, therefore the dimension of $V(\xi, \eta)$ cannot depend on the variables ξ, η. In this way a chiral model with values in a Grassmannian manifold can be considered as a reduction of a principal chiral model, see [31, 11, 18, 26]. The non–linear σ–model or n–field equation [21] arises when one considers one-dimensional subspaces characterized by a unitary vector $\mathbf{n}(\xi, \eta)$, which satisfies the equations

$$\mathbf{n}_{\xi\eta} + (\mathbf{n}_\xi \cdot \mathbf{n}_\eta)\mathbf{n} = 0.$$

3 The zero–curvature condition

Let us define

$$\omega(\lambda) = \frac{u}{1-\lambda} d\xi + \frac{v}{1+\lambda} d\eta, \qquad (3.1)$$

where $\lambda \in \overline{\mathbb{C}} \backslash \{\pm 1\}$ is a complex spectral parameter. Then, ω is a map

$$\omega : \overline{\mathbb{C}} \backslash \{\pm 1\} \to \Lambda(\mathbb{C}^2, \mathfrak{g}),$$

assigning to each λ a $\mathfrak{g}$–valued 1–form. We can consider ω as a connection, then its curvature Ω is defined by

$$\Omega := d\omega - \frac{1}{2}[\omega, \omega].$$

In [31] (cf. [30, 2]) it was observed that the vanishing condition for the curvature Ω of the connection (3.1) is equivalent to the principal chiral field equations (2.1,2.2) for the left currents u and v (this was already noticed, for the non–linear σ–model, in [21]). This follows easily from the identity

$$\frac{1}{1-\lambda}\frac{1}{1+\lambda} = \frac{1}{2}\left(\frac{1}{1-\lambda} - \frac{1}{1+\lambda}\right).$$

There is an equivalent zero–curvature formulation for the principal chiral model proposed in [26] which can be obtain from the one described above. Performing the following homographic transformation

$$\tilde{\lambda} = \frac{\lambda+1}{\lambda-1},$$

with

$$\lambda = \frac{\tilde{\lambda}+1}{\tilde{\lambda}-1},$$

the connection ω is transformed into

$$\omega(\tilde{\lambda}) = \frac{1}{2}\left((1-\tilde{\lambda})ud\xi + (1-\tilde{\lambda}^{-1})vd\eta\right),$$

and the vanishing condition for its curvature is equivalent to the equations (2.1,2.2).

4 The factorization technique

In this section we solve the principal chiral model in terms of a Birkhoff factorization in a Lie group, which can be considered as a matrix Riemann–Hilbert problem. This approach to the subject is not new, it goes back to [33] and was applied for the first time to the principal chiral model in [31], see also [18, 11, 10].

Notice that

$$\omega|_{\lambda=0} = ds \cdot s^{-1},$$

and that the zero–curvature condition for ω allows us to write

$$\omega = d\psi \cdot \psi^{-1}, \qquad (4.1)$$

for some $\psi(\lambda, \xi, \eta) \in G$. The function ψ is called an extended solution, or extended chiral field, corresponding to s. From this one concludes that $s = \psi|_{\lambda=0} \bmod G$. The notation of [26] is $E_{\tilde{\lambda}} = \psi(\lambda)$ and one has $s \cdot E_{-1} \in G$. Given an extended solution ψ the function $\psi \cdot g$ is also an extended solution for any $g(\lambda) \in G$. We shall introduce the undressing of a given solution and then give the dressing technique.

4.1 Undressing solutions

Given a solution to the principal chiral model we present the so called undressing of such a solution. Consider two small circles Γ_μ in the Riemman sphere surrounding the points $\mu = \pm 1$, such that ψ is regular at Γ_μ. Then, we have the following Birkhoff factorization problems defined in a neighbourhood of $\lambda = \pm 1$

$$\psi = \psi_{\mu+}^{-1} \cdot \psi_{\mu-}, \quad \mu = \pm 1, \tag{4.2}$$

where $\psi_{\mu-}$ is the holomorphic extension of $\psi|_{\Gamma_\mu}$ to the exterior of Γ_μ being normalized by the identity at $\lambda = \infty$, and $\psi_{\mu+}$ is the holomorphic extension of $\psi|_{\Gamma_\mu}$ to the interior of Γ_μ. If there is a solution to these factorization problems when $\xi, \eta = 0$, that is $\psi|_{\xi,\eta=0}(\lambda) = g(\lambda)$ can be factorized as in (4.2), then the pair of factorization problems (4.2) has a unique solution if ξ, η are close enough to the origin.

Defining the two zero–curvature 1–forms

$$\omega_\mu = d\psi_{\mu-} \cdot \psi_{\mu-}^{-1}, \quad \mu = \pm 1 \tag{4.3}$$

Eq.(4.2) implies

$$\omega_\mu = d\psi_{\mu+} \cdot \psi_{\mu+}^{-1} + \operatorname{Ad}\psi_{\mu+}\omega. \tag{4.4}$$

We introduce the notation

$$\omega_\mu = U_\mu d\xi + V_\mu d\eta,$$

then Eqs.(4.3,4.4) gives

$$U_{-1} = \partial_\xi \psi_{-1-} \cdot \psi_{-1-}^{-1} = \partial_\xi \psi_{-1+} \cdot \psi_{-1+}^{-1} + \operatorname{Ad}\psi_{-1+}\frac{u}{1-\lambda}.$$

From the first identity one concludes that the function U_{-1} is holomorphic outside the circle Γ_{-1}, moreover

$$U_{-1}|_{\lambda=\infty} = 0,$$

and from the second it follows that U_{-1} is holomorphic inside Γ_{-1}. Then, Liouville's theorem implies that $U_{-1} = 0$. Similarly $V_1 = 0$. The zero–curvature condition over the 1–forms

$$\omega_1 = U_1 d\xi, \quad \omega_{-1} = V_{-1} d\eta,$$

gives

$$\partial_\eta U_1 = 0, \quad \partial_\xi V_{-1} = 0,$$

therefore

$$U_1(\xi, \lambda) = \frac{u_0(\xi)}{1-\lambda},$$

where we have used

$$U_1 = \partial_\xi \psi_{1-} \cdot \psi_{1-}^{-1} = \partial_\xi \psi_{1+} \cdot \psi_{1+}^{-1} + \mathrm{Ad}\psi_{1+} \frac{u}{1-\lambda}.$$

In a similar way

$$V_{-1}(\eta, \lambda) = \frac{v_0(\eta)}{1+\lambda}.$$

The map sending $u(\xi, \eta), v(\xi, \eta)$ to $u_0(\xi), v_0(\eta)$ can be understood as an undressing transformation.

4.2 The dressing procedure

We shall now proceede to give the dressing technique. Define over Γ_μ the functions ψ_μ by the equations

$$\partial_\xi \psi_1 \cdot \psi_1^{-1} = \frac{u_0(\xi)}{1-\lambda}, \qquad \partial_\eta \psi_{-1} \cdot \psi_{-1}^{-1} = \frac{v_0(\eta)}{1+\lambda}.$$

Note that ψ_μ do not need to be holomorphic inside Γ_μ. Let be $\Gamma = \Gamma_1 \cup \Gamma_{-1}$, and consider the following factorization problem

$$\psi_\mu = \psi_{\mu+}^{-1} \cdot \psi, \qquad \mu = \pm 1, \tag{4.5}$$

where $\psi_{\mu+}$ has a holomorphic extension to the interior of Γ_μ and ψ can be analytically continued outside Γ and it is normalized by the identity at ∞. For small enough values of the coordinates ξ, η this problem has a unique solution if $\psi\mid_{\xi,\eta=0}$ can be factorized as in (4.5). We set

$$\omega = d\psi \cdot \psi^{-1},$$

then Eq.(4.5) gives

$$\omega = d\psi_{\mu+} \cdot \psi_{\mu+}^{-1} + \mathrm{Ad}\psi_{\mu+}\omega_\mu, \quad \mu = \pm 1, \tag{4.6}$$

where

$$\omega_\mu(\xi, \eta, \lambda) := \begin{cases} u_0(\xi)/(1-\lambda)d\xi, & \mu = 1 \\ v_0(\eta)/(1+\lambda)d\eta, & \mu = -1. \end{cases} \tag{4.7}$$

Taking into account

$$\partial_\eta \psi_1 \cdot \psi_1^{-1} = \partial_\xi \psi_{-1} \cdot \psi_{-1}^{-1} = 0,$$

one deduces

$$\omega(\lambda) = \frac{u}{1-\lambda}d\xi + \frac{v}{1+\lambda}d\eta.$$

Hence, we have a dressing transformation $u_0, v_0 \to u, v$, giving a solution to the principal chiral model $s = \psi\mid_{\lambda=0} \mathrm{mod} G$.

The infinitesimal version of the factorization problem (4.5) is rooted in the adele ring of the field of rational functions over $\mathbb{C}$, in [6] this was used to construct Lie algebras in connection with τ-functions. In [25, 23] there is a clear exposition, in relation to classical r-matrices, of this construction that, adapted to the principal chiral model, we present here. For each $\mu = \pm 1$ consider the loop algebra $L_{\mu,\text{pol}}\mathfrak{g}$ of polynomial loops over the circle Γ_μ with values in the Lie algebra $\mathfrak{g}$, see [22]. Denote by $\mathcal{D} = \{-1, 1\}$, and define

$$L_\mathcal{D}\mathfrak{g} := L_{-1,\text{pol}}\mathfrak{g} \oplus L_{1,\text{pol}}\mathfrak{g}.$$

Following [22] we denote by $L^+_{\mu,\text{pol}}\mathfrak{g}$ the set of those polynomials loops with holomorphic extension to the interior of Γ_μ, and we introduce

$$L^+_\mathcal{D}\mathfrak{g} := L^+_{-1,\text{pol}}\mathfrak{g} \oplus L^+_{1,\text{pol}}\mathfrak{g}.$$

Then, if $L^-_{\text{pol},1}(\Gamma, \mathfrak{g})$ is the set of polynomial loops over $\Gamma = \Gamma_{-1} \cup \Gamma_1$ with a holomorphic extension to the exterior of this curve, normalized by the identity at ∞, we have the spliting

$$L_\mathcal{D}\mathfrak{g} = L^+_\mathcal{D}\mathfrak{g} \oplus L^-_{\text{pol},1}(\Gamma, \mathfrak{g}).$$

This follows from the Mittag–Leffler theorem for $\mathbb{C}P^1$, and it can be extended to smooth loops using the Sochocki–Plemej's techniques. Thus, we have a solution to the modified classical Yang–Baxter equation, therefore a classical r-matrix, which is given by the difference of the projectors associated to the above spliting. This spliting, in the smooth case, is the Lie algebra version of the related factorization problem (4.5) in the corresponding group.

Dressing the initial data over the characteristics

The above method can be adapted to dress the initial data over the characteristics. For this purpose we normalize ψ in the following manner

$$\psi_\mu \mid_{\xi,\eta=0} = \text{id},$$

thus (4.5) implies

$$\psi \mid_{\xi,\eta=0} = \text{id}. \tag{4.8}$$

Then, we can identify the function ψ_μ of the factorization problem (4.5) with the function $\psi_{\mu-}$ appearing in (4.2). Noting that

$$\partial_\xi \psi \cdot \psi^{-1} = \frac{u}{1-\lambda},$$

consequence of (4.1), and taking into account the normalization condition (4.8) we conclude that $\psi\mid_{\eta=0}$ is regular at $\lambda = -1$. Then (4.2) implies

$$\psi \mid_{\eta=0} = \psi_{1-} \mid_{\eta=0},$$

The Principal chiral model as an integrable system

therefore
$$u_0(\xi) = u(\xi, 0).$$
Similarly
$$v_0(\eta) = v(0, \eta).$$
The dressing of these initial data is done with the use of the factorization problem (4.5) where we must insert the following information for ψ_μ:
$$\partial_\xi \psi_1 \cdot \psi_1^{-1} = \frac{u(\xi, 0)}{1 - \lambda}, \quad \partial_\eta \psi_{-1} \cdot \psi_{-1}^{-1} = \frac{v(0, \eta)}{1 + \lambda},$$
and impose the normalization condition
$$\psi_\mu \mid_{\xi, \eta = 0} = \mathrm{id}.$$
From the initial data over the characteristics we have recovered the solution to the principal chiral model via the factorization problem (4.5); in fact $\psi \mid_{\lambda = 0}$ is a chiral field.

Dressing the vacuum

We shall be interested now in dressing a vacuum solution to the principal chiral model instead of dressing the initial conditions. Suppose that
$$[u_0(\xi), v_0(\eta)] = 0.$$
Then u_0, v_0 is a solution to the principal chiral model which is called a vacuum solution. The zero–curvature 1–form ω_0 associated to a vacuum solution is
$$\omega_0(\xi, \eta, \lambda) = \frac{u_0(\xi)}{1 - \lambda} d\xi + \frac{v_0(\eta)}{1 + \lambda} d\eta.$$
Define ψ_0 by
$$\omega_0 := d\psi_0 \cdot \psi_0^{-1},$$
and such that
$$\psi_0 \mid_{\xi, \eta = 0} = \mathrm{id}.$$
Assuming the initial data
$$\psi_\mu \mid_{\xi, \eta = 0} = g,$$
we can write
$$\psi_\mu = \tilde{\psi}_\mu \cdot g,$$

where $\tilde\psi_\mu$ has a holomorphic extension to the exterior of Γ_μ and is normalized by $\tilde\psi_\mu\mid_{\xi,\eta=0} = \text{id}$. Observe that
$$\psi_0 = \tilde\psi_1 \cdot \tilde\psi_{-1} = \tilde\psi_{-1} \cdot \tilde\psi_1.$$
Then, the factorization problem (4.5) gives
$$\mathsf{X} := \psi \cdot g^{-1} \cdot \psi_0^{-1} = \psi_{1+} \cdot \tilde\psi_{-1}^{-1} = \psi_{-1+} \cdot \tilde\psi_1^{-1},$$
therefore X is holomorphic on the interior of Γ. Hence the factorization problem (4.5) for the vacuum solution is
$$\psi_0 \cdot g = \mathsf{X}^{-1} \cdot \psi,$$
where X is holomorphic inside the curve Γ and ψ is holomorphic outside Γ with $\psi\mid_{\lambda=\infty} = \text{id}$, $\psi\mid_{\lambda=0} = s$ being a chiral field.

Dressing a known solution

The dressing of a vacuum solution can be extended to a dressing of any known solution. Suppose that a chiral field is given, say s_0, in principle not a vacuum, with extended solution ψ_0; then the factorization problem
$$\psi_0 = \mathsf{X}^{-1} \cdot \psi, \tag{4.9}$$
where X is holomorphic inside the curve Γ and ψ is holomorphic outside Γ with $\psi\mid_{\lambda=\infty} = \text{id}$, gives a new extended solution ψ, $s = \psi\mid_{\lambda=0}$ being a chiral field. This can be proved as follows: The factorization problem (4.9) implies
$$d\mathsf{X} \cdot \mathsf{X}^{-1} + \text{Ad}\mathsf{X}\,\omega_0 = \omega, \tag{4.10}$$
ω_0 being the zero–curvature 1–form associated to the known solution s_0 and $\omega = d\psi \cdot \psi^{-1}$. To be an extended solution ψ needs ω to be meromorphic in λ with poles, which are simple, at $\mathcal{D} = \{-1, 1\}$. The analyticity of X on the interior of Γ and of ψ on the exterior of this curve imply that ω is as in (3.1) with
$$u = \text{Ad}\mathsf{X}\mid_{\lambda=1} u_0, \qquad v = \text{Ad}\mathsf{X}\mid_{\lambda=-1} v_0. \tag{4.11}$$
When X admits a meromorphic extension to $\mathbb{C}$ with no pole at $\lambda = 0$ we have the following formula giving the dressed chiral field
$$s = \mathsf{X}\mid_{\lambda=0} \cdot s_0. \tag{4.12}$$

5 Conservation laws

Solving the factorization problem (4.5) one finds solutions to the equations of the principal chiral model once $u_0(\xi)$ and $v_0(\eta)$ are fixed. There is another aspect of this factorization problem of interest for us. One can recover an infinite family of explicit local conservation laws from it. A local conservation law is an expression like

The Principal chiral model as an integrable system

$$\partial_\xi f = \partial_\eta g,$$

where the functions f, g are polynomials in u, v and its $\partial_\xi, \partial_\eta$-derivatives.
From (4.5) one obtains the condition

$$d\psi_{\mu+} \cdot \psi_{\mu+}^{-1} + P_{\mu+}\mathrm{Ad}\psi_{\mu+}\omega_\mu = \begin{cases} v/(1+\lambda)d\eta, & \mu = 1 \\ u/(1-\lambda)d\xi, & \mu = -1, \end{cases} \quad (5.1)$$

Here the projectors $P_{\mu+}$ are defined in terms of a Fourier decomposition by

$$P_{\mu+}\sum_n (1 + \mathrm{sgn}(\mu)\lambda)^n X_n = \sum_{n \geq 0}(1 + \mathrm{sgn}(\mu)\lambda)^n X_n, \qquad \mu = \pm 1.$$

We shall deal with a simple Lie algebra $\mathfrak{g}$, and a semi–simple vector $u_0(\xi) \in \mathfrak{g}$. Therefore, we have the linear splitting $\mathfrak{g} = \mathfrak{k}_\xi \oplus \mathfrak{m}_\xi$, where $\mathfrak{k}_\xi := \mathrm{Ker}\,\mathrm{ad}u_0(\xi)$ is a subalgebra and $\mathfrak{m}_\xi := \mathrm{Im}\,\mathrm{ad}u_0(\xi)$ is a subspace with $[\mathfrak{k}_\xi, \mathfrak{m}_\xi] \subset \mathfrak{m}_\xi$. One can write $\mathfrak{k}_\xi = \mathfrak{z}_\xi \oplus \mathfrak{l}_\xi$, where $\mathfrak{z}_\xi$ is the centre and $\mathfrak{l}_\xi$ is a semi–simple Lie algebra. Henceforth in this section we shall not indicate explicitly the ξ–dependence and we shall assume that $[u_0, \partial_\xi u_0] = 0$. Then, the maximal abelian subalgebra $\mathfrak{h}$ containing the semisimple vector u_0 is a Cartan subalgebra. Let Δ_+ be the associated positive root system, and define $u_0^\perp := \{\alpha \in \Delta_+ : \alpha(u_0) = 0\}$ and $\Xi_+ := \Delta_+ \backslash u_0^\perp$. Thus, the operator $\mathcal{U}_0$ defined as the restriction of $\mathrm{ad}u_0$ to $\mathfrak{m}$ is an invertible operator; in fact we have the formula

$$\mathcal{U}_0^{-1} = q(\mathcal{U}_0),$$

where

$$q(z) = \frac{1}{z}\left(1 - \prod_{\alpha \in \Xi_+}\left(1 - \frac{z^2}{\alpha(u_0)^2}\right)\right).$$

All the assumptions made above are satisfied when we dress a vacuum solution such that u_0 and v_0 takes its values in some Cartan subalgebra $\mathfrak{h}$ of $\mathfrak{g}$. For example when $G = SU(N)$ this case is generic.

To analyse Eq.(5.1) when $\mu = 1$ we propose the factorization

$$\psi_{1+} = \vartheta \cdot a \cdot \varphi, \qquad (5.2)$$

where

$$a = \psi_{1+}|_{\lambda=1};\ \ \ln\varphi(\xi,\eta,\lambda) \in \mathfrak{k},\ \varphi|_{\lambda=1} = \mathrm{id};\ \ \ln\vartheta(\xi,\eta,\lambda) \in \mathrm{Ad}\mathfrak{m},\ \vartheta|_{\lambda=1} = \mathrm{id}.$$

From

$$P_{1-}\mathrm{Ad}\psi_{1+}\frac{u_0}{1-\lambda} = \frac{u}{1-\lambda},$$

that follows from Eq.(4.5), and where we have use the resolution of the identity given by $\mathrm{id} = P_{\mu+} + P_{\mu-}$, we conclude

$$u = \mathrm{Ad}\,a\,u_0. \qquad (5.3)$$

The ξ–derivative of (5.3) gives
$$u_\xi = [a_\xi \cdot a^{-1}, u] + \mathrm{Ad}\, a\, \partial_\xi u_0,$$
and so
$$a_\xi \cdot a^{-1} = \mathcal{U}^{-1} u_\xi.$$
Here $\mathcal{U}$ is as $\mathcal{U}_0$ and $\mathcal{U}^{-1}$ as $\mathcal{U}_0^{-1}$ but replacing u_0 by u. The ξ–component of Eq.(5.1) with the factorization (5.2) reads
$$\vartheta_\xi \cdot \vartheta^{-1} + \mathrm{Ad}\vartheta(a_\xi \cdot a^{-1} + \mathrm{Ad}\, a\, \varphi_\xi \cdot \varphi^{-1}) = -P_{1+}\mathrm{Ad}\vartheta\frac{u}{1-\lambda}. \tag{5.4}$$

We shall use the following Fourier expansions:
$$\ln \vartheta = \sum_{n>0}(1-\lambda)^n \Theta_n,$$
$$\mathrm{Ad}\, a\, \varphi_\xi \cdot \varphi^{-1} = \sum_{n>0}(1-\lambda)^n \Phi_n,$$
and the formulae
$$\partial g \cdot g^{-1} = \sum_{n\geq 0} \frac{1}{(n+1)!}(\mathrm{ad}\ln g)^n \partial \ln g,$$
$$\mathrm{Ad}g X = \sum_{n\geq 0} \frac{1}{n!}(\mathrm{ad}\ln g)^n X.$$

Then, splitting Eq.(5.4) in terms of the above Fourier decomposition we find expressions for Θ_n and Φ_n as finite polynomials in $\partial_\xi^m u$. (If $\deg u = \deg \partial_\xi = 1$ then these polynomials are homogeneous of degree $n+1$.)

The η–component of (5.1) is
$$\vartheta_\eta \cdot \vartheta^{-1} + \mathrm{Ad}\vartheta(a_\eta \cdot a^{-1} + \mathrm{Ad}\, a\, \varphi_\eta \cdot \varphi^{-1}) = \frac{v}{1+\lambda}.$$

Thereby $a_\eta \cdot a^{-1} = v$. Using the formula
$$\frac{1}{1+\lambda} = \sum_{n\geq 0}\frac{1}{2^{n+1}}(1-\lambda)^n,$$
and denoting by
$$\mathrm{Ad}\, a\, \varphi_\eta \cdot \varphi^{-1} =: \sum_{n>0}(1-\lambda)^n \tilde{\Phi}_n,$$

we get for the coefficients $\tilde{\Phi}_n$ expressions as finite polynomials in the variables $\partial_\xi^m u, \partial_\eta^m u, v$.

Finally, $d\varphi \cdot \varphi^{-1}$ is a zero–curvature 1–form with values in $\mathfrak{k}$, projecting it, parallel to $\mathfrak{l}$, into the centre $\mathfrak{z}$ we obtain a 1–form $k := \sum_{n>0}(1-\lambda)^n(k_n^\xi d\xi + k_n^\eta d\eta)$ that satisfies

The Principal chiral model as an integrable system 159

$$\partial_\eta k_n^\xi = \partial_\xi k_n^\eta.$$

We have local expressions for Ada k_n^ξ, Ada k_n^η which are the projection of $\Phi_n, \tilde{\Phi}_n$ into Ada $\mathfrak{z}$. Suppose $S = $ Ada H where $H \in \mathfrak{z}$, then if B is the Cartan–Killing form of $\mathfrak{g}$ we have

$$a_n := \mathrm{B}(\mathrm{Ad}a\ k_n^\xi, S) = \mathrm{B}(k_n^\xi, H),$$
$$b_n := \mathrm{B}(\mathrm{Ad}a\ k_n^\eta, S) = \mathrm{B}(k_n^\eta, H).$$

Therefore

$$\partial_\eta a_n = \partial_\xi b_n, \quad n > 0. \tag{5.5}$$

Now, the image of u is the homogeneous space G/K where K is the corresponding Lie group to $\mathfrak{k}$; thus it is possible to parametrize this homogeneous space by u. The function S has its image in the same space — this follows from $H \in \mathfrak{z}$, so that it can be parametrized by u. An example is $H = u_0$, that is $S = u$.

Hence (5.5) is an infinity family of local conservations laws. Proceeding in an analogous way with Eq.(5.1) for $\mu = -1$, and supposing that $v_0(\eta)$ has analogous properties to those of u_0, we obtain other infinite family of explicit local conservations laws for the principal chiral model. (In this case we use the Fourier decomposition in terms of $(1 + \lambda)^n$.)

In [21] there was found explicitly an infinite family of local conservation laws for the non-linear σ–model. For the principal chiral model in [19] there is an infinite family of local conservation laws, but there is a differential obstruction to finding them explicitly. In [8] there was found an explicit infinite set of local conservation laws for an integrable system obtained from the principal chiral model by a gauge transformation. Finally in [4, 5, 7] one can find a doubly infinite family of local conservations laws for this model, the exposition in those papers is related to the construction of ψ but it is difficult to find its relation to the dressing method explained here.

We have given a derivation of a double family of conservation laws based in the factorization problem connected with dressing procedure. These conservations laws can be constructed explicitly once u_0, v_0 are fixed. Observe that these conservation laws are relevant in the hyperbolic case; if we are dealing with an evolution system the conservations laws give, when appropriate asymptotic conditions are imposed, constants of motion. In the elliptic case the interpretation is not so inmediate. We must remark that the factorization (5.2) was introduced, in relation with conservation laws, in [14].

6 The reality condition for $U(N)$–chiral fields

We shall give in this section the reality condition for the extended solution $\psi(\lambda)$ of a chiral field s defined in the appropriate real form of $\mathbb{C}^2$ and with values in $U(N)$. From $s^{-1} = s^\dagger$ it follows that

$$u^\dagger = (s_\xi \cdot s^{-1})^\dagger = s \cdot (s^{-1})_\xi = -s_\xi \cdot s^{-1},$$
$$v^\dagger = (s_\eta \cdot s^{-1})^\dagger = s \cdot (s^{-1})_\eta = -s_\eta \cdot s^{-1}.$$

In the Minkowski case, $\xi, \eta \in \mathbb{R}^{1,1}$, then the ξ, η–left currents of s take their values in the Lie algebra $\mathfrak{u}(N)$

$$u^\dagger + u = 0, \quad v^\dagger + v = 0.$$

For the Euclidean case we have $\bar{\xi} = \eta$, therefore

$$u^\dagger + v = 0,$$

and so the x, y–left currents of the chiral field s are maps into $\mathfrak{u}(N)$.

Observe that for the 1–form defined in (3.1) we have

$$\omega(\lambda)^\dagger = -\frac{u^\dagger}{1-\bar{\lambda}}d\bar{\xi} - \frac{v^\dagger}{1+\bar{\lambda}}d\bar{\eta}. \tag{6.1}$$

When one considers the Minkowski case the Eq.(6.1) reads

$$\omega(\lambda)^\dagger + \omega(\bar{\lambda}) = 0,$$

therefore

$$\psi(\lambda)^\dagger \cdot \psi(\bar{\lambda}) = \mathrm{id}. \tag{6.2}$$

This is the reality condition for the extended solution ψ when we consider the Lorentzian relativistic case. For harmonic maps from $\mathbb{R}^2$ into $U(N)$ one has $\bar{\xi} = \eta$ and so

$$\omega(\lambda)^\dagger = -\frac{v}{1-\bar{\lambda}}d\eta - \frac{u}{1+\bar{\lambda}}d\xi,$$

thereby

$$\omega(\lambda)^\dagger + \omega(-\bar{\lambda}) = 0,$$

hence for the extended solution defined in (4.1) one finds the following reality condition

$$\psi(\lambda)^\dagger \cdot \psi(-\bar{\lambda}) = \mathrm{id}. \tag{6.3}$$

Equation (6.2) implies that over the curve $\gamma_m = \mathbb{R} = \{\lambda \in \overline{\mathbb{C}} : \lambda = \bar{\lambda}\}$ the extended solution ψ, which by construction is holomorphic outside the curve Γ, can be consider as a $U(N)$–loop based at $\lambda = \infty$. When one performs the homographic transformation

$$\tilde{\lambda} = \frac{i\lambda + 1}{i\lambda - 1},$$

so that

The Principal chiral model as an integrable system 161

$$\lambda = i\frac{\tilde{\lambda}+1}{\tilde{\lambda}-1},$$

the curve γ_m is transformed into the circle S^1, and $\infty \to 1$, hence the extended solution is an element of the based loop group $\Omega U(N)$, see [22].

From Eq.(6.3) it follows that the extended solution, in the Euclidean case, is a smooth map from the curve $\gamma_e = i\mathbb{R} = \{\lambda \in \overline{\mathbb{C}} : \lambda = -\bar{\lambda}\}$ to $U(N)$ based at ∞. With the homographic transformation

$$\tilde{\lambda} = \frac{\lambda+1}{\lambda-1},$$

with inverse

$$\lambda = \frac{\tilde{\lambda}+1}{\tilde{\lambda}-1},$$

we get that the extended solution $E_{\tilde{\lambda}}$ is an element of the based loop group $\Omega U(N)$ [26].

7 Soliton type $U(N)$–chiral fields over Minkowski space

We present here the soliton type solutions of the principal chiral model over $\mathbb{R}^{1,1}$ with values in $U(N)$. These solutions were found for the first time in [31, 32]. We shall dress a known chiral field s_0 with extended solution ψ_0. For this purpose we need the function X that solves the factorization problem (4.9). The functions ψ, ψ_0 must satisfy the reality condition (6.2), therefore X also satisfies this condition. The M-soliton appears when one considers the function X, which is holomorphic inside the curve Γ, having a mereomorphic extension to the Riemann sphere with poles, which are simple, at the set $\mathcal{P}_m := \{\lambda_m, \bar{\lambda}_m\}_{m=1}^M$, and the known solution is a vacuum. The consideration of conjugate poles is motivated by the extension of the method to the real classical groups $O(p,q), Sp(N)$. Hence X can be written as

$$X(\lambda) = \mathrm{id} + \sum_{m=1}^M \frac{A_m}{\lambda-\lambda_m} + \frac{B_m}{\lambda-\bar{\lambda}_m}.$$

The reality condition (6.2) implies

$$X^{-1}(\lambda) = \mathrm{id} + \sum_{m=1}^M \frac{A_m^\dagger}{\lambda-\bar{\lambda}_m} + \frac{B_m^\dagger}{\lambda-\lambda_m},$$

which holds if and only if

$$A_n B_n^\dagger = 0, \tag{7.1}$$
$$A_n A_n^\dagger + B_n B_n^\dagger = 0, \tag{7.2}$$

where

$$\mathcal{A}_n := \mathrm{id} + \sum_{m=1}^{M} \frac{A_m}{\bar\lambda_n - \lambda_m} + \sum_{m \neq n} \frac{B_m}{\bar\lambda_n - \bar\lambda_m},$$

$$\mathcal{B}_n := \mathrm{id} + \sum_{m \neq n} \frac{A_m}{\lambda_n - \lambda_m} + \sum_{m=1}^{M} \frac{B_m}{\lambda_n - \bar\lambda_m}.$$

With the factorization

$$A_n = X_n F_n^\dagger, \quad B_n = Y_n G_n^\dagger, \tag{7.3}$$

where X_n, F_n are matrices with N lines and K_n columns and Y_n, G_n are matrices with N lines and S_n columns, Eq.(7.1) reads

$$F_n^\dagger G_n = 0, \tag{7.4}$$

and Eq.(7.2) is equivalent to

$$\mathcal{B}_n G_n = -X_n \alpha_n, \quad \mathcal{A}_n F_n = Y_n \alpha_n^\dagger, \tag{7.5}$$

which can be written as

$$G_n + \sum_{m \neq n} \frac{X_m F_m^\dagger G_n}{\lambda_n - \lambda_m} + \sum_{m=1}^{M} \frac{Y_m G_m^\dagger G_n}{\lambda_n - \bar\lambda_m} = -X_n \alpha_n, \tag{7.6}$$

$$F_n + \sum_{m=1}^{M} \frac{X_m F_m^\dagger F_n}{\bar\lambda_n - \lambda_m} + \sum_{m \neq n} \frac{Y_m G_m^\dagger F_n}{\bar\lambda_n - \bar\lambda_m} = Y_n \alpha_n^\dagger. \tag{7.7}$$

The pole set $\mathcal{P}_m$ is chosen such that it does not contain the points $1, -1$. Then, recalling that ω, ω_0 have their poles, which are simple, localised at $1, -1$, we see that there are no double poles in the left hand side of (4.10), this implies

$$d\mathcal{A}_n \, B_n^\dagger + \mathcal{A}_n \omega_0 \,|_{\lambda=\lambda_n}\, B_n^\dagger = 0,$$
$$d\mathcal{B}_n \, A_n^\dagger + \mathcal{B}_n \omega_0 \,|_{\lambda=\bar\lambda_n}\, A_n^\dagger = 0,$$

that using Eq.(7.3) reads

$$dF_n^\dagger \, G_n + F_n^\dagger \omega_0 \,|_{\lambda=\lambda_n}\, G_n = 0,$$
$$dG_n^\dagger \, F_n + G_n^\dagger \omega_0 \,|_{\lambda=\bar\lambda_n}\, F_n = 0.$$

These equations, taking into account (7.3) and using the freedom $X_n \to X_n f_n^{-1}$, $F_n \to F_n f_n$, $Y_n \to Y_n g_n^{-1}$ and $G_n \to G_n g_n$ for invertible square matrices f_n and g_n of order K_n and S_n respectively, have as solution

$$G_n(\xi, \eta) = \psi_0(\xi, \eta, \lambda_n) \cdot G_n^0, \tag{7.8}$$
$$F_n(\xi, \eta) = \psi_0(\xi, \eta, \bar\lambda_n) \cdot F_n^0, \tag{7.9}$$

for some constant matrices F_n^0, G_n^0 satisfying (7.4).

The absence of first order poles at $\mathcal{P}_m$ in the expression (4.10) gives the equations

$$dA_n \, \mathcal{A}_n^\dagger + A_n \omega_0 \mid_{\lambda=\lambda_n} \mathcal{A}_n^\dagger + dB_n \, \mathcal{B}_n^\dagger + B_n \omega_0 \mid_{\lambda=\lambda_n} \mathcal{B}_n^\dagger + A_n \left.\frac{d\omega_0}{d\lambda}\right|_{\lambda=\lambda_n} \mathcal{B}_n^\dagger = 0.$$

Notice that

$$dA_n \, \mathcal{A}_n^\dagger + A_n \omega_0 \mid_{\lambda=\lambda_n} \mathcal{A}_n^\dagger = dX_n \alpha_n Y_n^\dagger,$$

that follows from (7.3,7.9) and the second equation in (7.5). Taking the exterior derivative in the first equation of (7.5) and recalling (7.8) one finds

$$dB_n \, \mathcal{B}_n^\dagger + B_n \omega_0 \mid_{\lambda=\lambda_n} \mathcal{B}_n^\dagger = -(dX_n \alpha_n + X_n d\alpha_n) Y_n^\dagger.$$

Thus,

$$X_n (d\alpha_n - F_n^\dagger \left.\frac{d\omega_0}{d\lambda}\right|_{\lambda=\lambda_n} G_n) Y_n^\dagger = 0,$$

which is satisfied if

$$d\alpha_n = (F_n^0)^\dagger \left.\frac{d\omega_0}{d\lambda}\right|_{\lambda=\lambda_n} G_n^0. \tag{7.10}$$

Then, a chiral field with X having its poles at $\mathcal{P}_m$ (an M-soliton) is fixed by M constant rectangular $K_n \times N$ matrices F_n^0 and M constant rectangular $S_n \times N$ matrices G_n^0 that satisfy the constraint (7.4), M constant rectangular $K_n \times S_n$ matrices α_n^0, M complex numbers λ_n and a known solution (vacuum) s_0 with ξ, η-left currents u_0, v_0. With these data we reconstruct F_n and G_n from Eqs.(7.9,7.8) and α_n from Eq.(7.10) and the initial condition $\alpha_n \mid_{\xi,\eta=0} = \alpha_n^0$. Solving the linear system (7.6,7.7) we get X_n, Y_n that when introduced in (7.3) gives the functions A_n, B_n in terms of which X is defined. Using (4.12) one can construct the chiral field s in terms of s_0 and $X\mid_{\lambda=0}$. The ξ, η-left currents are computed, with the help of Eq.(4.11), in terms of $X\mid_{\lambda=\pm 1}$ and the undressed currents u_0, v_0.

8 $U(N)$ harmonic maps over Euclidean space

As in the previous section we dress a known solution s_0 with extended solution ψ_0. For this purpose we construct the function X that satisfies the reality condition (6.3), which is holomorphic on the interior of Γ and has a meromorphic extension to the exterior of Γ the set of its poles, which are simple, being $\mathcal{P}_e := \{\lambda_m, -\bar{\lambda}_m\}_{m=1}^M$; as previously the choice of this pole set is motivated by the extension of the method to the real classical groups $O(p,q), Sp(N)$. Hence we write

$$X(\lambda) = \mathrm{id} + \sum_{m=1}^{M} \frac{A_m}{\lambda - \lambda_m} + \frac{B_m}{\lambda + \bar{\lambda}_m}.$$

The reality condition (6.3) implies

$$X^{-1}(\lambda) = \mathrm{id} - \sum_{m=1}^M \frac{A_m^\dagger}{\lambda + \bar\lambda_m} + \frac{B_m^\dagger}{\lambda - \lambda_m},$$

which holds if and only if

$$A_n B_n^\dagger = 0, \tag{8.1}$$
$$A_n A_n^\dagger - B_n B_n^\dagger = 0, \tag{8.2}$$

where

$$\mathcal{A}_n := \mathrm{id} - \sum_{m=1}^M \frac{A_m}{\bar\lambda_n + \lambda_m} - \sum_{m\neq n} \frac{B_m}{\bar\lambda_n - \bar\lambda_m},$$

$$\mathcal{B}_n := \mathrm{id} + \sum_{m\neq n} \frac{A_m}{\lambda_n - \lambda_m} + \sum_{m=1}^M \frac{B_m}{\lambda_n + \bar\lambda_m}.$$

With the factorization (7.3) the Eq.(8.1) is equivalent to the condition (7.4) and Eq.(8.2) is equivalent to

$$\mathcal{B}_n G_n = X_n \alpha_n, \quad \mathcal{A}_n F_n = Y_n \alpha_n^\dagger.$$

These equations can be written explicitly as

$$G_n + \sum_{m\neq n} \frac{X_m F_m^\dagger G_n}{\lambda_n - \lambda_m} + \sum_{m=1}^M \frac{Y_m G_m^\dagger G_n}{\lambda_n + \bar\lambda_m} = X_n \alpha_n, \tag{8.3}$$

$$F_n - \sum_{m=1}^M \frac{X_m F_m^\dagger F_n}{\bar\lambda_n + \lambda_m} - \sum_{m\neq n} \frac{Y_m G_m^\dagger F_n}{\bar\lambda_n - \bar\lambda_m} = Y_n \alpha_n^\dagger. \tag{8.4}$$

As previously, assuming that the pole set $\mathcal{P}_e$ is chosen such that it does not contains the points $1, -1$, we obtain from Eq.(4.10)

$$G_n(\xi, \eta) = \psi_0(\xi, \eta, \lambda_n) \cdot G_n^0, \tag{8.5}$$
$$F_n(\xi, \eta) = \psi_0(\xi, \eta, -\bar\lambda_n) \cdot F_n^0, \tag{8.6}$$

for some constant matrices F_n^0, G_n^0 satisfying (7.4).

The absence in the expression (4.10) of first order poles at $\mathcal{P}_e$ gives

$$X_n(d\alpha_n + F_n^\dagger \left.\frac{d\omega_0}{d\lambda}\right|_{\lambda=\lambda_n} G_n)Y_n^\dagger = 0,$$

which is satisfied if

$$d\alpha_n = -(F_n^0)^\dagger \left.\frac{d\omega_0}{d\lambda}\right|_{\lambda=\lambda_n} G_n^0. \tag{8.7}$$

Then, a $U(N)$ harmonic map, with extended solution such that X has $2M$ poles located at $\mathcal{P}_e$, is fixed by M constant rectangular $K_n \times N$ matrices F_n^0 and M constant rectangular $S_n \times N$ matrices G_n^0 that satisfy the constraint (7.4), M constant rectangular $K_n \times S_n$ matrices α_n^0, M complex numbers λ_n, and a known solution s_0. With these data we reconstruct F_n and G_n from Eqs.(8.6,8.5), and α_n from Eq.(8.7) and the initial condition $\alpha_n\mid_{\xi,\eta=0} = \alpha_n^0$. Solving the linear system (8.3,8.4) we get X_n, Y_n which when introduced in (7.3) gives the functions A_n, B_n in terms of which X is defined. The harmonic map s is obtained from the expressions for s_0 and $X\mid_{\lambda=0}$ using (4.12). The left currents can be computed with the use of (4.11).

9 The reality condition for $O(p,q), Sp(N)$ chiral fields

We are interested now in chiral fields with values in the real classical groups $O(p,q)$, $Sp(N)$, the pseudorthogonal and the sympletic groups. Recall that such groups are the set of real matrices g of order $p+q$ for $O(p,q)$ and $2N$ for $Sp(N)$ such that

$$g^{-1} = \mathcal{J} g^t \mathcal{J}^{-1}, \tag{9.1}$$

where

$$\mathcal{J} := \begin{pmatrix} I_p & 0 \\ 0 & -I_q \end{pmatrix},$$

for $O(p,q)$ and

$$\mathcal{J} := \begin{pmatrix} 0 & -I_N \\ I_N & 0 \end{pmatrix}$$

for $Sp(N)$. Observe that $\mathcal{J}^{-1} = \mathcal{J}^t = \mathcal{J}$ in the pseudo-orthogonal case and $\mathcal{J}^{-1} = \mathcal{J}^t = -\mathcal{J}$ in the symplectic case. Therefore Eq.(9.1) can equally be written as

$$g^{-1} = \mathcal{J}^{-1} g^t \mathcal{J}.$$

The Lie algebras of these groups, denoted by $\mathfrak{o}(p,q)$ and by $\mathfrak{sp}(N)$ respectively, are the set of those real matrices satisfying

$$A^t \mathcal{J} + \mathcal{J} A = 0.$$

The chiral field s takes its values in one of these real groups, then it must be real $\bar{s} = s$, and (9.1) must be fulfilled. The ξ, η–left currents satisfies

$$\bar{u} = \overline{s_\xi \cdot s^{-1}} = s_{\bar\xi} \cdot s^{-1},$$
$$\bar{v} = \overline{s_\eta \cdot s^{-1}} = s_{\bar\eta} \cdot s^{-1}.$$

In the Minkowski case $\xi, \eta \in \mathbb{R}^{1,1}$, and so

$$\bar{u} = u, \quad \bar{v} = v.$$

In the Euclidean case we have $\bar\xi = \eta$, thus

$$\bar u = v.$$

Then, for the 1-form defined in (3.1) one has

$$\overline{\omega(\lambda)} = \omega(\bar\lambda),$$

in the Minkowski case, which implies the following reality condition for the extended solution

$$\overline{\psi(\lambda)} = \psi(\bar\lambda). \tag{9.2}$$

In the Euclidean case the reality condition is different, now

$$\overline{\omega(\lambda)} = \omega(-\bar\lambda),$$

thereby

$$\overline{\psi(\lambda)} = \psi(-\bar\lambda). \tag{9.3}$$

In both cases it must be also satisfied

$$\mathcal{J}\omega(\lambda) + \omega(\lambda)^t \mathcal{J} = 0,$$

which follows from $s(\xi,\eta) \in O(p,q), Sp(N)$. This implies the additional condition for the extended solution

$$\psi(\lambda)^{-1} = \mathcal{J}\psi(\lambda)^t \mathcal{J}^{-1}, \tag{9.4}$$

or equivalently

$$\psi(\lambda)^{-1} = \mathcal{J}^{-1}\psi(\lambda)^t \mathcal{J}.$$

To summarize, the extended solution ψ must satisfy the reality conditions (9.2,9.4) in the Minkowski case and (9.3,9.4) in the Euclidean one.

10 Soliton type $O(p,q), Sp(N)$–chiral fields over Minkowski space

Chiral fields from $\mathbb{R}^{1,1}$ into the groups $O(p,q)$ and $Sp(N)$ can be constructed following closely §7, the construction that we shall report here was first found in [32]. A relevant difference with the $U(N)$ case is that X needs to have conjugate poles. From (9.2) and (4.9), one has

$$\overline{\mathsf{X}(\lambda)} = \mathsf{X}(\bar\lambda),$$

which implies

The Principal chiral model as an integrable system 167

$$\mathsf{X}(\lambda) = \mathrm{id} + \sum_{m=1}^{M} \frac{A_m}{\lambda - \lambda_m} + \frac{\bar{A}_m}{\lambda - \bar{\lambda}_m},$$

and (4.9,9.4) gives

$$\mathsf{X}^{-1}(\lambda) = \mathrm{id} + \sum_{m=1}^{M} \frac{\mathcal{J} A_m^t \mathcal{J}^{-1}}{\lambda - \lambda_m} + \frac{\mathcal{J} A_m^\dagger \mathcal{J}^{-1}}{\lambda - \bar{\lambda}_m},$$

that holds if

$$A_n \mathcal{J} A_n^t = 0, \tag{10.1}$$
$$A_n \mathcal{J} \bar{A}_n^t + \bar{A}_n \mathcal{J} A_n^t = 0, \tag{10.2}$$

with

$$\mathcal{A}_n := \mathrm{id} + \sum_{m \neq n} \frac{A_m}{\lambda_n - \lambda_m} + \sum_{m=1}^{M} \frac{\bar{A}_m}{\lambda_n - \bar{\lambda}_m}.$$

Writing

$$A_n = X_n F_n^\dagger, \tag{10.3}$$

where X_n, F_n are matrices with N rows and K_n columns, Eq.(10.1) reads

$$F_n^t \mathcal{J} F_n = 0 \tag{10.4}$$

and Eq.(10.2) is equivalent to

$$\mathcal{A}_n \mathcal{J} \bar{F}_n = -X_n \alpha_n,$$

which can be written as

$$\mathcal{J} \bar{F}_n + \sum_{m \neq n} \frac{X_m F_m^\dagger \mathcal{J} \bar{F}_n}{\lambda_n - \lambda_m} + \sum_{m=1}^{M} \frac{\bar{X}_m \bar{F}_m^t \mathcal{J} \bar{F}_n}{\lambda_n - \bar{\lambda}_m} = -X_n \alpha_n, \tag{10.5}$$

where $\alpha_n^t = -\mathcal{J}^2 \alpha_n$, notice that $\mathcal{J}^2 = \mathrm{id}$ in the $O(p,q)$ case and $\mathcal{J}^2 = -\mathrm{id}$ in the $Sp(N)$ case. From Eq.(4.10) one finds

$$d\mathcal{A}_n \mathcal{J} A_n^t + \mathcal{A}_n \omega_0 \big|_{\lambda=\lambda_n} \mathcal{J} A_n^\dagger = 0,$$

which gives

$$\mathcal{J} \bar{F}_n(\xi, \eta) = \psi_0(\xi, \eta, \bar{\lambda}_n) \cdot \mathcal{J} \bar{F}_n^0, \tag{10.6}$$

for some constant matrix F_n^0 satisfying (10.4), and

$$d\mathcal{A}_n \mathcal{J} A_n^t + \mathcal{A}_n \omega_0 \big|_{\lambda=\lambda_n} \mathcal{J} A_n^t + d\mathcal{A}_n \mathcal{J} A_n^t + \mathcal{A}_n \omega_0 \big|_{\lambda=\lambda_n} \mathcal{J} A_n^t +$$
$$\mathcal{A}_n \frac{d\omega_0}{d\lambda}\bigg|_{\lambda=\lambda_n} \mathcal{J} A_n^t = 0$$

which, as one can see, implies

$$d\alpha_n = (F_n^0)^\dagger \left.\frac{d\omega_0}{d\lambda}\right|_{\lambda=\lambda_n} J\bar{F}_n^0. \tag{10.7}$$

Thus, a chiral field such that the associated function X has its poles at $\mathcal{P}_m$ (M-soliton) is fixed by M constant rectangular $K_n \times N$ matrices F_n^0 that satisfy the constraint (10.4), M constant $K_n \times K_n$ matrices α_n^0 satisfying $(\alpha_n^0)^t = -\mathcal{J}^2 \alpha_n^0$, M complex numbers λ_n, and a known solution (vacuum solution) s_0. With these data we reconstruct F_n from Eq.(10.6), and α_n from Eq.(10.7) and the initial condition $\alpha_n|_{\xi,\eta=0} = \alpha_n^0$. Solving the linear system (10.5) we get X_n which when introduced in (10.3) gives the functions A_n in terms of which X is defined. The solution s is recovered with the use of (4.12), and the associated ξ,η–left currents u,v can be computed from (4.11).

11 Euclidean harmonic maps into $O(p,q), Sp(N)$

The results of the previous section can be extended to Euclidean spaces, therefore we can construct harmonic maps from $\mathbb{R}^2$ into the Lie groups $O(p,q), Sp(N)$. The reality condition (9.3) implies

$$\overline{X(\lambda)} = X(-\bar{\lambda}),$$

thus

$$X(\lambda) = \mathrm{id} + \sum_{m=1}^{M} \frac{A_m}{\lambda - \lambda_m} - \frac{\bar{A}_m}{\lambda + \bar{\lambda}_m},$$

and from Eqs.(4.9,9.4) we get

$$X^{-1}(\lambda) = \mathrm{id} + \sum_{m=1}^{M} \frac{\mathcal{J} A_m^t \mathcal{J}^{-1}}{\lambda - \lambda_m} - \frac{\mathcal{J} A_m^\dagger \mathcal{J}^{-1}}{\lambda + \bar{\lambda}_m}.$$

This, together with (4.10), gives us exactly the same formulae as for the Minkowski case but now

$$\mathcal{A}_n := \mathrm{id} + \sum_{m \neq n} \frac{A_m}{\lambda_n - \lambda_m} - \sum_{m=1}^{M} \frac{\bar{A}_m}{\lambda_n + \bar{\lambda}_m},$$

and the system (10.5) is transformed into

$$\mathcal{J}\bar{F}_n + \sum_{m \neq n} \frac{X_m F_m^\dagger \mathcal{J}\bar{F}_n}{\lambda_n - \lambda_m} - \sum_{m=1}^{M} \frac{\bar{X}_m F_m^t \mathcal{J}\bar{F}_n}{\lambda_n + \bar{\lambda}_m} = -X_n \alpha_n. \tag{11.1}$$

The Principal chiral model as an integrable system 169

Then, a harmonic map from $\mathbb{R}^2$ into $O(p,q), Sp(N)$, with extended solution such that X has its poles at $\mathcal{P}_e$ is fixed by M constant rectangular $K_n \times N$ matrices F_n^0 that satisfy the constraint (10.4), a set of M constant $K_n \times K_n$ matrices α_n^0 satisfying $(\alpha_n^0)^t = -\mathcal{J}^2 \alpha_n^0$, a set of M complex numbers λ_n, and a known solution s_0. With these data we reconstruct F_n from Eq.(10.6), and α_n from Eq.(10.7) and the initial condition $\alpha_n |_{\xi,\eta=0} = \alpha_n^0$. Solving the linear system (11.1) we get X_n which when introduced in (10.3) gives the functions A_n in terms of which X is defined. Then, the harmonic map s is determined by (4.12), and its ξ, η-left currents by (4.11).

12 Simple examples

In this section we present simple examples of the above construction of chiral fields. Let us start with the unitary group $U(N)$; in this case we are allowed to suppose that X has only one pole, say at $\lambda = \alpha$. When we consider the Minkowski space $\mathbb{R}^{1,1}$ we find

$$\mathsf{X}(\lambda) = \mathrm{id} + \frac{\alpha - \bar{\alpha}}{\lambda - \alpha} F(F^\dagger F)^{-1} F^\dagger,$$

where F is a $N \times K$ matrix such that $F(\xi, \eta) = \psi_0(\xi, \eta, \bar{\alpha}) \cdot F^0$ and $\det F^\dagger F \neq 0$. The rank of F^0 is K, its K columns giving a basis for the K-dimensional subspace $V^0 \subset \mathbb{C}^N$, therefore the set of K columns of $F(\xi, \eta)$ is a basis for the K-dimensional subspace $V(\xi, \eta) := \psi_0(\xi, \eta, \bar{\alpha}) V^0$. The operator

$$\pi := F(F^\dagger F)^{-1} F^\dagger,$$

is the Hermitian projector into V in terms of which the dressed chiral field is expressed as

$$s = (\pi^\perp + \frac{\bar{\alpha}}{\alpha} \pi) \cdot s_0,$$

where $\pi^\perp := \mathrm{id} - \pi$.

A similar analysis in the Euclidean case gives

$$\mathsf{X}(\lambda) = \mathrm{id} + \frac{\alpha + \bar{\alpha}}{\lambda - \alpha} \pi, \tag{12.1}$$

and the harmonic map is

$$s = (\pi^\perp - \frac{\bar{\alpha}}{\alpha} \pi) \cdot s_0,$$

which when $\alpha \in \mathbb{R}$ gives

$$s = (\pi^\perp - \pi) \cdot s_0.$$

We have $A = (\alpha + \bar{\alpha})\pi$ and $\mathcal{A} = \pi^\perp$, therefore

$$d\pi \ \pi^\perp + \pi \omega |_{\lambda = \alpha} \pi^\perp = 0.$$

Recall that $\alpha \notin \mathcal{D} = \{-1, 1\}$, but when $\alpha \to 1$ the equation collapses to

$$\pi u_0 \pi^\perp = 0,$$
$$\partial_\xi \pi \, \pi^\perp + \pi 2 v_0 \pi^\perp = 0.$$

If s_0 is a finite energy uniton one says that π is a flag factor for it [3, 29].

A number of remarks are in order at this point. In the Euclidean case we see that the above formula, in the limit case $\alpha \to 1$ (now X is not holomorphic at $\lambda = 1$), reproduces the singular Bäcklund transformation, or adding a uniton, of §12 of [26]. The function X that generates this transformation is [26]

$$\mathsf{X}(\lambda) = \pi^\perp + \tilde{\lambda}\pi = \pi^\perp + \frac{\lambda+1}{\lambda-1}\pi = \mathrm{id} + \frac{2}{\lambda-1}\pi.$$

Moreover, as one can see, the simple factors $f(\tilde{\lambda})$ of §5 of [26] can be written as

$$f(\tilde{\lambda}) = \boldsymbol{\xi}_{\tilde{\alpha}}(\tilde{\lambda}) \mathsf{X}(\lambda),$$

where $\boldsymbol{\xi}_{\tilde{\alpha}}(\tilde{\lambda})$ is defined in formula (18) of [26] and X is given in (12.1). Therefore, the simple factor f factorizes into a product of an element that belongs to the centre of $U(N)$ and the Zakharov–Mikhailov function X; notice that in [31, 32] the group is $SU(N)$ and no centre appears. So Uhlenbeck's Bäcklund transformations [26] are nothing other than the dressing given in the papers [31, 32].Even more, Theorem 5.4 of [26] regarding the factorization of any dressing into simple factors was already present in, for example, Chapter III, §2 of the book [18]. Nevertheless, Part II of [26] cannot be deduced from the known results in integrable systems theory; finite energy unitons or flag factors are highly nontrivial objects.

For the real groups the simple examples are as follows. We need a matrix with N rows and K columns say F^0 that satisfies

$$(F^0)^t J F^0 = 0, \qquad \det(F^0)^t J \bar{F}^0 \neq 0,$$

denote by $JF = \psi_0 \mid_{\lambda=\alpha} JF^0$. Then, in the Minkowski case the function X, that has its poles located at $\alpha, \bar{\alpha}$, is

$$\mathsf{X}(\lambda) = \mathrm{id} - \frac{\bar{\alpha}-\alpha}{\lambda-\alpha} JF(F^\dagger JF)^{-1} F^\dagger - \frac{\alpha-\bar{\alpha}}{\lambda-\bar{\alpha}} J\bar{F}(F^t J\bar{F})^{-1} F^t,$$

The rank of JF^0 is K, giving its K columns a basis for the K–dimensional subspace $V^0 \subset \mathbb{C}^N$, therefore the set of K columns of $JF(\xi, \eta)$ is a basis for the K–dimensional subspace $V(\xi, \eta) := \psi_0(\xi, \eta, \alpha) V^0$. The operator

$$\pi := JF(F^\dagger JF)^{-1} F^\dagger,$$

is the Hermitian projector into V. Observe that

$$\bar{\pi} = J \pi^t J^{-1},$$

and that

$$\pi\bar{\pi} = 0.$$

The operator $P := \pi + \bar{\pi}$ is an Hermitian projector with $\bar{P} = P$ and $P^t = \mathcal{J} P \mathcal{J}^{-1}$. The chiral field is

$$s = \left(P^\perp + \frac{\bar{\alpha}}{\alpha}\pi + \frac{\alpha}{\bar{\alpha}}\bar{\pi}\right) \cdot s_0.$$

In the Euclidean case we have

$$\mathsf{X}(\lambda) = \mathrm{id} - \frac{\bar{\alpha} + \alpha}{\lambda - \alpha}\pi - \frac{\alpha + \bar{\alpha}}{\lambda - \bar{\alpha}}\bar{\pi},$$

and the harmonic map is

$$s = \left(P^\perp - \frac{\bar{\alpha}}{\alpha}\pi - \frac{\alpha}{\bar{\alpha}}\bar{\pi}\right) \cdot s_0$$

This dressing transformation when $\alpha \in \mathbb{R}$ is

$$s = (P^\perp - P) \cdot s_0.$$

Is this construction connected with a possible uniton solution with values in the real groups?

Acknowledgments

The author would like to acknowledge Dr. A.P. Fordy and Dr. J.C. Wood for the invitation to the meeting *Harmonic Maps and Integrable Systems* and financial support from the CICYT proyecto PB89–0133. Finally the author is indebted to Dr.F.Guil for many conversations.

Bibliography

[1] E. Beggs, *Solitons in the chiral equation*, Commun. Math. Phys.**128** (1990), 131-139.

[2] M. Bordemann, M. Forger, J. Laartz and U. Schäper, *2-dimensional nonlinear sigma models: zero curvature and Poisson structure*, this volume.

[3] F.E. Burstall and J.H. Rawnsley, *Twistor Theory for Riemannian Symmetric Spaces*, Lect. Notes in Math. **1424**, Springer, Berlin, 1990.

[4] I.V. Cherednik, *Local conservations laws of principal chiral fields*, Teo. Mat. Fiz. **38** (1979), 179-185; English transl.: Theor. Math. Phys. **38** (1979), 120.

[5] I. Cherednik, *Algebraic aspects of two–dimensional chiral fields*, Itogi Nauchn. i Tekhn. Inform., Moscow, Ser. Sov. Prob. Mat. **17** (1981), 175-218, 220; English transl.: J. Sov. Math. **21** (1983), 601.

[6] I. Cherednik, *Definition of functions for generalized affine Lie algebras*, Funk. Anal. Prilozh. **17** (1983), 93-95; English transl.: Func. Anal. Appl. **17** (1983), 243-245.

[7] I. Cherednik, *Elliptic curves and soliton matrix differential equations*, Itogi Nauchn. i Tekhn. Inform., Moscow, Alg. Top. Geom. **22** (1984), 205-265, 267; English transl.: J.Sov.Math. **38** (1987), 1989.

[8] L. Dickey, *Symplectic structure, Lagrangian and involutiveness of first integrals of the principal chiral model*, Commun. Math. Phys.**87** (1983), 505-513.

[9] B.A. Dubrovin, A.T. Fomenko and S.P. Novikov, *Modern Geometry – Methods and Applications: The Geometry and Topology of Manifolds*, Springer, Berlin, 1985.

[10] B.A. Dubrovin, I.Krichever and S.P. Novikov, *Integrable Systems I*, in: *Encyclopaedia of Mathematical Science 4 (Dynamical Systems IV)*, V.I. Arnold and S.P. Novikov (eds.), Springer, Berlin, 1990.

[11] L. Faddeev and L. Takhtajan, *Hamiltonian Methods in the Theory of Solitons*, Springer, Berlin, 1987.

[12] M.A. Guest and Y.Ohnita, *Group actions and deformations of harmonic maps*, J. Math. Soc. Japan. (to appear).

[13] F .Guil and M. Mañas, *The Homogeneous Heisenberg subalgebra and equations of AKNS type*, Lett. Math .Phys. **19** (1990), 89-95.

[14] F. Guil and M. Mañas, *Two-dimensional integrable systems and self-dual Yang-Mills equations*, Preprint FT/UCM/18/92.

[15] N. Hitchin, *The self-duality equations on a Riemann surface*, Proc. London Math .Soc. **55** (1987), 59-126.

[16] D.A. Korotkin, *Finite-gap solutions of the $SU(1,1)$ and $SU(2)$ duality equations and their axisymmetric steady-state reductions*, Mat. Sbornik **181** (1990), 923-933; English transl.: Math. USSR Sbornik **70** (1991), 355.

[17] D.A. Korotkin, *Self-dual Yang-Mills fields and deformations of algebraic curves*, Commun. Math .Phys. **134** (1990), 397-412.

[18] S.P. Novikov, S. Manakov, L. Piatveskii and V.E. Zakharov, *Theory of Solitons (The Inverse Scattering Method)*, Consultants Bureau, New York, 1984.

[19] A. Ogielski, M. Prasad, A. Sinha and L. Chau, *Bäcklund transformations and local conservation laws for principal chiral fields*, Phys. Lett. **91B** (1980), 387-391.

[20] B. Piette and W.J. Zakrzewski, *General solutions of the $U(3)$ and $U(4)$ chiral σ Models in two dimensions*, Nucl. Phys .**B300** (1988), 207-222.

[21] K. Pohlmeyer, *Integrable Hamiltonian systems and interactions through quadratic constraints*, Commun. Math .Phys. **46** (1976), 207-221.

[22] A. Pressley and G .Segal, *Loop groups*, Oxford University Press, Oxford, 1986.

[23] A.G. Reiman and M.A. Semenov-Tyan-Shaskii, *Lie algebras and Lax equations with spectral parameter dependence in an elliptic curve*, Zap. Nauchn. Sem. Leningrad Otdel Mat. Inst. Steklov **150** (1986) 104-118,221; English transl.: J. Sov. Math. **46** (1989), 1631.

[24] G. Segal, *Loop Groups and Harmonic Maps*, in: *Advances in Homotopy Theory*, London Math Soc .Lect. Notes **139**, B. Salamon, B. Steer and W. Sutherland (eds.), Cambridge University Press, Cambridge, 1989, 153.

[25] M. Semenov-Tyan-Shanskii, *Classical r-Matrices, Lax Equations, Poisson-Lie Groups and Dressing Transformations*, in: *Field Theory, Quantum Gravity and Strings II*, Lect. Notes Phys. **280**, H.de Vega and N. Sànchez (eds.), Springer, Berlin, 1987, 174.

[26] K. Uhlenbeck, *Harmonic maps into Lie groups (Classical solutions of the chiral model)*, J. Diff. Geom. **30** (1989), 1-50.

[27] K. Uhlenbeck, *On the connection between harmonic maps and the self-dual Yang–Mills and the sine–Gordon Equation*, J. Geom. Phys. **8** (1992), 283-316.

[28] R. Ward, *Classical solutions of the chiral model, unitons, and holomorphic vector bundles*, Commun. Math. Phys. **128** (1990), 319-332.

[29] J.C. Wood, *Explicit construction and parametrization of harmonic two-spheres in the unitary group*, Proc. London Math. Soc. **58** (1989), 608-624.

[30] J.C. Wood, *Harmonic maps into symmetric spaces and integrable systems*, this volume.

[31] V.E. Zakharov and A.V. Mikhailov, *Relativistically invariant two-dimensional model of field theory which is integrable by means of the inverse scattering problem method*, Zh. Eksp. Teor .Fiz. **74** (1978), 1953-1973; English transl.: Sov. Phys. JETP **47** (1978) 1017-1027.

[32] V.E. Zakharov and A.V. Mikhailov, *On the integrability of classical spinor models in two-dimensional space-time*, Commun. Math. Phys. **74** (1980), 21-40.

[33] V.E. Zakharov and A.V. Shabat *Integration of nonlinear equations of Mathematical-Physics by inverse scattering II*, Funk. Anal. Prilozh. **13** (1978), 13-22; English transl.: Func. Anal. Appl. **13** (1979), 166-174.

2-dimensional nonlinear sigma models: zero curvature and Poisson structure

M. Bordemann, M. Forger, J. Laartz and U. Schäper

1 Introduction

In this chapter some aspects of two-dimensional nonlinear sigma models are discussed. The solutions to the field equations are (pseudo-)harmonic maps of two-dimensional Minkowski space into some homogeneous Riemannian manifolds.
In Section 2 we shall give a brief introduction to the physics of these models and discuss the conditions under which a zero curvature ansatz built out of the conserved currents is possible: Theorem (2.1) which is an extended version of old work of the physicists K. Pohlmeyer [29] and H. Eichenherr & M. Forger [11] shows under weak technical assumptions that such an ansatz works if and only if the target manifold is a (pseudo-)Riemannian symmetric space. We have included the rather technical, mainly Lie algebraic proof because the result is new (although not very surprising). The use of symmetric spaces in this context had recently been rediscovered in the mathematics literature (see, for example [31]) and applied to harmonic tori, see for example the work of Burstall-Ferus-Pedit-Pinkall [8]. At the end of this section we give a brief incomplete historical outline of some papers in mathematical physics between 1976 and 1981 dealing with the integrability of two-dimensional sigma models.
Section 3 deals with a symplectic formulation of these models à la Faddeev-Takhtajan [14] or Maillet [26]: The phase space in question consists of smoothly embedded curves into the cotangent bundle of the target manifold with a "canonical" symplectic form. We show that for a certain class of functions of these curves including the above-mentioned currents, Poisson brackets are well-defined and can be computed. The last part of the section compiles recent work of the authors: The computation of the Poisson algebra of the currents [17] and –included in Theorem(3.1) – an explicit calculation of an r-matrix for these models [7] which in finite-dimensional settings is responsible for the Poisson-commutativity of the conserved charges of the system.

2 Zero curvature condition and dual symmetry

Nonlinear sigma models were introduced by Gell-Mann and Levi in 1960 [19]. Their aim was to describe pion-nucleon physics in a low energy approximation to some Lagrangian field theory defined, say, in terms of a set of self-interacting scalar fields. These fields can be assembled into a single map ϕ from d-dimensional Minkowski space ($\mathbb{R}^d, \eta$) ($(\eta_{\mu\nu}) = diag(1, -1, \cdots, -1)$) into some (internal) finite-dimensional real vector space E with positive definite scalar product $\cdot$, and the Lagrangian is given by

$$\mathcal{L}[\phi] = \frac{1}{2}\eta^{\mu\nu}\partial_\mu\phi \cdot \partial_\nu\phi - V(\phi). \tag{2.1}$$

Here V is a smooth function of E into the nonnegative real numbers (a potential) which describes self-interactions as soon as it does not reduce to a quadratic polynomial. In a low-energy approximation this Lagrangian $\mathcal{L}$ is modified by requiring the original fields ϕ to be constrained to the set of minima M of the function V, this set is assumed to be a real connected submanifold of E (see [28] for a discussion of M if V is a polynomial invariant under the unitary action of a Lie group). The scalar product $\cdot$ on E induces a Riemannian metric h on M. The new Lagrangian $\mathcal{L}'$ looks like that of a free field theory, but the fields ϕ take values in a Riemannian manifold (M, h) which in general is curved:

$$\mathcal{L}'[\phi] = \frac{1}{2}\eta^{\mu\nu}h_{ij}(\phi)\partial_\mu\phi^i\partial_\nu\phi^j. \tag{2.2}$$

The resulting Euler-Lagrange equations are then the well-known conditions for ϕ to define a (pseudo-)harmonic map (cf. [32]):

$$\eta^{\mu\nu}(\partial_\mu\partial_\nu\phi^i + \Gamma^i_{jk}(\phi)\partial_\mu\phi^j\partial_\nu\phi^k) = 0 \tag{2.3}$$

(where the Γ^i_{jk} are the Christoffel symbols of the metric h). The Lagrangian (2.2) together with the quadruple $(\mathbb{R}^d, \eta, M, h)$ is called a *d-dimensional sigma model*. Suppose now that a connected Lie group G acts effectively on (M, h) by isometries. Let $\mathcal{G}$ denote its Lie algebra and for each ξ in $\mathcal{G}$ let ξ_M denote the fundamental field of the group action defined by $m \mapsto \frac{d}{dt}(exp(t\xi)m)|_{t=0}$. Moreover, define the following one-form α on M with values in the dual space $\mathcal{G}^*$ of $\mathcal{G}$:

$$\alpha(m)(v_m)(\xi) = -h(m)(v_m, \xi_M(m)) \tag{2.4}$$

(where m is a point in M and v_m is a tangent vector at m). The one-form α is often called the momentum map, see [32]. By this we can define what in physics is called a *current* as the pull-back $\phi^*\alpha$ of α to Minkowski space by the field ϕ:

$$j_\mu[\phi](\xi) = -h_{ij}(\phi)\partial_\mu\phi^i(\xi_M)^j(\phi). \tag{2.5}$$

A straightforward computation shows that if ϕ is a pseudo-harmonic map (i.e is a solution to the field equations) then this current will be conserved:

$$\partial_\mu j^\mu[\phi](\xi) = 0. \tag{2.6}$$

If in addition M is a *homogeneous space* under G, i.e. $M = G/H$ where H is the closed subgroup of G stabilizing one arbitrarily chosen fixed point o of M then that computation shows that on the other hand the conservation of the above current will imply that ϕ is pseudo-harmonic.

Now in case that (M, h) is not a flat manifold nothing much can be said about the "general" solution of the above field equation (2.3) if the dimension of spacetime is four. However, in two-dimensional Minkowski space there is an integrability procedure for certain classes (M, h):

Let us assume that M is connected and homogeneous under the G-action. Since it is a Riemannian manifold we know that M is *reductive*, i.e. there exists an $\mathrm{Ad}_G(H)$-invariant subspace $\mathcal{M}$ of the Lie algebra $\mathcal{G}$ which complements the subalgebra $\mathcal{H}$ (see [21], Ch. X, p.190, Vol.II). Indeed, it is known that the isotropy subgroup of o in the full isometry group is compact (see [21], Ch. VI, p. 190, Vol. I). Using the Haar measure for this subgroup one can find a positive scalar product on the Lie algebra of the full isometry group that is invariant under the adjoint action of the isotropy subgroup. The restriction of this scalar product to the Lie algebra $\mathcal{G}$ will then be $\mathrm{Ad}_G(H)$ invariant because the subgroup H is contained in the above isotropy subgroup. The orthogonal complement $\mathcal{M}$ of $\mathcal{H}$ in $\mathcal{G}$ will then do the job. Now G is a principal bundle over M with structure group H. The tangent bundle TM of M can be regarded as an associated bundle over M with typical fibre $\mathcal{M}$, i.e $TM = G \times_H \mathcal{M}$ and the metric h is determined by its $\mathrm{Ad}_G(H)$invariant value h_o at the point o on $T_oM = \mathcal{M}$. For ξ in $\mathcal{G}$ denote by $\xi_\mathcal{H}$ and $\xi_\mathcal{M}$ its projection on to $\mathcal{H}$ and $\mathcal{M}$, respectively. Observing that the fundamental field $\xi_M(gH)$ is given by $\pi_E(g, (\mathrm{Ad}_G(g^{-1})\xi)_\mathcal{M})$ (π_E denoting the natural map $G \times \mathcal{M} \to TM$) the current for $\phi(t, x) = g(t, x)H$ is given by

$$j_\mu[gH](\xi) = -h_o((g^{-1}\partial_\mu g)_\mathcal{M}, (\mathrm{Ad}_G(g^{-1})\xi)_\mathcal{M}). \tag{2.7}$$

For fixed ϕ this current can be regarded as a one-form on a two-dimensional spacetime with values in the dual space $\mathcal{G}^*$ of the Lie algebra $\mathcal{G}$. In order to make contact with the zero curvature trick one needs some linear map $\mathcal{G} \to \mathcal{G}^*$ to convert the above current into a Lie algebra-valued one-form on space-time. In order to do that we shall make the stronger assumption that $\mathcal{G}$ is *metrisable*, i.e. there exists an ad-invariant nondegenerate symmetric bilinear form K_0 on the Lie algebra $\mathcal{G}$ (i.e. $K_0([\xi, \eta], \zeta) = K_0(\xi, [\eta, \zeta])$ for all ξ, η, ζ in $\mathcal{G}$). We shall speak of the pair $(\mathcal{G}, K_0)$ as a *metrised Lie algebra*. Not every Lie algebra is metrisable (see for example [5] for a discussion) but, for instance, all semisimple or abelian Lie algebras are. Moreover we shall assume that the subalgebra $\mathcal{H}$ and the subspace $\mathcal{M}$ are orthogonal with respect to K_0 in which case we shall speak of a *metrised triple* $(\mathcal{G}, \mathcal{H}, \mathcal{M}, K_0)$. If the form K_0 is not specified we say that the triple $(\mathcal{G}, \mathcal{H}, \mathcal{M})$ is *metrisable*.

Note that a quite general class of homogeneous spaces induces an ad-invariant form on $\mathcal{G}$: the manifold M is called a *naturally reductive homogeneous space* with respect to G and H if and only if the scalar product h_o on $\mathcal{M}$, in addition to being

$\mathrm{Ad}_G(H)$-invariant, satisfies the condition $h_o([X,Y]_{\mathcal{M}}, Z) = h_o(X, [Y, Z]_{\mathcal{M}})$ for all X, Y, Z in $\mathcal{M}$ ([21], Ch. X, p. 202, Vol.I). This is equivalent to the condition that each geodesic emanating from o with velocity vector X in $\mathcal{M}$ is of the form $\tau \mapsto \pi(exp(\tau X))$ where π denotes the canonical projection $G \to M = G/H$. Consider the normal subgroup G' of G corresponding to the ideal $\mathcal{G}' = \mathcal{M} + [\mathcal{M}, \mathcal{M}]$ of $\mathcal{G}$. Since $\mathcal{M}$ is contained in $\mathcal{G}'$ this G'-action is automatically transitive, so we have $M = G'/(H \cap G')$. Then it can be shown (see [22] for the compact and [15] for the general case) that there exists a nondegenerate symmetric $\mathrm{ad}_{\mathcal{G}'}$-invariant bilinear form K_0 on $\mathcal{G}'$ such that $\mathcal{M}$ and $\mathcal{H}$ are orthogonal and the restriction of K_0 to $\mathcal{M}$ equals h_o. In this manner one obtains a large class of examples, notably all pseudo-Riemannian symmetric spaces (e.g. compact Lie groups, spheres, Grassmannians,...) and all reductive homogeneous spaces whose group G is semisimple and for which the restriction of (some multiple of) the Killing form to $\mathcal{M}$ is positive or at least nondegenerate. However, a compact homogeneous Kähler manifold is naturally reductive with respect to a semisimple group of Kähler isometries if and *only if* it is Hermitian symmetric (see for example [6] for a proof). Thus for example, nonsymmetric flag manifolds with their natural $SU(N)$-invariant Kähler structure are *not* naturally reductive.

In any case, we shall suppose from now on that the triple $(\mathcal{G}, \mathcal{H}, \mathcal{M})$ is metrisable. Consequently, after having chosen a bilinear form K_0 such that the triple $(\mathcal{G}, \mathcal{H}, \mathcal{M}, K_0)$ is metrised, we can uniquely express the metric h_o at o by means of a K_0-symmetric nondegenerate linear endomorphism S of $\mathcal{M}$:

$$h_o(X, Y) = K_0(SX, Y) \qquad \text{for all } X, Y \in \mathcal{M}. \tag{2.8}$$

Identifying $\mathcal{G}^*$ with $\mathcal{G}$ via K_0 we arrive at a Lie algebra-valued version of the current (2.7) in the following, slightly more familiar form:

$$j_\mu[gH] = -\mathrm{Ad}(g)(S(g^{-1}\partial_\mu g)_{\mathcal{M}}). \tag{2.9}$$

Clearly, we can write the current as the pull-back $\phi^* A$ where A is the Lie algebra-valued one-form on $M = G/H$ given by

$$A(gH)(\pi_E(g, X)) = \mathrm{Ad}(g)SX \qquad \text{for } g \in G \text{ and } X \in \mathcal{M}. \tag{2.10}$$

The basic idea for a certain zero-curvature representation of the field equations is contained in following condition [29, 11]:

Under the above assumptions, we say that (M, h) *admits a zero-curvature representation for pseudo-harmonic maps by dual symmetry* if and only if the following holds: There exist smooth real-valued functions a, b on some open dense subset of the real axis and a bilinear form K on $\mathcal{G}$ making the triple $(\mathcal{G}, \mathcal{H}, \mathcal{M}, K)$ metrised such that for any smooth map ϕ of two-dimensional Minkowski space to M the following is satisfied: the new current (depending on the form K and a spectral parameter λ):

$$L_\mu(\lambda)[\phi] = a(\lambda) j_\mu[\phi] + b(\lambda)(*j)_\mu[\phi]. \tag{2.11}$$

(where $*j$ denotes the Hodge dual of j, i. e. $(*j)_0 = j_1$ and $(*j)_1 = j_0$) has curvature zero:

$$(dL(\lambda))_{\mu\nu} + [L_\mu(\lambda), L_\nu(\lambda)] = 0. \tag{2.12}$$

for all values of λ if and only if ϕ is pseudo-harmonic.

However, this approach does not work in general: it imposes restrictions on the target manifold G/H: The following theorem which is a slight generalization of a theorem of Eichenherr and Forger [11] deals with the general case:

Theorem 2.1 *Let $(M = G/H, h)$ be a connected simply connected homogeneous reductive pseudo-Riemannian manifold of dimension greater or equal than 2 where the Lie group G is connected and simply connected and acts almost effectively on M. Suppose that*

1. *the triple $(\mathcal{G}, \mathcal{H}, \mathcal{M})$ is metrisable and that*

2. *any given two-plane in any tangent space of M is the image of two-dimensional Minkowski space under the tangent map of a suitable pseudo-harmonic map to M.*

Then the following two statements are equivalent:

1. *(M, h) admits a zero-curvature representation for pseudo-harmonic maps by dual symmetry. If this is the case the functions a and b have to satisfy the equation*

$$a(\lambda)^2 - b(\lambda)^2 = ca(\lambda)$$

for some fixed real number c.

2. *(M, h) is isometric to the cartesian product $(M_1 \times M_2, h_1 \times h_2)$ where (M_2, h_2) is a connected simply connected normal Lie subgroup of G with a bi-invariant pseudo-Riemannian metric h_2 and $(M_1 = G_1/H, h_1)$ is a pseudo-Riemannian symmetric space on which another connected simply connected normal Lie subgroup G_1 of G acts transitively. Moreover, G is isomorphic to $G_1 \times G_2$, their Lie algebras $\mathcal{G}_1$ and $\mathcal{G}_2$ are orthogonal with respect to K_0, H is contained in G_1, and M_1 is different from a point if and only if the subalgebra $\mathcal{H}$ is nonzero.*

Proof: Since the curvature F of $L(\lambda)$ is a two-form on a two-dimensional manifold there is only one nonzero component, say F_{01} (where 0 stands for the time and 1 for the space coordinate). In order to compute the $(dL(\lambda))_{01}$ part of F_{01} we have to know the exterior derivative dA of the Lie algebra-valued one-form A defined above. Using the obvious formula $\Phi_g^* A = \text{Ad}(g)A$ where Φ_g denotes the canonical group action on M we can compute dA in terms of fundamental fields. Extending S to a linear endomorphism of $\mathcal{G}$ by defining it to be zero on the subalgebra $\mathcal{H}$ we can rewrite this formula to get (letting gX denote the tangent vector $\pi_E(g, X)$ for g in G and X in $\mathcal{M}$)

$$dA(gH)(gX, gY) = -\mathrm{Ad}(g)([SX, Y] + [X, SY] - S[X, Y]).$$

Writing now $X = (g^{-1}\partial_0 g)_\mathcal{M}$ and $Y = (g^{-1}\partial_1 g)_\mathcal{M}$ where the field $\phi(t, x)$ is given by $g(t, x)H$ we get

$$\begin{aligned}F_{01} &= b(\lambda)(d * j)_{01}[\phi] + \\ &+ \mathrm{Ad}(g)a(\lambda)(-[SX, Y] - [X, SY] + S[X, Y]) + \\ &+ \mathrm{Ad}(g)(a(\lambda)^2 - b(\lambda)^2)[SX, SY]\end{aligned}$$

Now, in view of Hypothesis 2 and Condition (2.6) we see that (M, h) admits a zero curvature representation for harmonic maps if and only if $b(\lambda)$ is nonzero and the terms after $\mathrm{Ad}(g)$ vanish for all X, Y in $\mathcal{M}$. Without any harm this condition can be extended on all of $\mathcal{G}$ since S vanishes on the subalgebra $\mathcal{H}$ and is $\mathrm{ad}_\mathcal{G}(\mathcal{H})$-invariant: $[V, SX] = S[V, X]$. Therefore the crucial condition for $\mathcal{G}$ is the following algebraic equation:

$$a(\lambda)(-[S\xi, \eta] - [\xi, S\eta] + S[\xi, \eta]) + (a(\lambda)^2 - b(\lambda)^2)[S\xi, S\eta] = 0$$

for all ξ, η in $\mathcal{G}$ and λ in some open dense subset of the real axis.

Now suppose that Statement 1 holds. We have to distinguish three cases:

Firstly, suppose that $a(\lambda) = 0$. Since $b(\lambda)$ must be nonzero, it follows that $[S\xi, S\eta]$ vanishes for all ξ, η in $\mathcal{G}$. S being a linear isomorphism of $\mathcal{M}$ this entails the relation $[\mathcal{M}, \mathcal{M}] = 0$. Moreover, as can easily be computed, S has to be a derivation of $\mathcal{G}$, i.e. $S[\xi, \eta] = [S\xi, \eta] + [\xi, S\eta]$ for all ξ, η in $\mathcal{G}$. But since S is symmetric with respect to K and $\mathrm{ad}(\xi)$ antisymmetric it follows that we must have $[S, \mathrm{ad}(\xi)] = \mathrm{ad}(S\xi)$ and $-[\mathrm{ad}(\xi), S] = -\mathrm{ad}(S\xi)$ at the same time showing that $\mathrm{ad}(S\xi) = 0$ which implies that $[\mathcal{G}, \mathcal{M}] = 0$. But then $\mathcal{H}$ is an ideal of $\mathcal{G}$ and has to vanish because of the assumed almost effective group action. It follows that $G = M$ is an abelian simply connected Lie group.

Secondly, suppose that $a(\lambda) = \pm b(\lambda)$. Then $a(\lambda) \neq 0$ and S has to be a derivation of $\mathcal{G}$. By the second part of the reasoning of the first case (which was independent of the relation $[\mathcal{M}, \mathcal{M}] = 0$) we conclude that $[\mathcal{M}, \mathcal{G}] = 0$ showing again that $G = M$ is an abelian simply connected Lie group.

Thirdly, suppose that $a(\lambda) \neq 0$ and that $(a(\lambda)^2 - b(\lambda)^2)/a(\lambda) = c \neq 0$ where c has to be a real constant. It follows that the map $\Phi = 1 - cS$ is a homomorphism of the Lie algebra $\mathcal{G}$, i.e. $\Phi[\xi, \eta] = [\Phi\xi, \Phi\eta]$. Consider the Fitting decomposition of $\mathcal{G}$ with respect to Φ, i.e. $\mathcal{G} = \mathcal{G}_n \oplus \mathcal{G}_u$ where the subspace $\mathcal{G}_n$ (the Fitting 0-space) is defined to be the union of all the kernels $\mathrm{Ker}\Phi^k$ ($k \geq 1$) and the subspace $\mathcal{G}_u$ (the Fitting 1-space) is equal to the intersection over all images $\mathrm{Im}\Phi^k$ ($k \geq 1$). $\mathcal{G}_n$ is obviously an ideal of $\mathcal{G}$ and since Φ is symmetric w.r.t K it follows that $\mathcal{G}_u$ is the orthogonal space of $\mathcal{G}_n$ whence it is also an ideal of $\mathcal{G}$ by the ad-invariance of K. Clearly, the subalgebra $\mathcal{H}$ is contained in $\mathcal{G}_u$, hence $\mathcal{G}_n$ is contained in the subspace $\mathcal{M}$. Now, by Fitting's Lemma the restriction Φ_u of Φ to $\mathcal{G}_u$ is a linear isomorphism whereas the restriction Φ_n of Φ to $\mathcal{G}_n$ is nilpotent. Since Φ_n is a K-symmetric homomorphism of $\mathcal{G}_n$ it satisfies both $\Phi_n \mathrm{ad}(\xi) = \mathrm{ad}(\Phi_n \xi)\Phi_n$ and the transposed condition $-\mathrm{ad}(\xi)\Phi_n = -\Phi_n \mathrm{ad}(\Phi_n \xi)$. But repeated application of these two equations gives $\Phi_n \mathrm{ad}(\xi) = \Phi_n \mathrm{ad}(\Phi_n^{2k} \xi)$ and $\mathrm{ad}(\xi)\Phi_n = \mathrm{ad}(\Phi_n^{2k} \xi)$ for an arbitrary positive integer k. Since Φ_n is nilpotent it follows that both equations are equal to zero. In particular S_n (the restriction of S to $\mathcal{G}_n$ commutes with all $\mathrm{ad}(\xi)$, ξ in

2-dimensional nonlinear sigma models: zero curvature and Poisson structure 181

$\mathcal{G}_n$ whence the metric $K(S\xi, \eta)$ on $\mathcal{G}_n$ is $\mathrm{Ad}(G_n)$-invariant where G_n denotes the normal simply connected subgroup of G generated by the ideal $\mathcal{G}_n$. Since Φ_u is a K-symmetric automorphism of $\mathcal{G}_u$ it again satisfies both $\Phi_u \mathrm{ad}(\xi) = \mathrm{ad}(\Phi_u \xi)\Phi_u$ and the transposed condition $-\mathrm{ad}(\xi)\Phi_u = -\Phi_u \mathrm{ad}(\Phi_u \xi)$ from which one can easily derive the equation $(1 - \Phi_u^2)[\xi, \eta] = 0 = [(1 - \Phi_u^2)\xi, \eta]$. Consider again a Fitting decomposition of $\mathcal{G}_u$ with respect to $\chi = 1 - \Phi_u^2$. By the same reasoning as above we conclude that the Fitting 1-space $\mathcal{G}_{uu}$ is a central ideal of $\mathcal{G}_u$ which is orthogonal to the Fitting 0-space $\mathcal{G}_{un}$ which in turn is an ideal of $\mathcal{G}_u$. Clearly, the subalgebra $\mathcal{H}$ is contained in $\mathcal{G}_{un}$ whence $\mathcal{G}_{uu}$ is again contained in the subspace $\mathcal{M}$. Consider the equation $\Phi_{un}^2 = 1 - \chi_n$ where Φ_{un} and χ_n denote the restrictions of the corresponding maps to $\mathcal{G}_{un}$. Observe that the linear map $(1 - \chi_n)^{-1/2}$ makes sense as a polynomial in the nilpotent map χ_n with coefficients of the Taylor series of that square root. Clearly $(1-\chi_n)^{-1/2} = 1 - \chi'$ commutes with Φ_{un} where χ' is a nilpotent map which depends polynomially on χ_n. Since $1 - \chi'$ is equal to the identity on and inside commutators the linear map $\Phi' = \Phi_{un}(1 - \chi')$ is an involutive automorphism of the ideal $\mathcal{G}_{un}$. Clearly, the subalgebra $\mathcal{H}$ is contained in the set of fixed points of Φ'. Suppose that an element X of the subspace $\mathcal{M}_{un} = \mathcal{M} \cap \mathcal{G}_{un}$ is contained in the set of fixed points of Φ'. Then, setting $1 - \chi'' = (1 - \chi')^{-1}$ (where χ'' is another nilpotent polynomial of χ_n) we get the equation $cSX = \chi''X$. Since S commutes with χ'' we would get $(cS)^k X = (\chi'')^k X = 0$ for large enough powers k which is absurd since S is invertible on $\mathcal{M}$. Hence the subalgebra $\mathcal{H}$ is the set of fixed points of Φ' and the subspace $\mathcal{M}_{un}$ must be equal to the set of eigenvectors belonging to the eigenvalue -1 of Φ' because Φ' is involutive (i.e. $\Phi'^2 = 1$) and leaves $\mathcal{M}_{un}$ invariant since it is a polynomial in S. Therefore we get the characteristic commutation relation $[\mathcal{M}_{un}, \mathcal{M}_{un}] \subset \mathcal{H}$ for symmetric spaces. If $\mathcal{H}$ is zero then $\mathcal{M}_{un}$ will be an abelian ideal of $\mathcal{G}$.

Now we can define the following: If the subalgebra $\mathcal{H}$ is zero then the subgroup H is equal to $\{1\}$ because it is connected and we set M_1 equal to the set containing one point and $(M_2, h_2) = (G, h')$ where h' is a bi-invariant metric on G induced by the $\mathrm{Ad}(G)$-invariant scalar product h_0 on $\mathcal{M} = \mathcal{G}$: this is well-defined because $\mathcal{M}$ is an orthogonal direct sum of the central ideal $\mathcal{M}_{un} \oplus \mathcal{G}_{uu}$ plus the ideal $\mathcal{G}_n$ on which the restriction of h has already been shown to be Ad-invariant.

If the subalgebra $\mathcal{H}$ is nonzero we set (M_1, h_1) equal to $(G_1/H, h_{un})$ where $G_1 = G_{un}$ is the connected normal Lie subgroup of G generated by the ideal $\mathcal{G}_{un}$ and h_{un} is the pseudo-Riemannian metric on M_1 induced by the restriction of h_o to the subspace $\mathcal{M}_{un}$. Since $\mathcal{G}_{un}$ has already been shown to be a symmetric Lie algebra it follows that (M_1, h_1) is a pseudo-Riemannian symmetric space. Let M_2 be the connected normal subgroup of G generated by the ideal $\mathcal{M}_2 = \mathcal{G}_{uu} \oplus \mathcal{G}_n$. By an analogous reasoning as above it follows that the bi-invariant metric h_2 on M_2 induced by the restriction of h_o to $\mathcal{M}_2$ is well-defined. Since G was assumed to be simply connected it follows from the decomposition of its Lie algebra into the direct sum of ideals that it is isomorphic to the Cartesian product of G_1 and G_2.

Conversely, suppose that Statement 2 holds. We have to find an $\mathrm{Ad}(G)$-invariant nondegenerate symmetric bilinear form K on $\mathcal{G}$ such that the λ-dependent current satisfies the zero curvature condition. Define K'' on $\mathcal{G}_2 \times \mathcal{G}_2$ equal to the value of h_2 at the unit element of G_2. Consider now the ideal $\mathcal{I} = \mathcal{M}_1 \oplus [\mathcal{M}_1, \mathcal{M}_1]$ of $\mathcal{G}_1$ where $\mathcal{M}_1$ is defined to be the intersection of $\mathcal{M}$ with $\mathcal{G}_1$. Its orthogonal space in $\mathcal{G}_1$ with respect to K_0 is an ideal of $\mathcal{G}$ contained in $\mathcal{H}$, hence it follows that $\mathcal{G}_1$ is equal to $\mathcal{I}$. Define the standard form K' on $\mathcal{I}$ by $K'(X, Y) = (h_1)_o(X, Y)$ for all X, Y in $\mathcal{M}_1$, $K'(V, X) = 0$ for all V in $\mathcal{H}$, X in $\mathcal{M}_1$, and $K'([X, Y], [X', Y']) = (h_1)_o(X, [Y, [X', Y']])$ for all

X, Y, X', Y' in $\mathcal{M}_1$. A lengthy but straightforward computation shows that K' is a well-defined, symmetric, nondegenerate, Ad-invariant bilinear form on $\mathcal{G}_1$. Define K to be the orthogonal sum $K' + 2K''$. Clearly, K is an Ad(G)-invariant nondegenerate scalar product on $\mathcal{G}$. Choosing $c = 2$ and observing that the map S is equal to $\pi_{\mathcal{M}_1} + \frac{1}{2}\pi_{\mathcal{G}_2}$ ($\pi_{\mathcal{M}_1}$ and $\pi_{\mathcal{G}_2}$ denoting the orthogonal projections of the corresponding subspaces) the zero curvature condition is quickly checked. **Q.E.D.**

Remark: This proof simplifies drastically if one assumes that S is proportional to $\pi_{\mathcal{M}}$; this was done in the original proof of Eichenherr and Forger [11].

Choosing the function $a(\lambda)$ equal to $2/(1 - \lambda^2)$ and $b(\lambda)$ equal to $2\lambda/(1 - \lambda^2)$ (in case $c = 2$) we get the following expressions for the λ-dependent current which we shall use in the next section:

$$L_0 = \frac{2}{1 - \lambda^2}(j_0 + \lambda j_1) \tag{2.13}$$

$$L_1 = \frac{2}{1 - \lambda^2}(j_1 + \lambda j_0). \tag{2.14}$$

We shall not discuss the construction of conserved charges and Bäcklund transformations. The interested reader is referred to the paper of K. Uhlenbeck [31] and to the contribution of M. Mañas in this book [27]).

We conclude this section with some (of course incomplete) historical remarks concerning the development of the above zero curvature representation in the physics literature at the end of the seventies (see also Forger's review [15]):
In 1976, K. Pohlmeyer published a paper in Comm. Math. Phys. ([29]) which initiated several further investigations in the theory of nonlinear sigma models and harmonic maps: firstly, previous joint investigations by H. Lehmann, K. Pohlmeyer, and G. Roepstorff had re-established the following fact which – in spite of being known in the mathematical theory of surfaces of constant curvature – seemed to be unknown to the physicists working in field theory at that time: the $O(3)$-sigma model (where the fields q take values in the 2-sphere in $\mathbb{R}^3$) admits a reduction to the *Sine-Gordon equation*: if $u = t + x$ and $v = t - x$ denote the light cone coordinates, and (,) denotes the Euclidean scalar product in $\mathbb{R}^3$, then the field equations allow for a reparametrisation in u and v such that $(\partial_u q, \partial_u q) = 1 = (\partial_v q, \partial_v q)$, $\cos \alpha = (\partial_u q, \partial_v q)$ leading to the equation $\partial_u \partial_v \alpha + \sin \alpha = 0$. Secondly, trying to "lift back" the known conserved charges of the Sine-Gordon equation to the sigma model, Pohlmeyer observed that he first had to find a one-parameter family of transformations on the solutions of the sigma model that did *not* change the above angle $\cos \alpha = (\partial_u q, \partial_v q)$. In this way he discovered the linear system or equivalently the zero curvature representation by dual symmetry of the $O(N)$-sigma model where the fields take values in the $(N-1)$-dimensional sphere by using the $O(N)$-currents $j_\mu = q\partial_\mu q^T - \partial_\mu q \; q^T$, compare ([29], eqs. (V.5.1) - (V.6.2)). Taking the solutions $\mathcal{R}(\lambda)$ of the linear system (cf. eqs. (2.13) ff) above) $\partial_\mu \mathcal{R}(\lambda) = \mathcal{R}(\lambda)L_\mu(\lambda)$ (which are $O(N)$-valued functionals on the fields q) and conjugating by these symmetry transformations a certain generalized "lifted version" of the $\gamma = 1$-Sine-Gordon-Bäcklund transformation ([29], eqs. (VI.4.1$^+$) - (VI.4.4$^+$)), he arrives at the desired

one parameter family of Bäcklund transformations of the sigma model ([29], eq. (VI.6). Along with this family he easily gets an infinite family of *local* conserved currents (see [29], eqs. (VI.8) - (VI.9.3)) which appear as the coefficients in a certain power series in λ around the point $\lambda = 1$ (the parameter γ appearing in [29] is equal to $(\lambda-1)/(\lambda+1)$). On the other hand, the family $\mathcal{R}(\lambda)$ itself gives rise to an another infinite family of *nonlocal* conserved charges that take the form of iterated integrals and appear as coefficients in a similar power series in λ around $\lambda = \infty$: this had been worked out in 1978 by M. Lüscher and K. Pohlmeyer [23] again for the $O(N)$ sigma model.

After the publication of Pohlmeyer's 1976 paper several people arrived at more and more generalized statements: In his dissertation [10] H. Eichenherr affirmed Pohlmeyer's conjecture that the above zero curvature representation and the analysis of nonlocal charges was also correct for sigma models with values in complex projective space (the so-called $\mathbb{C}P^n$-models). In the same year 1978, V. E. Zakharov and A. V. Mikhailov showed that the above zero curvature representation worked for sigma models with values in arbitrary Grassmannians and also for the so-called principal or chiral models where the target space is an arbitrary reductive Lie group ([33], eq. (4.8)). One year later, in 1979, H. Eichenherr and M. Forger gave their classification theorem [11]: Among those 2-dimensional sigma models with values in a naturally reductive homogeneous space it is precisely the class having values in a symmetric space that allow for the zero curvature representation by dual symmetry. Their proof is also applied to Euclidean sigma models which are parametrized by a Riemannian $\mathbb{R}^2$. The necessary differential geometric and Lie algebraic background for this is contained in M. Forger's Dissertation [16]. Their 1980 paper [12] related the principal models to the ones with values in a symmetric space $M = G/H$ by using the Cartan immersion $M \to G : gH \mapsto g\sigma(g^{-1})$ where σ denotes the involutive automorphism of G such that H is contained in G_σ, the group of fixed points of σ, whose identity component is in turn contained in H (see for example [20], p. 209 and p. 276). The formulation of the above-mentioned higher local conservation laws for sigma models with values in general symmetric spaces was also done by Eichenherr and Forger in 1981 [13].

3 Current Algebra and r-matrices

In this section we shall consider the *initial value problem* or *Cauchy problem* for sigma models in order to allow for a Hamiltonian formulation. We shall see that the notion of Liouville integrability used for Hamiltonian systems can also be (partly) ascribed to sigma models.

Since our parameter space is two-dimensional Minkowski space and not some Riemannian manifold like the two-dimensional sphere there is a natural framework available: Fix a spacelike hyperplane (for example the x-axis in some fixed inertial frame). The values of the field ϕ on this axis, $\phi(x) = \phi(0, x)$, form a curve in the target manifold (M, h) which we assume to be a smooth embedding for simplicity.

Since the field equation (2.3) is second order in time we also have to fix its first time derivative $\dot{\phi} = \partial \phi(t,x)/\partial t \mid_{t=0}$. This curve in the tangent bundle TM of M (which we also assume to be smooth) projects onto $\phi(x)$, i.e. $\tau_M(\dot{\phi}(x)) = \phi(x)$ with τ_M denoting the obvious bundle projection $TM \to M$. These two initial conditions uniquely determine the field $\phi(t,x)$ locally as a solution of the field equation (2.3).

Now we have to specify our symplectic manifold (i.e. the set of initial conditions) in more detail: Using the Riemannian metric h as a bundle isomorphism $h^\flat : TM \to T^*M : v \mapsto (w \mapsto h(v,w))$ we can form the smooth curve $\pi(x) = h^\flat(\dot{\phi}(x))$ of Minkowski space into the cotangent bundle T^*M of M. Clearly, $\tau_M^*(\pi(x)) = \phi(x)$ where τ_M^* denotes the obvious bundle projection $T^*M \to M$. Consider the set

$$\mathcal{C} = \{\psi : \mathbb{R} \to T^*M \mid \psi \text{ is a smooth embedding}\} \tag{3.1}$$

of smoothly embedded curves in T^*M projecting onto smoothly embedded curves in M. In order to define the tangent bundle of $\mathcal{C}$ and a "canonical symplectic form" we have to ensure the convergence of certain integral expression to come: Decompose the tangent space T_pT^*M at p in T_mM^* (m in M) into its vertical subspace $V_pT^*M = \{p'^{ver}(p) = \frac{d}{dt}(p + tp') \mid_{t=0} \mid p' \in T_mM^*\}$ and its horizontal subspace $H_pT^*M = \{v^{hor}(p) = \frac{d}{dt}\tau_{c(t)}^*p \mid_{t=0} \mid v = \dot{c}(0) \in T_mM\}$ ($\tau_{c(t)}^*$ denoting the parallel transport in the cotangent bundle along a curve $c(t)$ in M induced by the Levi-Cività connection of the metric h) and define the Sasaki metric h^{Sas} on T^*M by declaring the horizontal and vertical subspace to be mutually orthogonal and setting $h^{Sas}(p)(p'^{ver}, p''^{ver}) = h(m)(h^\sharp p', h^\sharp p'')$ and $h^{Sas}(p)(v^{hor}(p), w^{hor}(p)) = h(m)(v,w)$ ($h^\sharp$ denoting the inverse of $h^\flat$). For ψ in $\mathcal{C}$ define the tangent space $T_\psi \mathcal{C}$ to be the space of all smooth curves $\chi : \mathbb{R} \to TT^*M$ which project onto ψ (i. e. $\tau_{T^*M}(\chi) = \psi$) and are L^2 with respect to the integrated Sasaki metric, i. e. $\int dx h^{Sas}(\psi(x))(\chi(x), \chi(x)) < \infty$. The tangent bundle $T\mathcal{C}$ of $\mathcal{C}$ can then formally be defined as the disjoint union of all these tangent spaces. In order to allow for the existence of the currents to be made more precise further down we can make the space $\mathcal{C}$ slightly smaller by demanding the derivatives of ψ w. r. t. x to be L^2 w. r. t. the Sasaki metric, i. e. $\int dx h^{Sas}(\psi(x))(\frac{d\psi}{dx}(x), \frac{d\psi}{dx}(x)) < \infty$. The following two-form on $T_\psi \mathcal{C}$ will then exist, i. e. the integral converges:

$$\Omega(\psi)(\chi, \chi') = \int dx \omega(\psi(x))(\chi(x), \chi'(x)) \tag{3.2}$$

(ω denoting the canonical symplectic form on T^*M) because ω can be expressed by the Sasaki metric as can easily be computed:

$$\omega(p)(v^{hor}(p) + p'^{ver}(p), w^{hor}(p) + p''^{ver}(p)) = p'(w) - p''(v)$$
$$= h^{Sas}(p)((h^\sharp p')^{hor}(p) - (h^\flat v)^{ver}(p), w^{hor}(p) + p''^{ver}(p))$$

Using the lifted fields $\chi(\psi) = X \circ \psi$ where X is a suitable vector field on T^*M one can compute that $d\Omega$ vanishes. A standard bump function argument shows that Ω is weakly nondegenerate whence Ω could be called a symplectic form on $\mathcal{C}$.

Having a symplectic form at hand we would like to compute Poisson brackets of suitable real valued functionals on $\mathcal{C}$. But since this is an infinite dimensional manifold the symplectic form Ω is, in general, not invertible and one cannot expect to have well-defined Poisson brackets for all pairs of functionals on $\mathcal{C}$ even if they are in some sense C^∞. We would also like to avoid a detailed analysis of possible topologies and differentiable structures on mapping spaces like $\mathcal{C}$. Therefore we shall resign ourselves to giving the definition of the Poisson bracket of pairs of functionals on $\mathcal{C}$ which are of a particular, frequently used kind:

Recall that the initial value $\phi(x)$ of the field ϕ is nothing but the projection $\tau_M^*(\psi(x))$ of the curve ψ in T^*M discussed above. Let F, H be two smooth functions on T^*M, let B, B' be two one-forms on M, and let X, Y be two real valued smooth functions with compact support defined on the real axis. Define the following functionals on $\mathcal{C}$:

$$\langle F, X \rangle(\psi) = \int dx\, F(\psi(x))X(x) \tag{3.3}$$

$$\langle B, Y \rangle(\psi) = \int dx\, B(\phi(x))(\frac{d\phi}{dx})Y(x). \tag{3.4}$$

Again using the above-mentioned lifted fields, we can compute their Hamiltonian vector fields:

$$\hat{X}_{\langle F,Y \rangle}(\psi) = (x \mapsto X_F(\psi(x))Y(x)\ ) \tag{3.5}$$

$$\hat{X}_{\langle B,Z \rangle}(\psi) = (x \mapsto -Z(x)((i_{\frac{d\psi}{dx}}d(\tau_Q^*)^*B)(\psi(x)))^{ver}(\psi(x))$$
$$- \frac{dZ}{dx}(x)(((\tau_Q^*)^*B)(\psi(x)))^{ver}(\psi(x))\ ). \tag{3.6}$$

where Y and Z are smooth real-valued functions with compact support on the real axis and X_F denotes the Hamiltonian vector field on T^*M, i.e. $dF = \omega(X_F,\ )$. Their Poisson brackets can be computed from these formulae:

$$\{\langle F, X \rangle, \langle H, Y \rangle\}(\psi) = \langle \{F, H\}, XY \rangle(\psi) \tag{3.7}$$

$$\{\langle F, X \rangle, \langle B, Y \rangle\}(\psi) = \langle -i_{T\tau_M^*X_F}dB, XY \rangle(\psi)$$
$$+ \langle B(T\tau_M^*X_F), X\frac{dY}{dx} \rangle(\psi) \tag{3.8}$$

$$\{\langle B, X \rangle, \langle B', Y \rangle\}(\psi) = 0. \tag{3.9}$$

Here $\{F, H\} = dF(X_F)$ denotes the Poisson bracket of functions on T^*M, and XY denotes the usual pointwise product of real valued functions. It is clear that the prescription above defines a closed antisymmetric bracket on the real vector space spanned by the functionals (3.3, 3.4). The verification of the Jacobi identity is a straight forward computation. It should be mentioned that the Poisson brackets (3.7ff) have been derived from the simpler rule used by physicists (compare [14], page 322, eq. (5.11))

$$\{\phi^i(x), \pi_j(y)\} = \delta^i_j \delta(x-y) \;,\; \{\phi^i(x), \phi^j(y)\} = 0 = \{\pi^i(x), \pi^j(y)\} \qquad (3.10)$$

where (ϕ^i, π_j), $(1 \le i, j \le n = \dim M)$ denote the components of ψ in a local bundle chart of T^*M, and δ denotes the delta distribution for functions on the real axis. This rule can be shown to be independent of the chosen bundle chart [17]. Using eq. (3.10) on can presumably write down consistent Poisson brackets for a larger class of functionals which are local in the sense that they may depend on a finite number of arbitrarily high (covariant) derivatives of the curve ψ. But for the current algebra to be discussed further down it suffices to stay in the Lie algebra of the above functionals (3.3, 3.4). However, we emphasize that the heuristic value of the rule (3.10) is restricted: for instance, the naive application of (3.10) to Poisson brackets of nonlocal functionals like certain charges, defined as a suitable parallel transport over the real axis leads to ugly violations of the Jacobi identity (see [9], [26], [7]). At this point it seems that a careful analysis of the topology and differentiable structure of the mapping space $\mathcal{C}$ and the smoothness of the Poisson structure and the considered functionals becomes indispensable.

The above Poisson brackets (3.7ff) can now be applied to the current components $j_0[\psi](\xi)(x) = F(\psi(x))$ and $j_1[\psi](\xi)(x) = B(\phi(x))(d\phi/dx(x))$. Introducing the field

$$j[\psi](\xi, \eta)(x) = h(\phi(x))(\xi_M(\phi(x)), \eta_M(\phi(x))) \qquad (3.11)$$

(where ψ is given by $\pi = h^\flat \phi$) we arrive at the following formulae for the Poisson brackets: Here $(T_a)_{a=1,\ldots,\dim\mathcal{G}}$ denotes a basis for the symmetry Lie algebra $\mathcal{G}$, $f^a_{bc} T_a = [T_b, T_c]$ are the usual structure constants. Moreover, $j_{\mu,a} = j_\mu[\psi](T_a)(x)$ and $j_{ab}(x) = j[\psi](T_a, T_b)(x)$, and $\delta'(x) = \partial_x \delta(x)$ denotes the derivative of the δ-distribution (see [17] for more details of the computation):

$$\{j_{0,a}(x), j_{0,b}(y)\} = -f^c_{ab} j_{0,c}(x) \delta(x-y) \qquad (3.12)$$

$$\{j_{0,a}(x), j_{1,b}(y)\} = -f^c_{ab} j_{1,c}(x) \delta(x-y) + j_{ab}(y) \delta'(x-y) \qquad (3.13)$$

$$\{j_{1,a}(x), j_{1,b}(y)\} = 0 \qquad (3.14)$$

$$\{j_{0,a}(x), j_{bc}(y)\} = -\left(f^d_{ab} j_{cd}(x) + f^d_{ac} j_{bd}(x)\right) \delta(x-y) \qquad (3.15)$$

$$\{j_{1,a}(x), j_{bc}(y)\} = 0 \qquad (3.16)$$

$$\{j_{ab}(x), j_{cd}(y)\} = 0 \,. \qquad (3.17)$$

Integrating this over dx and dy together with the smooth real valued functions $X(x)$ and $Y(y)$ we can turn this into the form described above. Note that these formulae are valid without the assumption of M being homogeneous. The occurrence of the δ'-term makes these models "nonultralocal". As we shall see further down, the field j in front of the δ'-term is independent of ϕ if (G, h) is a semisimple Lie group. In this case, the δ'-term can be treated as a central extension of the current Lie algebra above with the δ'-term omitted.

We return now to the situation where (M, h) is a Riemannian symmetric space, and we assume for simplicity that its isometry group is simple. The zero-curvature

2-dimensional nonlinear sigma models: zero curvature and Poisson structure 187

representation of the pseudo-harmonic map equation guaranteed by Theorem 2.1 can equally be expressed by the fact that spatial and temporal covariant derivatives commute: $[\partial_0 + L_0(\lambda), \partial_1 + L_1(\lambda)] = 0$. For the initial value problem we can re-express this commutator in Lax form by writing

$$D(x, \lambda) = \partial_x + L_1(x, \lambda) \qquad (3.18)$$

and getting the well-known Lax-equation:

$$\frac{\partial D}{\partial t} = [D, L_0]. \qquad (3.19)$$

Here D is thought of as a function of the initial value space $\mathcal{C}$.

Before we proceed, we shall recall some general facts about completely integrable systems in Lax form:
A *completely integrable Hamiltonian system* (P, ω, H) is a smooth real-valued function H on a finite-dimensional symplectic manifold (P, ω) such that there exist $n = \dim P/2$ smooth real-valued functions $F_1, \ldots, F_n$ such that i) $\{H, F_i\} = 0$ for all $1 \leq i \leq n$ (the F_i are conservation laws), ii) $\{F_i, F_j\} = 0$ for all $1 \leq i, j \leq n$ (the F_i are in involution), iii) $dF_1 \wedge \cdots \wedge dF_n(x) \neq 0$ for almost all x (with respect to the symplectic volume) in P (the F_i are functionally independent), and iv) there exists a smooth real-valued function f of $\mathbb{R}^n$ such that $H = f(F_1, \ldots, F_n)$ (H is a function of the F_i). The functions $F_1, \ldots, F_n$ may then serve as new momentum (or action) variables almost everywhere, and by algebraic manipulations involving nothing more than integrations over chosen paths new local configuration (or angle) variables $Q^1, \ldots, Q^n$ can be constructed yielding a canonical (i.e. symplectic) transformation of P such that the Hamiltonian equations of motion become trivial (see for example [1], Ch. 5 for details). The Hamiltonian system (P, ω, H) is said to be in *Lax form* if and only if there exist smooth functions D and L_0 of P into a finite-dimensional real Lie algebra $\mathcal{G}$ such that i) the Hamiltonian equations of motion imply the Lax equation $dD/dt = [D, L_0]$ and ii) D is an injective immersion (hence the Lax equation implies the equations of motion). It follows that the "trace-polynomials" $tr D^k$ are constants of motion for all positive integers k (here a faithful representation of $\mathcal{G}$ as a matrix Lie algebra is assumed; more generally, one can take Ad-invariant functions on $\mathcal{G}$, see [4] for generalizations). Moreover, it can be shown under the assumption that D maps into regular adjoint orbits that these trace-polynomials are in involution if and only if there exists a so-called *classical r-matrix* (see [3]). This is a smooth map $d: P \to \mathcal{G} \otimes \mathcal{G}$ such that the following equation is satisfied (the notation is explained in more detail further down):

$$\{D \overset{\otimes}{,} D\} = [d_{12}, D \otimes 1] - [d_{21}, 1 \otimes D]. \qquad (3.20)$$

Introducing components in the above Lie algebra basis (i.e. $D = D^a T_a$, $d = d^{ab} T_a \otimes T_b$, $[T_b, T_c] = f^a_{bc} T_a$) this means

$$\{D^a, D^b\} = f^a_{cd} d^{cb} D^d - f^b_{cd} d^{ca} D^d. \qquad (3.21)$$

A convenient and commonly used index-free notation is achieved by taking the universal enveloping algebra $U\mathcal{G}$ of $\mathcal{G}$ and an N-fold tensor power therof, $U\mathcal{G}^{\otimes N}$. This vector space is an associative unital algebra in which commutators are defined as usual. For each positive integer r smaller or equal than N and integers $1 \leq i_{\sigma(1)} < \cdots < i_{\sigma(r)} \leq N$ (where σ is a permutation of $\{1,\ldots,r\}$) there is a natural linear map $(\ )_{i_1,\ldots,i_r} : \mathcal{G}^{\otimes r} \to U\mathcal{G}^{\otimes N}$ by mapping $\xi_1 \otimes \cdots \otimes \xi_r$ to $1 \otimes \cdots \otimes 1 \otimes \xi_{\sigma(1)} \otimes 1 \otimes \cdots \otimes 1 \otimes \xi_{\sigma(r)} \otimes 1 \otimes \cdots \otimes 1$ (where $\xi_{\sigma(i)}$ sits at the $i_{\sigma(i)}^{\text{th}}$ place for $1 \leq i \leq r$). For instance, ξ_2 means $1 \otimes \xi$ in $U\mathcal{G}^{\otimes 2}$ for ξ in $\mathcal{G}$ and d_{31} means $d^{ab}T_b \otimes 1 \otimes T_a$ in $U\mathcal{G}^{\otimes 3}$ for d in $\mathcal{G}^{\otimes 2}$. We shall make use of this notation for the rest of the section. The compatibility equation coming from the Jacobi identity for Poisson brackets is then computed as follows: Setting

$$\text{YB}(d)_{123} = [d_{12}, d_{13}] + [d_{12}, d_{23}] - [d_{13}, d_{32}] \tag{3.22}$$

one has

$$[D_1, \text{YB}(d)_{123} + \{D_2, d_{13}\} - \{D_3, d_{12}\}] + \text{ cycl.} = 0. \tag{3.23}$$

In some situations, d is antisymmetric, independent of P, and satisfies the Classical Yang-Baxter equation

$$\text{YB}(d)_{123} = 0. \tag{3.24}$$

In this case one can construct a Lie algebra structure on the so-called "double" $\mathcal{G} \oplus \mathcal{G}^*$ which allows the construction of more explicit solutions by a factorisation procedure (see for example [30]) for details. But in general, the r-matrix will depend on the phase space.

Returning to our sigma model, we shall soon see that these structures also appear in an infinite-dimensional setting:
The symplectic manifold for the sigma model will be the space of curves $\mathcal{C}$ (3.1 in the cotangent bundle of M which we had defined above. The Hamiltonian of the sigma model can be identified with the *energy* functional

$$\mathcal{E}(\psi) = \frac{1}{2}\int dx\, h^{-1}(\psi(x), \psi(x)) + \frac{1}{2}\int dx\, h(T\tau_M^* \frac{d\psi}{dx}(x), T\tau_M^* \frac{d\psi}{dx}(x)) \tag{3.25}$$

where h^{-1} denotes the usual induced fibre metric in the cotangent bundle. We define the Lie algebra $\hat{\mathcal{G}}$ to be the semidirect sum of the Lie algebra of all differential operators and the tensor product of the Lie algebra of all smooth functions on the real line taking values in the finite-dimensional Lie algebra $\mathcal{G}$ with the commutative associative algebra of all rational functions of the spectral parameter. Then the operator $D(x, \lambda)$ (3.18) and the current component $L_0(x, \lambda)$ regarded as functionals of ψ are maps of the phase space $\mathcal{C}$ to the "big" Lie algebra $\hat{\mathcal{G}}$ which satisfy the Lax equation because of equation (3.19). It is now interesting whether one can also construct a classical r-matrix d for this system, i. e. one has to compute the fundamental Poisson bracket (3.20). Before we show that this is indeed the case we have to introduce some notation: With the aid of the Killing form Kil we can

2-dimensional nonlinear sigma models: zero curvature and Poisson structure

identify the Lie algebra $\mathcal{G}$ with its dual space $\mathcal{G}^*$. For example, the identity map of $\mathcal{G}$ can thus be identified with the Casimir $C = Kil^{ab} T_a \otimes T_b$ in $\mathcal{G} \otimes \mathcal{G}$ by raising the second index. On the other hand, raising the first index of the bilinear form (-valued field) j (compare eq. (3.11)) leads to the endomorphism (-valued field)

$$j(gH) = \text{Ad}(g) \, \pi_\mathcal{M} \, \text{Ad}(g)^{-1} , \qquad (3.26)$$

where $\pi_\mathcal{M}$ denotes the orthogonal projection on the subspace $\mathcal{M}$. Note that this expression is well-defined on cosets $gH = \phi(x)$ since $\text{Ad}(h)$ commutes with $\pi_\mathcal{M}$ for all h in the subgroup H. In case M is equal to G this projection is equal to the identity whence j is independent of M.

An important quantity will be the field

$$\sigma(gH) = \text{Ad}(g) (1 - 2\pi_\mathcal{M}) \, \text{Ad}(g)^{-1} \qquad (3.27)$$

which is a sort of field dependent involutive automorphism of the Lie algebra. Raising of the second index will give the equivalent equation $\sigma = C - 2j$. After some lengthy computations we arrive at the following

Theorem 3.1 *Let*

$$D(x, \lambda) = \partial_x + L_1[\psi](\lambda)(x) \qquad (3.28)$$

$$d(x, y, \lambda, \mu) = \frac{2\mu}{1-\mu^2} \left(\frac{\mu C}{\lambda - \mu} + \frac{\sigma(\phi(x))}{1 - \lambda\mu} \right) \delta(x - y) . \qquad (3.29)$$

$$\qquad (3.30)$$

Then we have the following equations related to Poisson brackets:

$$\{D_1(x, \lambda), D_2(y, \mu)\} =$$
$$[d_{12}(x, y, \lambda, \mu), D_1(x, \lambda)] - [d_{21}(y, x, \mu, \lambda), D_2(y, \mu)] \qquad (3.31)$$

$$\{D_1(x, \lambda), d_{23}(y, z, \mu, \nu)\} =$$
$$[\frac{2\lambda}{1-\lambda^2}(C_{12}\delta(x-y) + C_{13}\delta(x-z)), d_{23}(y, z, \mu, \nu)] \qquad (3.32)$$

$$\{d_{12}(x, y, \lambda, \mu), d_{34}(z, v, \nu, \rho)\} = 0 \qquad (3.33)$$

$$\text{YB}(d)(\lambda, \mu, \nu)\delta(x-y)\delta(y-z)$$
$$+ [\frac{2\mu}{1-\mu^2} C_{12}\delta(x-y) + \frac{2\mu}{1-\mu^2} C_{13}\delta(x-z), d_{13}(x, z, \lambda, \nu)]$$
$$- [\frac{2\nu}{1-\nu^2} C_{13}\delta(x-z) + \frac{2\nu}{1-\nu^2} C_{23}\delta(y-z), d_{12}(x, y, \lambda, \mu)]$$
$$= 0 . \qquad (3.34)$$

For a detailed proof we refer the reader to the paper [7]. Apart from identities stemming from the symmetric space property such as, for example, $\mathrm{ad}(j_\mu) = j\mathrm{ad}(j_\mu) + \mathrm{ad}(j_\mu)j$ or $\partial_\mu j = j\mathrm{ad}(j_\mu) - \mathrm{ad}(j_\mu)j$, the computation is elementary, but lengthy. Here $\mathrm{YB}(d)(\lambda, \mu, \nu)$ is defined as in (3.22), the only modification being the appropriate assignment of the spectral parameters (λ, μ, ν) to the numbers $(1, 2, 3)$.

The first equation (3.31) can indeed be seen as an infinite-dimensional analogue of the fundamental Poison bracket equation (3.20) mentioned above. However, as the equations show the tensor product of the big Lie algebra $\hat{\mathcal{G}}$ with itself is bigger than the algebraically defined tensor product: it includes rational functions in two variables and also distributions in two variables. The second and the third equation (3.32, 3.33) show that the space spanned by the D operator and the d-matrix close under the Poisson bracket. The last equation (3.34) is a slightly stronger version of the Yang-Baxter type equation (3.23) mentioned above since there is no cyclic sum any more. However, the r-matrix d does not satisfy the classical Yang-Baxter equation (3.24): Avan and Talon have shown [2] in a more general context that all those solutions of (3.24) that are linear combinations of the Casimir C and the involutive automorphism σ with coefficients depending on the spectral parameter are of the particular form $g(\mu)(\frac{C_{12}}{f(\lambda)-f(\mu)} \pm \epsilon \frac{\sigma_{12}}{f(\lambda)\mp f(\mu)})$; here ϵ is 0 or 1, and f and g are arbitrary nonconstant functions lying in some "nice" field of functions (such as for instance rational or meromorphic, see also [7] for details). Further computations show (see [7]) that the Jacobi identity for the algebra spanned by the D's and d's is indeed satisfied.

To end this chapter we should like to mention some open problems connected to this Hamiltonian formulation of sigma models: firstly, a nice direct analogue of the Lie algebra $\mathcal{G}$ with some nondegenerate trace normally appearing in the known examples of finite-dimensional Lax pairs is still lacking. Secondly, the conserved quantities or charges usually associated with integrable sigma models are monodromy matrices, i. e. parallel transports from $-\infty$ to $+\infty$ w. r. t. the covariant derivative $D(x, \lambda)$ (particular solutions of the associated linear system). Although it can easily be proved that these functions of the field ϕ (via the dependence of $D(x, \lambda)$ on ϕ) are constants of time, they do not seem to have well-defined Poisson brackets, at least when these brackets are computed by the physicist's rule (3.10). Being integrals, hence nonlocal quantities, they do not fall under the class of functions mentioned above (3.3, 3.4). The problem is a violation of the Jacobi identity, first noticed by Lüscher and Pohlmeyer [23] and described in more detail by de Vega, Eichenherr, and Maillet[9]. In a subsequent paper [26] Maillet was able to give a regularisation procedure for the Poisson bracket of the charges such that the Jacobi identity was satisfied. However, his prescription is "multi step" insofar as each multiple Poisson bracket requires a separate regularisation that cannot be reduced to the regularisation of multiple Poisson brackets with a smaller number of factors. Hence, one does not really achieve a full Poisson algebra structure. On the other hand, all "one step" regularisation procedures do unfortunately not work so far (see for example [7] for a no-go theorem).

Bibliography

[1] R. Abraham and J. Marsden, *Foundations of Mechanics*, Addison-Wesley, 1978.

[2] J. Avan and M. Talon, *Rational and trigonometric constant non-antisymmetric r-matrices*, Phys. Lett. B **241** (1990), 77-82.

[3] O. Babelon and C.-M. Viallet, *Hamiltonian structures and Lax equations*, Phys. Lett. **237 B** (1990), 411-416.

[4] M. Bordemann, *Generalized Lax pairs, the modified classical Yang-Baxter-equation, and affine geometry of Lie groups*, Comm. Math. Phys. **135** (1990), 201-216.

[5] M. Bordemann, *Nondegenerate bilinear forms on nonassociative algebras*, Universität Freiburg preprint THEP 92/3, April 1992.

[6] M. Bordemann, M. Forger and H. Römer, *Homogeneous Kähler manifolds: Paving the way to new supersymmetric sigma models*, Comm. Math. Phys. **102** (1986), 605-647.

[7] M. Bordemann, M. Forger, J. Laartz and U. Schäper, *The Lie-Poisson structure of integrable classical non-linear sigma models*, Comm. Math. Phys. **152** (1993), 167-190.

[8] F.E. Burstall, D. Ferus, F. Pedit and U. Pinkall, *Harmonic tori in symmetric spaces and commuting Hamiltonian systems on loop algebras*, Ann. of Math. **138** (1993), 173-212.

[9] H.J. de Vega, H. Eichenherr and J.-M. Maillet, *Classical and quantum algebras of non-local charges in sigma models*, Comm. Math. Phys. **92** (1984), 507-524.

[10] H. Eichenherr, *SU(N)-invariante nichtlineare sigma-modelle*, Universität Heidelberg, Fakultät für Physik und Astronomie, Dissertation SS 78-2 (1978); published as: *SU(N)-invariant nonlinear σ-models*, Nuclear Physics B **146** (1978), 215-223.

[11] H. Eichenherr and M. Forger, *On the dual symmetry of the nonlinear sigma models*, Nucl. Phys. B **155** (1979), 381-393.

[12] H. Eichenherr and M. Forger, *More about nonlinear sigma models on symmetric spaces*, Nucl. Phys. B **164** (1980), 528-535 and B **282** (1987), 745-746 (erratum).

[13] H. Eichenherr and M. Forger, *Higher local conservation laws for nonlinear sigma models on symmetric spaces*, Comm. Math. Phys. **82** (1981), 227-255.

[14] L.D. Faddeev and L.A. Takhtajan, *Hamiltonian Methods in the Theory of Solitons*, Springer-Verlag, Berlin 1987.

[15] M. Forger, *Nonlinear sigma models on symmetric spaces*. In: *Nonlinear Partial Differential Operators and Quantization Procedures*, Proceedings, Clausthal, Germany 1981, ed. S.I. Andersson and H.D. Doebner, Lect. Notes in Math. **1037**, Springer-Verlag, Berlin 1983.

[16] M. Forger, *Differential Geometric Methods in Nonlinear σ-models and Gauge Theories*, Dissertation, FU Berlin, Fachbereich Physik, 1979.

[17] M. Forger, J. Laartz and U. Schäper, *Current algebra of classical non-linear sigma models*, Comm. Math. Phys. **146** (1992), 397-402.

[18] L. Freidel and J.M. Maillet, *Quadratic algebras and integrable systems*, preprint LPTHE-24/91; *On classical and quantum integrable field theories associated to Kac-Moody current algebras*, preprint LPTHE-25/91.

[19] M. Gell-Mann and M. Lévy, *The axial vector current in beta decay*, Nuovo Cimento **16** (1960), 705-726.

[20] S. Helgason, *Differential Geometry, Lie Groups, and Symmetric Spaces*, Academic Press, New York 1978.

[21] S. Kobayashi and K. Nomizu, *Foundations of Differential Geometry*, Interscience, New York, 1963 (Vol.I) and 1969 (Vol.II).

[22] B. Kostant, *On differential geometry and homogeneous spaces II*, Proc. Nat. Acad. Sci. **42** (1956), 354-357.

[23] M. Lüscher and K. Pohlmeyer, *Scattering of massless lumps and nonlocal charges in the two-dimensional nonlinear σ-model*, Nuclear Physics **B 137** (1978), 46-54.

[24] J.-M. Maillet, *Kac-Moody algebra and extended Yang-Baxter relations in the $O(N)$ non-linear sigma model*, Phys. Lett. **162 B** (1985), 137-142.

[25] J.-M. Maillet, *Hamiltonian structures for integrable classical theories from graded Kac-Moody algebras*, Phys. Lett. **167 B** (1986), 401-405.

[26] J.-M. Maillet, *New integrable canonical structures in two-dimensional models*, Nucl. Phys. **B 269** (1986), 54-76.

[27] M. Mañas, *The principal chiral model as an integrable system*, this volume.

[28] L. Michel, *Minima of Higgs-Landau polynomials*, CERN preprint Ref.Th.2716-CERN (1979).

[29] K. Pohlmeyer, *Integrable Hamiltonian systems and interactions through quadratic constraints*, Comm. Math. Phys. **46** (1976), 207-221.

[30] M.A. Semenov-Tyan-Shanskii, *What is a classical r-matrix?*, Funk. Anal. i Prilozh. **17** (1983), 17-33; English transl.: Funct. Anal. Appl. **17** (1983), 259-272.

[31] K. Uhlenbeck, *Harmonic maps into Lie groups (Classical solutions of the chiral model)*, J. Diff. Geom. **30** (1989), 1-50.

[32] J.C. Wood, *Harmonic maps into symmetric spaces and integrable systems*, this volume.

[33] V.E. Zakharov and A.V. Mikhailov, *Relativistically invariant two-dimensional models of field theory which are integrable by means of the inverse scattering problem method*, Zh. Eksp. Teor. Fiz. **74** (1978), 1953-1973; Engl. transl.: Sov. Phys. JETP **47**(6) (1978), 1017-1027.

Sigma models in 2+1 dimensions

R.S. Ward

1 Introduction

Finite-energy harmonic maps from $\mathbf{R}^2$ into simple manifolds such as $\mathrm{CP}^1 \cong S^2$ or $\mathrm{SU}(2) \cong S^3$ are thoroughly understood. The relevant equations constitute an integrable system, the general solution of which can be written down explicitly. Such solutions may be regarded as static multi-soliton configurations in two-dimensional space. In the language of physics, these "harmonic-map" systems are known as sigma models or chiral models.

When one adds time-dependence, and deals with functions which depend on time t as well as on the spatial coordinates x and y, then the situation is much less clearly understood. The solitons are now capable of evolving, moving, colliding and so forth. The time-evolution is determined by a set of nonlinear partial differential equations, which correspond (roughly speaking) to the harmonic map equations on $\mathbf{R}^{2+1}$ (three-dimensional Minkowski space-time). But there are various different choices of equation that one can make, leading to different systems.

This article begins by reviewing the static case. We then see that the simplest way of introducing time-dependence results in solitons that are unstable, and so these "simple" systems are not really satisfactory. Nevertheless, one can still study the collision of solitons in this case. Of greater interest are a number of modified systems, in which the solitons are stable; three of these are reviewed. The dynamics of solitons in these systems is rich and varied, and has as yet only partly been explored. The main results of what has been done are summarized in the final section of this review.

The term "soliton" is used throughout in a rather broad sense, to mean "lump of energy (or solution of nonlinear equation) which is localized in space". The solitons described near the beginning of this review are not even stable. Later on, we deal with stable solitons; but their interactions are in general inelastic, so they are certainly not solitons in the strictest sense of the word.

2 Static Solitons

Let us begin by establishing some notation. The space-time $\mathbf{R}^{2+1}$ has coordinates $x^\mu = (x^0, x^1, x^2) = (t, x, y)$; the spatial coordinates are $x^j = (x^1, x^2) = (x, y)$, and the time coordinate is $x^0 = t$. The (flat) metric on $\mathbf{R}^{2+1}$ is $\eta_{\mu\nu} = \text{diag}(-1, 1, 1)$, and its inverse is denoted $\eta^{\mu\nu}$; these are used to raise and lower indices. The Einstein summation convention on repeated indices applies throughout. A partial derivative with respect to x^μ is denoted ∂_μ. So, for example, the wave equation is written

$$\eta^{\mu\nu}\partial_\mu\partial_\nu\phi = 0, \quad \text{or} \quad \partial^\mu\partial_\mu\phi = 0. \tag{2.1}$$

The spatial metric corresponds to the Kronecker delta δ_{jk}.

The O(3) σ-model, or $\mathbf{CP}^1$ model, deals with functions ("fields") which take values on the Riemann sphere. By stereographic projection, we may represent such a field as a complex-valued function $W(x^\mu)$ which is allowed to take the value ∞. In this section, we shall be dealing only with static fields, so W depends only on the spatial coordinates (x, y).

The most important quantity associated with a field is its potential energy E_P, which is the spatial integral of the norm-squared of the field (using the standard metric on the Riemann sphere). In terms of W, this energy is

$$E_P = \int_{\mathbf{R}^2} (1 + |W|^2)^{-2} \delta^{jk} (\partial_j W)(\partial_k W^*) \, dx \, dy, \tag{2.2}$$

where W^* denotes the complex conjugate of W. In general, when W becomes time-dependent, E_P will also depend on time t.

Since the expression (2.2) is invariant under conformal transformations of the spatial domain $\mathbf{R}^2$, it makes sense to require that W extends smoothly to the conformal compactification S^2; in other words, for any t, that W extends to a smooth map from S^2 to $\mathbf{CP}^1 \cong S^2$. From now on, we shall assume that W satisfies this regularity property. A consequence is that the energy E_P is finite.

Another consequence is that the field configurations are classified by an integer N, which is the degree of the extended map $S^2 \to S^2$. Roughly speaking, N is the number of solitons, where antisolitons are counted negatively. This review will deal with three cases: a single soliton ($N = 1$), two-soliton interactions ($N = 2$), and soliton-antisoliton interactions ($N = 0$).

The energy E_P, being positive-definite, is of course bounded below by zero. But the well-known "Bogomolny argument" shows that one can improve this bound: in fact

$$E_P \geq 2\pi |N|. \tag{2.3}$$

If $N > 0$, then equality in (2.3) is attained if and only if W is a rational meromorphic function of $z = x + iy$. (The case $N < 0$ corresponds to rational functions of z^*.) Such rational functions are harmonic maps from $\mathbf{R}^2$ to $\mathbf{CP}^1$, and all finite-energy harmonic maps are of this type. If one plots the energy density (the integrand of 2.2) as a function of x and y, one sees (generically) a picture consisting of N

lumps (solitons) on the xy-plane. In other words, meromorphic rational functions of degree N represent static N-soliton solutions of the CP^1 model, depending on $2N+1$ complex parameters. The model (and in particular the energy E_P) is invariant under rotations of the target sphere, so three real parameters should be subtracted from this tally, leaving $4N - 1$ real moduli.

For example, in the case $N = 1$, we get a 1-soliton solution depending on three real parameters: we can take

$$W(x, y) = \lambda(z - c), \qquad (2.4)$$

where λ is a nonzero real number and c is a complex number. Examination of the energy density for (2.4) reveals that the soliton is located at the point $z = c$, and its width is (proportional to) λ^{-1}.

It is also worth looking at the energy density of the $N = 2$ solution

$$W(x, y) = \lambda(z^2 - b^2).$$

If $|b|$ is large, then this looks like two solitons, located at $z = \pm b$. As b is reduced in magnitude, the two solitons approach each other and merge, forming not a single lump, but rather a ring of radius $(3\lambda^2)^{-1/4}$.

Finally, let us turn briefly to the SU(2) chiral model; see [1] for more details. Here the field J takes values in SU(2), and its potential energy is

$$E_P = -\tfrac{1}{8} \int_{\mathbf{R}^2} \mathrm{tr}\left[\delta^{jk} J^{-1}(\partial_j J) J^{-1}(\partial_k J)\right] dx\, dy. \qquad (2.5)$$

Static solitons correspond to finite-energy critical points of this functional, and these turn out to be embeddings of the CP^1 solitons discussed previously, where the CP^1 is embedded as a totally geodesic submanifold ("equator") of SU(2). Without loss of generality, we can take this embedding to be

$$J = \frac{\mathrm{i}}{1 + |W|^2} \begin{pmatrix} 1 - |W|^2 & 2\mathrm{i}W \\ -2\mathrm{i}W^* & -1 + |W|^2 \end{pmatrix}. \qquad (2.6)$$

3 Soliton Instability in the Basic Model

From now on, the CP^1 field W (or chiral field J) are allowed to depend on time t as well as on the two space variables x and y. The time evolution of W is determined by the Euler-Lagrange equations for the Lagrangian

$$L = \int_{\mathbf{R}^2} (1 + |W|^2)^{-2} \eta^{\mu\nu} (\partial_\mu W)(\partial_\nu W^*)\, dx\, dy. \qquad (3.1)$$

Note that L has the form $E_P - E_K$, where E_K is the kinetic energy

$$E_K = \int_{\mathbf{R}^2} (1 + |W|^2)^{-2} |\partial_0 W|^2\, dx\, dy. \qquad (3.2)$$

The total energy $E = E_P + E_K$ is constant in time, provided W satisfies the Euler equations for (3.1): these equations are

$$\partial^\mu \partial_\mu W = 2W^*(1 + |W|^2)^{-1}(\partial^\mu W)(\partial_\mu W), \tag{3.3}$$

which is a second-order hyperbolic partial differential equation for W. One specifies initial data (W and $\partial_0 W$) at $t = 0$, and (3.3) then determines W for all t.

The most important question about the 1-soliton (2.4) is whether or not it is stable. Since the energy in the $N = 1$ sector is bounded below by 2π, and (2.4) saturates this bound, the only ways that W can change correspond (roughly speaking) to λ and c changing. Changing c simply corresponds to the soliton moving in the plane; this can indeed happen, and is not regarded as an instability. So the question is: if one starts with a soliton located at $z = 0$ (ie. $W = \lambda z$) and perturbs its shape, does the solution stay close to the initial configuration for all t?

This question has been studied numerically, using two very different numerical procedures for solving the evolution equation (3.3). One way involved imposing radial symmetry in the xy-plane, and discretizing (3.3) in such a way that the topological lower bound (2.3) on the energy was maintained [2]. The other way used a more traditional Runge-Kutta approach to simulating (3.3) [3]. The results in the two cases were the same: the width of the soliton tends to change, essentially linearly, with time. For example, it may shrink, so that after a finite time interval it becomes an infinite spike (a singular solution). Or the soliton may spread out, with this expansion continuing indefinitely.

Let us turn briefly to the SU(2) chiral model. The Lagrangian for $J(t, x, y)$ is

$$L = -\tfrac{1}{8} \int_{\mathbf{R}^2} \mathrm{tr}\left[\eta^{\mu\nu} J^{-1}(\partial_\mu J) J^{-1}(\partial_\nu J)\right] dx\, dy, \tag{3.4}$$

and the corresponding equation of motion is

$$\eta^{\mu\nu}\partial_\mu(J^{-1}\partial_\nu J) = 0. \tag{3.5}$$

This contains the CP^1 model, in the sense described at the end of the previous section. So this chiral model inherits the instability of the CP^1 case (and may have other instabilities as well).

The conclusion, therefore, is that the "solitons" introduced hitherto are unstable in size. This instability arises from the conformal invariance of the models in xy-space; and in order to stabilize the solitons, one has to break the conformal invariance. We shall see in section 5 how this can be done.

4 Dynamics of CP^1 Solitons

Even though CP^1 solitons are unstable, one can still investigate how they scatter off each other. The instability is rather weak, and it can be arranged for the collisions to occur before the size of the solitons has changed very much. Such collisions have been studied in two ways: by an approximation in which the dynamics corresponds to geodesic flow on the moduli space of static solutions [4, 5]; and by direct numerical solution of the equations of motion (3.3) [6]. The results of these two very different approaches are in broad agreement.

The geodesic approximation involves reducing the problem to one with a finite (indeed, small) number of degrees of freedom. One replaces the infinite-dimensional configuration space of the field theory with the moduli space M of static solutions, which for two solitons ($N = 2$) is 7-dimensional. The potential energy E_P is identically equal to 4π, and the kinetic energy E_K defines a metric on M. The geodesics on M correspond approximately to evolutions of the 2-soliton system, and the approximation is a good one if the speeds are small (ie. if E_K is small compared to $E_P = 4\pi$).

The studies referred to above reveal a rich structure in the dynamics of soliton-soliton interactions. The reason for this is that the scattering of solitons is strongly influenced by their sizes, and also by internal "phases" (parameters that occur in W, but do not show up in the energy density). In a typical head-on collision of two solitons, they momentarily form a ring when they coincide, and then separate at right angles to their original direction of approach. In other words, the scattering angle is 90°. If the impact parameter is nonzero (ie. if the encounter is a glancing one, rather than head-on), then the scattering angle is less than 90°; and the angle tends to zero as the impact parameter increases (as one might expect). But other sorts of behaviour may also occur: for example, cases where the scattering angle is identically zero, irrespective of the impact parameter [4].

Finally, it might be mentioned that the collision of a soliton and an antisoliton has also been studied [6]. Here $N = 0$, and there is no "nearby" static solution, so the problem can only be attacked by numerical solution of the partial differential equation. Such numerical simulations indicate that the soliton and antisoliton attract each other and annihilate, leaving radiation.

5 Stabilizing the Solitons

In order to have stable solitons, the "basic" sigma model has to be modified. There are ways of doing this which involve the introduction of extra fields (in particular, gauge fields); see, for example, [7]. But our attention here will be restricted to three different ways which do not involve additional fields, namely Q-lumps, a Skyrme term, and integrability.

(i) *Q-lumps.* The CP^1 model admits a phase symmetry: the energy and the equation of motion are invariant under $W \mapsto W \exp(i\theta)$. Rotation in this internal phase gives (in effect) an "expansionary" force, which tries to increase the width of a soliton. In order to counteract this, one may add to E_P a term which provides a "shrinking" force. The integral of any function of W (not including derivatives of W) has this property. In [8] the term

$$\int_{\mathbf{R}^2} \alpha^2 |W|^2 (1+|W|^2)^{-2} \, dx \, dy \tag{5.1}$$

is added to the basic expression (2.2) for E_P, and the resulting model is investigated. Here α is a nonzero real constant.

Associated with the symmetry $W \mapsto W \exp(i\theta)$ there is a conserved quantity

$$Q = i \int_{\mathbf{R}^2} (1+|W|^2)^{-2} (W^* \partial_0 W - W \partial_0 W^*) \, dx \, dy. \tag{5.2}$$

The Bogomolny bound (2.3) on the energy is no longer valid, but a modified version of it is. (The extra term in E_P has to be carefully chosen in order for this to be so, motivating the particular choice 5.1). In the modified model, the total energy E is bounded below as follows:

$$E \geq 2\pi N + |\alpha Q|, \tag{5.3}$$

with equality if and only if

$$W(t,x,y) = W_0(z) \exp(i\alpha t). \tag{5.4}$$

Here W_0 is a rational meromorphic function of $z = x + iy$, of degree N. (We are taking N to be positive.) In order for Q and E to be finite, we need $N \geq 2$.

The result (see [8] for details) is that this modification of the CP^1 model admits "spinning solitons" called Q-lumps, with topological charge $N \geq 2$. The size of such a soliton can have any specified value (it is fixed by the initial conditions). And these solitons are stable, except for the fact that a "single" soliton of topological charge N can (and does) break up into N solitons which go their separate ways.

(ii) Skyrmions. As mentioned above, we can introduce a shrinking force by adding to the Lagrangian a term not involving derivatives of W (by contrast with the basic Lagrangian (3.1) which is quadratic in $\partial_\mu W$). In the case of Q-lumps, the compensating expansion force comes from an internal rotation. But another possibility is to add a so-called Skyrme term, which is quartic in $\partial_\mu W$: this also provides an expansion force.

The precise form of the Skyrme term is determined (up to a constant factor) by the requirements that the system be Lorentz-invariant, that the total energy be positive-definite, and that the equations of motion involve time derivatives of no more than second order. For the non-derivative term there is more choice: one could use (5.1), but other possibilities would work equally well. The choice made in [9, 10] is motivated by the requirement that the explicit static 1-soliton (2.4) should be a solution of the modified system: such a choice is

$$\int_{\mathbf{R}^2} 4\theta_2 (1+|W|^2)^{-4} \, dx \, dy, \tag{5.5}$$

where θ_2 is a positive constant. So the complete Lagrangian which describes the modified Skyrme-type CP^1 model is the sum of (3.1), (5.5), and the Skyrme term

$$\int_{\mathbf{R}^2} 2\theta_1 (1+|W|^2)^{-4} \{[(\partial_\mu W)(\partial^\mu W^*)]^2 - |(\partial_\mu W)(\partial^\mu W)|^2\} \, dx \, dy, \tag{5.6}$$

Sigma models in 2+1 dimensions

where θ_1 is another positive constant. The two coupling constants θ_1 and θ_2 can be set to have whatever (positive) values one wishes.

It is a straightforward matter to derive the Euler-Lagrange equations from this Lagrangian, although the result looks rather complicated (see [9]). One can then easily check that the 1-soliton expression $W = \lambda(z - c)$ is a solution if and only if

$$\lambda = [\theta_2/(2\theta_1)]^{1/4}. \tag{5.7}$$

In other words, the width λ^{-1} of the soliton is now fixed. One would expect this soliton to be stable (this was the purpose of adding the θ-terms), and numerical experiments confirm that this is indeed the case [9].

(iii) Integrable chiral model. One can achieve stability in a quite different way, by a modification which renders the model integrable (in a sense to be described below). More specifically, one can modify the chiral model (3.4, 3.5) in this way. However, a reduction to CP^1 such as (2.6) is not consistent with the modified equations of motion: there is no integrable version of the CP^1 model in 2+1 dimensions.

The modification consists of replacing (3.5) by

$$(\eta^{\mu\nu} + \varepsilon^{\mu\nu})\partial_\mu(J^{-1}\partial_\nu J) = 0, \tag{5.8}$$

where $\varepsilon^{\mu\nu}$ is a constant skew-symmetric tensor. This is a hyperbolic system of partial differential equations. Note, for example, that if J is restricted to be a diagonal matrix diag[exp(iϕ), exp(-iϕ)], then (5.8) reduces to the archetypal hyperbolic system, namely the wave equation (2.1). In the general (nonlinear) case, (5.8) can be rewritten in a form which makes its hyperbolic nature manifest, namely

$$\partial^\mu \partial_\mu J = (\text{polynomial in } J, J^{-1}, \partial_\mu J).$$

It is only for certain values of $\varepsilon^{\mu\nu}$ that (5.8) is an integrable system [11], namely if

$$\varepsilon^{\mu\nu}\varepsilon_{\mu\nu} = -2. \tag{5.9}$$

Indeed if (5.9) holds, then (5.8) is a dimensional reduction of the self-dual Yang-Mills equations in 2+2 dimensions [12, 13], and is integrable in the same sense. Any choice satisfying (5.9) is acceptable; let us set $\varepsilon^{20} = 1 = -\varepsilon^{02}$ (other components zero).

The equations (5.8) arise as consistency conditions for a pair of linear equations (this is what is meant by integrability in the present context). The linear equations are

$$\begin{aligned}(\zeta\partial_x - \partial_t - \partial_y)\psi &= A\psi, \\ (\zeta\partial_t - \zeta\partial_y - \partial_x)\psi &= B\psi.\end{aligned} \tag{5.10}$$

Here ζ is a complex parameter, and A, B, ψ are 2×2 matrix functions of x^μ, with ψ additionally dependent on ζ. In order for the overdetermined system (5.10) to have a solution ψ, the matrices A and B have to satisfy a pair of equations; these in turn imply that A and B have the form

$$A = J^{-1}(\partial_t + \partial_y)J, \quad B = J^{-1}\partial_x J,$$

for some J satisfying (5.8).

The total energy E of the field (which is constant in time) is given by exactly the same expression as for the unmodified ($\varepsilon^{\mu\nu} = 0$) case. In other words, it is the sum of the potential energy (2.5) and the analogous expression for the kinetic energy. The equations of motion (5.8) arise from a Lagrangian L, but this is not equal to $E_P - E_K$ (which would be the unmodified case 3.4). Rather, L has a form which one may think of as arising from torsion on the target space SU(2) [11], or from a background magnetic field on the configuration space of the system [12].

The expressions (2.4, 2.6) define a field J which represents a 1-soliton, and which is a static solution of the integrable equation (5.8) (as well as of the original chiral equation 3.5). As a solution of (3.5), this soliton is unstable (as remarked previously). The point of the modification is that it should ensure stability. As yet, there is no analytic proof of stability, but there is strong evidence for it from numerical experiments [14]. For technical reasons, this numerical work uses alternative forms of the equations (5.8); in particular, a Lorentz-invariant form in terms of gauge fields, which emphasizes the relation of this system to the self-dual Yang-Mills system [13].

6 Dynamics in the Stabilized Systems

If we take a pair of solitons in one of the stabilized models described above, and fire them towards each other, how do they scatter? A number of studies (analytic and numerical) have addressed this question (see [8, 9, 10, 12, 15] for some of these). The brief review which follows deals only with the case of head-on collisions; results on glancing encounters may be found in the references cited above. The outcome of a head-on collision is that the two solitons scatter through some angle, which is typically 0° (no scattering), 90°, or 180° (bounce-back).

The case of Q-lumps is, however, atypical. One can study a slow-motion approximation, in which the trajectories of the solitons correspond to geodesics on the moduli space of static solutions saturating the Bogomolny bound (5.3). This reveals [8] that the scattering angle depends on the impact speed, and lies strictly between 90° and 180°. Presumably this is a consequence of the "internal rotation" of the Q-lumps. It serves as another example of the way in which the internal structure of σ-model solitons can affect their dynamical behaviour.

Let us turn to the CP^1 model modified by the addition of (5.5) and the Skyrme term (5.6). In this case, collisions have been studied in a couple of different ways. One of these was a "collective-coordinate" approximation in which $W(t, x, y)$ was assumed to have the 2-soliton form

$$W = \lambda(z^2 - b^2), \tag{6.1}$$

where λ and b are real-valued functions of t. If λ and b are constant, then (6.1) is a (static) solution of the unmodified CP^1 model, but not of the modified model.

(Essentially, the extra terms introduce a repulsive force between the solitons: if two solitons are placed, at rest, on the plane, then they begin to move apart.) Now the field equations for W reduce to a set of ordinary differential equations for $\lambda(t)$ and $b(t)$, and these can be solved numerically. The results are as follows [10].

If the impact speed is below a certain critical value v_c, then the solitons scatter back-to-back (ie. the scattering angle is 180°). This is what one would expect: because of their mutual repulsion, a certain minimum energy is needed if they are to reach each other. For impact speeds above v_c, the solitons collide, momentarily merging into a ring; then two individual solitons re-form, and separate at right angles to the incoming directions. In other words, the scattering angle is 90°. An analytic argument gives an expression for v_c in terms of the product $\theta_1\theta_2$ of the two parameters occurring in the model, and the numerical result is in agreement with this.

The problem had previously been tackled by direct numerical simulation of the partial differential equation, and the results were the same as those described above [9]. Features which are suppressed by the collective-coordinate approximation (such as radiation, and internal oscillations of the solitons) show up. But the two studies agree on the basic nature of the scattering behaviour.

There have also been numerical studies of soliton-antisoliton collisions in the modified CP^1 model [15]. Here the topological charge N is zero, so there is (on the face of it) nothing to stop the soliton and antisoliton from annihilating each other when they collide. This is indeed what happens (although the dynamics can be quite complicated): after the collision, only radiation remains.

Finally, let us consider soliton collisions governed by the integrable chiral equation (5.8). In view of the integrability, we would expect to be able to write down at least some explicit solutions. This is indeed the case: if we take ψ in (5.10) to have n simple poles in ζ (their location not depending on x^μ), then the corresponding solution of (5.8) represents n solitons, each moving at constant velocity [12]. When any two of these solitons collide, there is no scattering (not even a phase shift). The solution is an explicit rational function of x^μ.

However, the situation is more complicated than this simple picture might suggest. The chiral field has internal degrees of freedom, and these can affect the scattering of solitons: one can get nontrivial scattering, despite the fact that the system is integrable. This was discovered in numerical experiments [16, 15]. (In principle, one should be able to use the techniques of inverse scattering to analyse the problem, but this does not appear to be straightforward.)

The numerical procedure was tested on the 2-soliton exact solution, and this confirmed its reliability (and the absence of scattering). Then different initial data (still representing two solitons fired at each other) was used: this time, the two solitons collided to form a ring, and then separated at 90°. In other words, one gets highly nontrivial scattering, similar in nature to that occurring in nonintegrable systems.

Numerical simulations of soliton-antisoliton collisions in the integrable system also reveal 90° scattering [16, 15]. In this case, the integrability preserves the solitons (unlike in nonintegrable models, where they annihilate), but permits nontrivial scattering.

Clearly the solitons of this integrable model have a rich and varied behaviour, the elucidation of which requires much more theoretical and numerical analysis.

Bibliography

[1] K K Uhlenbeck, *Harmonic maps into Lie groups (Classical solutions of the chiral model)*, J Diff Geom **30** (1989), 1–50.

[2] R A Leese, *Discrete Bogomolny equations for the non-linear $O(3)$ σ-model in 2+1 dimensions*, Phys Rev D **40** (1989), 2004–2013.

[3] R A Leese, *Soliton stability in the $O(3)$ σ-model in (2+1) dimensions*, M Peyrard & W J Zakrzewski, Nonlinearity **3** (1990), 387–412.

[4] R S Ward, *Slowly-moving lumps in the CP^1 model in (2+1) dimensions*, Phys Lett B **158** (1985), 424–428.

[5] R A Leese, *Low-energy scattering of solitions in the CP^1 model*, Nucl Phys B **344** (1990), 33–72.

[6] W J Zakrzewski, *Soliton-like scattering in the $O(3)$ σ model in (2+1) dimensions*, Nonlinearity **4** (1991), 429–475.

[7] M A Mehta, J A Davies & I J R Aitchison, *Stable soliton solution of the CP^1 model with a Chern-Simons term*, Phys Lett B **281** (1992), 86–89.

[8] R A Leese, *Q-lumps and their interactions*, Nucl Phys B **366** (1991), 283–311.

[9] R A Leese, M Peyrard & W J Zakrzewski, *Soliton scattering in some relativistic models in (2+1) dimensions*, Nonlinearity **3** (1990), 773–807.

[10] P M Sutcliffe, *The interaction of Skyrme-like lumps in (2+1) dimensions*, Nonlinearity **4** (1991), 1109–1121.

[11] R S Ward, *Integrablility of the chiral equations with torsion term*, Nonlinearity **1** (1988), 671–679.

[12] R S Ward, *Solition solutions in an integrable chiral model in (2+1) dimensions*, J Math Phys **29** (1988), 386–389.

[13] R S Ward, *Twistors in 2+1 dimensions*, J Math Phys **30** (1989), 2246–2251.

[14] P M Sutcliffe, *Yang-Mills-Higgs solitons in (2+1) dimensions*, Phys Rev D, to appear.

[15] B Piette, P M Sutcliffe & W J Zakrzewski, *Soliton antisoliton scattering in (2+1)-dimensions*, Int J Mod Phys C **3** (1992), 637–660.

[16] P M Sutcliffe, *Nontrivial soliton scattering in an integrable chiral model in (2+1)-dimensions*, J Math Phys **33** (1992), 2269–2278.

THE ALGEBRAIC APPROACH

Infinite dimensional Lie groups and the two-dimensional Toda lattice

I. McIntosh

1 Introduction.

This article is concerned with the system of nonlinear partial differential equations

$$\frac{\partial^2 w_j}{\partial x \partial t} = -\exp(w_{j-1}) + 2\exp(w_j) - \exp(w_{j+1}),$$

which has been called the two-dimensional Toda lattice (or the Toda field equations) of A-type. Initially I will assume all the variables to be complex-valued. In fact it will be more convenient to work with the equivalent system

$$\frac{\partial^2 \ln(v_j)}{\partial x \partial t} = -v_{j-1} + 2v_j - v_{j+1}, \qquad (1.1)$$

in which $v_j = \exp(w_j)$. There are three situations which I will consider: the infinite lattice, where $j \in \mathbf{Z}$; the periodic lattice, where $j \in \mathbf{Z}_{n+1}$; and the finite lattice, where $1 \leq j \leq n$ with $v_0 = v_{n+1} = 0$. Of course the last two are specialisations of the first for appropriate conditions on the v_j.

There has been, at one time or another, much attention given to the finite and periodic Toda lattice in the mathematical physics community, beginning with the observations of [18, 22, 23] that a two dimensional Toda lattice exists for every simple Lie algebra (the term 'A-type' above refers to the the type of simple Lie algebra the system (1.1) corresponds to). The coefficients of the v_j in (1.1) are equated with the row entries in the Cartan matrix, for either a finite rank Lie algebra or a Kač-Moody Lie algebra: these correspond to the finite and periodic lattices respectively. The periodic lattice equations have been shown to be formally 'completely integrable' [10, 26]. The finite lattice equations do not share this property and have been treated separately (see, for example, [19, 12]).

Our principal interest here is in the elliptic version of these equations which arises in the study of harmonic sequences of maps from a Riemann surface Σ into $\mathbf{CP}^n$. Without going into the details (see, for example, [5, 7]) the elliptic system

$$\partial\bar{\partial}\ln(v_j) = v_{j-1} - 2v_j + v_{j+1}, \qquad (1.2)$$

(where $\partial = \partial/\partial x$ and now all the v_j are real-valued and strictly positive functions of $x, \bar{x}$) arises from the following construction. Any harmonic map $f : \Sigma \to \mathbf{CP}^n$ admits (at least locally) a lift ϕ_0 which maps into $\mathbf{C}^{n+1}$. If we let π_0 denote the orthogonal projection of $\mathbf{C}^{n+1}$ onto the line generated by ϕ_0, then $\phi_1 = \pi_0^\perp(\partial\phi_0)$ is again the lift of a harmonic map into $\mathbf{CP}^n$ (called the ∂-transform or ∂'-Gauss map of f, see [28]). We proceed inductively to define $\phi_{j+1} = \pi_j^\perp(\partial\phi_j)$ for $j \geq 0$. Similarly, we define $\phi_{j-1} = \pi_j^\perp(\bar{\partial}\phi_j)$ for $j \leq 0$ and the total sequence of lifts $\{\phi_j : j \in \mathbf{Z}\}$ determines the harmonic sequence generated by the initial harmonic map f, see [28, 7]. It is shown in [5, 7] that if we set $v_j = |\phi_{j+1}|^2/|\phi_j|^2$ then the v_j satisfy (1.2), which is referred to in [5] as the 'unintegrated Plücker formulae' for harmonic sequences.

Initially the aim here is to describe a very large class of solutions to (1.2), in the case of the infinite lattice, which are globally real-analytic on the complex x-plane. These will be found inside a larger class of almost globally holomorphic solutions to (1.1). More importantly, this construction provides us with the lifts ϕ_j in the case where the lattice is periodic. The finite lattice case can also be treated, after a slight alteration, and in both cases we find harmonic maps into the full flag manifold U_{n+1}/T, where T is the maximal torus of diagonal matrices. The harmonic maps into $\mathbf{CP}^n$ follow by homogeneous projection from this flag manifold. The maps we find in the case of the finite lattice extend to give harmonic spheres in $\mathbf{CP}^n$: each belongs to the harmonic sequence generated by a smooth anti-holomorphic rational curve of degree n in $\mathbf{CP}^n$.

The space of solutions to (1.1) described here is essentially 'the dressing orbit of the trivial solution' in the sense of [27, 29, 21]. This requires a choice of infinite dimensional Lie group and initially we will choose the restricted general linear group of the Hilbert space $H = L^2(S^1, \mathbf{C})$ (in the sense of [24]). Here we are following the work of Segal & Wilson [25] who used this group to study the generalised KdV equations. This connection with KdV should not come as a surprise; it has been known for some time that the equations of the (generalised) modified KdV hierarchy can be thought of as symmetries of the two dimensional periodic Toda lattice (cf. [17]). To get solutions of (1.2) we impose a reality condition which turns the dressing construction into an example of the 'natural action' described by Guest & Ohnita [13, 14].

However, if we only work with GL_{res} we find that we cannot recover Krichever's famous θ-function solutions to (1.1) [16] which are solutions for the infinite lattice and also, with the right choice of θ-function, for the periodic lattice. These are very interesting because they offer the possibilty of finding doubly periodic harmonic maps of the x-plane into U_{n+1}/T i.e. harmonic tori (and it seems certain that all orthogonal periodic sequences of harmonic tori in $\mathbf{CP}^n$ arise in this way, given the results of [6]). In the case $n = 1$, where the periodic lattice equation (1.2) is simply the elliptic sinh-Gordon equation, these solutions have been thoroughly discussed by Bobenko [3, 4] (see also [15]). For the dressing construction to produce these solutions we must replace GL_{res} by a larger group. At present I only understand how to do this for the periodic Toda lattice (which appears to be the only interesting case if one is solely interested in harmonic tori). The periodic lattice solutions which come from GL_{res} all come from an embedding of a loop group into GL_{res}, where the domain for these loops is the unit circle $\{\zeta \in \mathbf{C} : |\zeta| = 1\}$. To get any θ-function solutions we must replace the unit circle by a pair of circles, one near $\zeta = 0$ and another near $\zeta = \infty$. Then we can obtain all θ-function solutions for which the θ-function arises from a Riemann surface whose branch points lie in between these circles (except for the branch points we are obliged to put at 0 and ∞). This is explained in the final section, which is based on the article [20].

To proceed with the dressing construction we must recall the definition of GL_{res} as well as two of its homogeneous spaces: the Grassmannian Gr and the flag space Fl. These play the principal role in passing between the group GL_{res} and solutions of (1.1). The aim is to construct not just the functions v_j but, at least for the periodic lattice, the 'lifts' ϕ_j. To achieve this we also need the notion of a *Baker function*; this is an H-valued function and there is one for each point in the Grassmannian Gr.

2 The Lie group GL_{res}, the Grassmannian Gr and the flag space Fl.

The Hilbert space $H = L^2(S^1, \mathbf{C})$ splits into the direct sum $H_+ \oplus H_-$, where H_+ (respectively, H_-) is the subspace of all functions $f(\zeta)$ in H whose Fourier series possess only non-negative (respectively, negative) powers of ζ. Consequently any linear operator A on H may be written in the block form

$$A = \begin{pmatrix} a & b \\ c & d \end{pmatrix} \tag{2.1}$$

where $a : H_+ \to H_+$, $b : H_+ \to H_-$, etc. Following [24] we define GL_{res} to be the group of all bounded invertible linear operators on H whose block form (2.1) is such that b and c are Hilbert-Schmidt operators. The space of all these has a norm with respect to which it is a Banach Lie group (*cf.* [24]). If A belongs to GL_{res} then it follows that a and d are Fredholm operators. From now on we will always work in the connected component of the identity in GL_{res}, which is where a (and hence d) has Fredholm index zero; we will simply call this component GL_{res}.

A subspace W of H will belong to Gr if it is closed and the orthogonal projections $pr_+ : W \to H_+$ and $pr_- : W \to H_-$ are, respectively, Fredholm and Hilbert-Schmidt (in fact we only want the case where the first map has Fredholm index zero). An equivalent definition of Gr is as the GL_{res}-orbit of H_+ *i.e.* $Gr = \{AH_+ : A \in GL_{res}\}$

where AH_+ denotes $\{Af : f \in H_+\}$. Hence Gr is a homogeneous space of GL_{res}, isomorphic to $GL_{res}/\mathbf{P}$ where $\mathbf{P}$ is the stabiliser of H_+.

Given any $A \in GL_{res}$ we can define a bi-infinite sequence $\{W_j\}_{j \in \mathbf{Z}}$ of elements of Gr by

$$W_j = \zeta^{-j} A \zeta^j H_+, \quad j \in \mathbf{Z}, \tag{2.2}$$

where we have identified ζ with the multiplication operator $f \mapsto \zeta f$ in GL_{res}. This gives us a bi-infinite flag

$$\ldots \subset \zeta W_1 \subset W_0 \subset \zeta^{-1} W_{-1} \subset \ldots.$$

Conversely, any bi-infinite sequence $\{V_j\}_{j \in \mathbf{Z}}$ of closed subspaces of H satisfying $V_{j+1} \subset V_j$ and $\zeta^{-j} V_j \in Gr$ provides us with a sequence $\{W_j\}$ of the type above. It follows that the collection of all of these is a homogeneous space of GL_{res}, isomorphic to $GL_{res}/\mathbf{B}$, where the subgroup $\mathbf{B}$ is defined by $A \in \mathbf{B}$ iff $\zeta^{-j} A \zeta^j H_+ = H_+$ for all $j \in \mathbf{Z}$. This group $\mathbf{B}$ is just the analogue of the Borel subgroup of upper triangular matrices for GL_{n+1}. We define this collection of bi-infinite flags to be the flag space Fl.

2.1 Baker functions.

For each W in Gr there is a unique function $\psi_W(x)$, called the Baker function for W, which takes values in W and has Fourier series of the form

$$\psi_W(x) = \exp(x\zeta)(1 + O(\zeta^{-1})). \tag{2.3}$$

This relies on the following results which are proven in [25]. Inside Gr there is an open dense subset, called the big cell, which consists of all W for which $pr_+ : W \to H_+$ is invertible. Clearly, for any W in the big cell the family $e^{-x\zeta}W$ will also be in the big cell for x suitably near 0. Therefore there is a locally defined function ψ_W uniquely characterised by $pr_+(e^{-x\zeta}\psi_W) = 1$. However, it was shown in [25] that for *any* W in Gr the family $e^{-x\zeta}W$ belongs to the big cell for almost all (complex) values of x, therefore the Baker function is defined almost globally. Indeed if we write

$$\psi_W(x) = \exp(x\zeta)(1 + a_W(x)\zeta^{-1} + \ldots) \tag{2.4}$$

it has been shown in [25] that $a_W(x)$ is a meromorphic function (in fact all the coefficients in this series are) and its poles are precisely at the x-values for which $e^{-x\zeta}W$ does not lie in the big cell.

3 Solutions of the Toda lattice equations.

Given any $A \in GL_{res}$ we will define a 1-parameter family in Fl by

$$W_j(t) = \zeta^{-j} A \zeta^j \exp(t\zeta^{-1}) H_+. \tag{3.1}$$

It follows at once from the definition that

$$\partial_t W_j(t) \subset \zeta^{-1} W_{j-1}(t). \tag{3.2}$$

For each t we have the Baker functions $\psi_j(x,t)$ taking values in $W_j(t)$ and one easily checks that this gives an analytic family of Baker functions. Moreover, from the definition and (3.2) we have:

$$\partial_x \psi_j(x,t) \in W_j(t) \quad ; \quad \partial_t \psi_j(x,t) \in \zeta^{-1} W_{j-1}(t). \tag{3.3}$$

Proposition 1 *There are functions $q_j(x,t)$ and $v_j(x,t)$ for which the Baker functions satisfy:*

$$\partial_x \psi_j + q_j \psi_j - \zeta \psi_{j+1} = 0; \tag{3.4}$$
$$\zeta \partial_t \psi_j - v_j \psi_{j-1} = 0. \tag{3.5}$$

The compatibility conditions for these equations imply that the functions v_j satisfy the Toda equations (1.1).

Proof. As in (2.4) we can write $\psi_j = \exp(x\zeta)(1 + a_j \zeta^{-1} + \ldots)$. The claim is that $q_j = a_{j+1} - a_j$ and $v_j = \partial_t a_j$. Indeed an examination of the Fourier series shows that the left hand side of (3.4) looks like

$$e^{x\zeta}(q_j + a_j - a_{j+1} + O(\zeta^{-1})), \tag{3.6}$$

while the left hand side of (3.5) looks like

$$e^{x\zeta}(\partial_t a_j - v_j + O(\zeta^{-1})). \tag{3.7}$$

However, all the terms in (3.6) belong to $W_j(t)$ and all the terms in (3.7) belong to $W_{j-1}(t)$. As explained earlier, for almost every x the space $e^{-x\zeta} W_j(t)$ belongs to the big cell and therefore $e^{-x\zeta} W_j(t) \cap H_- = \{0\}$. It follows that (3.6) and (3.7) are identically zero for almost all x, t with the above choices for q_j and v_j.

It remains to prove that v_j satisfy the Toda equations (1.1). The identity $\partial_x \partial_t \psi_j = \partial_t \partial_x \psi_j$, when combined with the linear independence of ψ_j and $\zeta^{-1} \psi_{j-1}$ in $W_j(t)$, implies that

$$\partial_t q_j = v_{j+1} - v_j \quad ; \quad \partial_x v_j = v_j(q_{j-1} - q_j). \tag{3.8}$$

The second equation is equivalent to $\partial_x \ln(v_j) = q_{j-1} - q_j$. When combined with the first equation we obtain the equations (1.1).

Remark 1. Notice that, if Γ_+ denotes the subgroup of all elements of **P** commuting with ζ, then any element of the coset $A\Gamma_+$ will yield the same flag (2.2). Also, it is clear from the definition that, for any j, $\psi_j a(\zeta)$ is the Baker function for aW_j whenever $a(\zeta) = 1 + O(\zeta^{-1})$. Let Γ_- denote the group of all these functions $a(\zeta)$ (thought of as a subgroup of GL_{res}) then any element of the double coset $\Gamma_- A\Gamma_+$ provides the same solution $\{v_j\}_{j \in \mathbf{Z}}$ for the Toda equations (1.1). In fact there is a one-to-one correspondence between these double cosets and these solutions of (1.1), a fact that the reader can easily verify.

3.1 Globally analytic solutions of the elliptic Toda lattice.

To obtain solutions of the elliptic Toda lattice (1.2) we must impose two reality conditions on the construction just described. The first is to set $t = -\bar{x}$ and the second is to choose $A = L^*L$ for any $L \in GL_{res}$, where L^* denotes the adjoint operator. In that case we have the following result.

Theorem 1 *Suppose $A = L^*L$ where $L \in GL_{res}$. Then with $t = -\bar{x}$ the functions $v_j(x, -\bar{x})$ given by the construction above are globally analytic, real-valued and strictly positive solutions to (1.2).*

The key to this result is that GL_{res} admits an Iwasawa decomposition i.e. every $L \in GL_{res}$ can be uniquely written as a product UDN where U is a unitary operator, D maps $\zeta^j \mapsto s_j \zeta^j$ where each s_j is a strictly positive real, and N maps $\zeta^j \mapsto \zeta^j(1 + O(\zeta))$. Let us denote the corresponding subgroups of GL_{res} by **U**, **D** and **N**, then this result is a corollary of the proof that $GL_{res}/\mathbf{B} \simeq Fl \simeq \mathbf{U}/\mathbf{T}$ where $\mathbf{T} = \mathbf{B} \cap \mathbf{U}$. The proof of this is quite straightforward; it follows exactly the lines of the proof for GL_{n+1} in which the role of **D** is played by the subgroup of diagonal positive definite matrices and **N** corresponds to the subgroup of upper triangular unipotents. Indeed this analogy is quite close if we identify each $f = \sum f_j \zeta^j$ with the infinite length column vector $(\ldots, f_1, f_0, f_{-1}, \ldots)^t$.

Proof. Using the Iwasawa decomposition we write $L \exp(-\bar{x}\zeta^{-1}) = UDN$, then we have

$$\exp(-x\zeta)L^*L\exp(-\bar{x}\zeta^{-1}) = N^*D^*DN = N^*D^2N \qquad (3.9)$$

Now we make two claims:

1. $e^{-x\zeta}W_j(-\bar{x})$ belongs to the big cell for all x,

2. $e^{x\zeta}N^* : \zeta^j \mapsto \zeta^j \psi_j$.

The first claim follows from the identity $e^{-x\zeta}W_j(-\bar{x}) = (\zeta^{-j}N\zeta^j)^*H_+$ (from (3.9), since $D^2N \in \mathbf{B}$ and as an operator $\zeta^* = \zeta^{-1}$). It is easy to see that, using the block form (2.1),

$$(\zeta^{-j}N\zeta^j)^* = \begin{pmatrix} a_j^* & 0 \\ b_j^* & d_j^* \end{pmatrix},$$

where a_j, b_j, d_j come from the block form for $\zeta^{-j}N\zeta^j$. It follows that the first claim is true precisely when a_j^* is invertible, which is always the case since $\zeta^{-j}N\zeta^j$ is invertible.

The second claim comes from the fact that $N^* : \zeta^j \mapsto \zeta^j(1 + O(\zeta^{-1}))$, since $\zeta^j \psi_j$ is characterised as the unique element of $\zeta^j W_j(-\bar{x})$ with Fourier series $\zeta^j e^{x\zeta}(1 + O(\zeta^{-1}))$.

For brevity, let us denote $e^{x\zeta}N^*$ by ψ. Then from the second claim we deduce that all the equations (3.5) are encapsulated in the single equation

$$\psi^{-1}\bar{\partial}\psi = -v \qquad (3.10)$$

where v is the operator defined by $\zeta^j \mapsto v_j \zeta^{j-1}$. Now we observe, from (3.9), the identity

$$\exp(x\zeta)N^* = A\exp(-\bar{x}\zeta^{-1})N^{-1}D^{-2}.$$

Placing the right hand side of this into (3.10) gives

$$v = D^2 N \zeta^{-1} N^{-1} D^{-2} - D^2 N \bar{\partial}(N^{-1}) D^{-2} + 2\bar{\partial}D D^{-1}.$$

An examination of this identity shows that v must equal $D^2 \zeta^{-1} D^{-2}$, which maps $\zeta^j \mapsto (s_{j-1}^2 s_j^{-2})\zeta^{j-1}$ (where $D : \zeta^j \mapsto s_j \zeta^j$). It follows that each v_j is real-valued and strictly positive, while the first claim proves that v_j is globally analytic. This finishes the proof of the theorem.

4   Harmonic maps into U_{n+1}/T and $\mathbf{CP}^n$.

While the Baker functions have led us to solutions of (1.2) it is unclear how any of these solutions are related to harmonic maps. In this section we will see that all the periodic lattice solutions which arise from this construction correspond to *equi*-harmonic maps into U_{n+1}/T (a map into a homogeneous space of U_{n+1} is equi-harmonic if it is harmonic with respect to every U_{n+1}-invariant metric). Moreover, after a slight alteration to the construction we will be able to conclude the same thing about solutions to the finite lattice equations.

The first step is to reformulate the previous proposition slightly.

Corollary 1 *Let $L \in GL_{res}$ and let UDN denote the Iwasawa decomposition of $L\exp(-x\zeta)^*$. Then U satisfies the equations*

$$\begin{aligned} U^{-1}\partial U &= -\partial D.D^{-1} + D^{-1}\zeta D \\ U^{-1}\bar{\partial}U &= D^{-1}\bar{\partial}D - D\zeta^{-1}D^{-1} \end{aligned} \qquad (4.1)$$

whose compatibility conditions are equivalent to the Toda lattice equations (1.2).

In fact the compatibility conditions for this system, which are explicitly

$$\partial\bar{\partial}\ln(s_j^2) = s_{j-1}^2 s_j^{-2} - s_j^2 s_{j+1}^{-2}, \qquad (4.2)$$

are gauge equivalent to the Toda equations (1.2), since

$$U = L\psi^{*-1}D^{-1}. \qquad (4.3)$$

Remark 2. We could show that the group GL_{res} acts on the space of all unitary operators U arising from this corollary: the action is just the 'natural action' (*cf.* [13, 14]) in which $L \circ U$ is the unitary factor in the Iwasawa decomposition of the product LU.

Now we will examine the periodic and finite lattice equations respectively.

4.1 The periodic lattice.

The periodic lattice corresponds to periodic flags in Fl i.e. those flags $\{W_j\}$ for which $W_{j+n+1} = W_j$. The collection of all such flags is denoted $Fl^{(n+1)}$. Clearly, for any such flag, $\zeta^{n+1}W_j \subset W_j$: the points in Gr with this property form a submanifold $Gr^{(n+1)}$. Both $Fl^{(n+1)}$ and $Gr^{(n+1)}$ are homogeneous spaces of the subgroup $GL_{res}^{(n+1)}$ of GL_{res} which is the commutant of ζ^{n+1}.

Pressley and Segal [24] point out that $GL_{res}^{(n+1)}$ is identifiable with the loop group $LG = L_{\frac{1}{2}}GL_{n+1}$ of so-called '$\frac{1}{2}$-differentiable' maps $S^1 \to GL_{n+1}$ (to be precise, those whose determinant has winding number zero). However for our purposes it is more useful to use another realisation of $GL_{res}^{(n+1)}$, called the *principal realisation*, as (somewhat surprisingly) the subgroup $L(G,\nu)$ of elements $g(\zeta)$ of LG satisfying $\sigma g(\zeta)\sigma^{-1} = g(\omega\zeta)$, where σ is the diagonal matrix $diag(1,\omega,\ldots,\omega^n)$ and $\omega = \exp(2\pi i/(n+1))$. We will call such loops ν-*equivariant*, where ν denotes the periodic automorphism $g \mapsto \sigma g \sigma^{-1}$ of GL_{n+1}.

The identification of $GL_{res}^{(n+1)}$ with $L(G,\nu)$ goes as follows. First we note that any $f \in H$ may be expanded in the form

$$f(\zeta) = f_0(\zeta^{n+1}) + \zeta f_1(\zeta^{n+1}) + \ldots + \zeta^n f_n(\zeta^{n+1}). \tag{4.4}$$

For $g \in L(G,\nu)$ we define $g \circ f$ to be that element of H got by adding together the entries of the vector

$$(f_0, \zeta f_1, \ldots, \zeta^n f_n)g^{-1}. \tag{4.5}$$

This identifies g with an element of GL_{res} which clearly commutes with ζ^{n+1}. Conversely, given $A \in GL_{res}^{(n+1)}$ we see from (4.5) that we can construct an element g of $L(G,\nu)$ by taking the j-th row of g^{-1} to be $\zeta^{-j}(a_{0j},\ldots,a_{nj})$ where $a_{0j} + \ldots + \zeta^n a_{nj}$ is the expansion of $A\zeta^j \in H$ according to (4.4).

Given this identification let us use Φ to denote the $L(G,\nu)$-valued function of $x, \bar{x}$ which represents U defined by (4.3). Since U is unitary it follows that Φ is too. It is a simple matter to check that the equations for U above imply

$$\begin{aligned}\Phi^{-1}\partial\Phi &= -\partial s.s^{-1} - \zeta s^{-1}\Lambda s \\ \Phi^{-1}\bar{\partial}\Phi &= s^{-1}\bar{\partial}s + \zeta^{-1}s\Lambda^t s^{-1}\end{aligned} \tag{4.6}$$

where s^{-1} is the diagonal, positive definite, matrix $diag(s_0,\ldots,s_n)$. Here $\Lambda \in gl_{n+1}$ is the cyclic matrix characterised by $\Lambda\delta_{j+1} = \delta_j$ where δ_j is the unit basis vector in $\mathbf{C}^{n+1}$ for the j-th coordinate (with j taken over $\mathbf{Z}_{n+1}$). Observe that $\exp(x\zeta\Lambda)$ corresponds to the multiplication operator $e^{-x\zeta}$ on H.

In the context of this equation the previous corollary now amounts to the following statement. Set $\Phi^{(0)} = \exp(-x\zeta\Lambda + \bar{x}\zeta^{-1}\Lambda^t)$ (which produces the trivial solution of (1.2) i.e. $v_j = 1$ for all j) and take any $g \in L(G,\nu)$. Then $g\Phi^{(0)}$ has Iwasawa decomposition Φsn and the pair Φ, s satisfy (4.6). Here all the factors in this decomposition are ν-equivariant: s is as above and n has Fourier series $I + O(\zeta)$.

Infinite dimensional Lie groups and the two-dimensional Toda lattice 213

Proposition 2 *Let Φ be any global solution of (4.6) and let Φ_α denote Φ evaluated at $\zeta = \alpha$, $|\alpha| = 1$. The map $\phi : \mathbf{R}^2 \to U_{n+1}/T$ defined by $\phi(x, \bar{x}) = \Phi_\alpha(x, \bar{x})T$ is equi-harmonic.*

This is a property of the equation (4.6) and not just the class of solutions we are currently studying. It is a corollary of a deeper result, due to Black [2], and follows from the fact that the map ϕ is a *primitive map* in the following sense of Burstall's [8]:

Let $\mathbf{g}$ denote the Lie algebra of U_{n+1} and let M denote U_{n+1}/T. Recall (from e.g. [9, 28]) that there is a surjective endomorphism of bundles

$$\begin{aligned} M \times \mathbf{g} &\longrightarrow TM \\ (x, \xi) &\longmapsto \tfrac{d}{dt}(\exp t\xi.x)|_0 \end{aligned} \quad (4.7)$$

whose kernel is the bundle $[\mathbf{t}]$ (where this denotes the vector bundle $U_{n+1} \times_T \mathbf{t}$, where T acts adjointly on its Lie algebra $\mathbf{t}$). Given a reductive splitting $\mathbf{g} = \mathbf{t} + \mathbf{m}$ the restriction of this endomorphism (4.7) to $[\mathbf{m}]$ is invertible and we denote its inverse by β. We can think of β as a $\mathbf{g}$-valued 1-form on M (it is called the *Maurer-Cartan form* for M).

The periodic automorphism ν on $\mathbf{g}$ induces a $\mathbf{Z}_{n+1}$-grading on the complexification $\mathbf{g}^C$ whose homogeneous subspaces we denote by $\mathbf{g}^C_k$. Each of these is simply the eigenspace of ν for the eigenvalue ω^k. In particular, $\mathbf{g}^C_0 = \mathbf{t}^C$ and $\mathbf{m}^C = \mathbf{g}^C_1 + \ldots + \mathbf{g}^C_n$. Burstall [8] calls a map $\phi : \mathbf{R}^2 \to M$ *primitive* (or ν-*primitive*) whenever $\phi^*\beta^{(1,0)}$ takes values in $[\mathbf{g}^C_1]$. This condition guarantees that ϕ is f-holomorphic with respect to a horizontal f-structure (as defined in [2]) which one can naturally associate to the automorphism ν, provided ν has order greater than 2 (i.e. $n \geq 2$). One of the main results of [2] implies that any such map whose domain is a Riemann surface is equi-harmonic.

In the case of the map $\phi = \Phi_\alpha T$ the definitions are such that ϕ is primitive precisely when $P_m(\Phi^{-1}\partial\Phi)$ takes values in $\mathbf{g}^C_1$, where P_m is the projection onto $\mathbf{m}^C$ along $\mathbf{t}^C$. But this is always the case if Φ satisfies (4.6), since $\nu(\Lambda) = \omega\Lambda$ and s takes values in T^C. (When $n = 1$ we can prove the result without this machinery.)

Proposition 3 *Let ϕ_k denote the k-th column of Φ_α defined above, and let $\langle\phi_k\rangle$ denote the line it generates in $\mathbf{C}^{n+1}$. Then the map $\langle\phi_k\rangle : \mathbf{R}^2 \to \mathbf{CP}^n$ it defines is equi-harmonic.*

This follows from another result in [2], that if the primitive map ϕ is post-composed with the homogeneous projection $U_{n+1}/T \to U_{n+1}/K$, where $K \supset T$ is any closed subgroup, the result is also an equi-harmonic map. For $\mathbf{CP}^n$ there are, of course, $n+1$ ways of doing this; one for each column of Φ_α. All these harmonic maps lie in the same orthogonal, periodic, harmonic sequence. Indeed one sees from (4.6) that the ∂-transform of the k-th column yields the $(k+1)$-st, up to scaling.

To compute Φ_α explicitly we see from (4.3) that we need the Baker functions together with the entries s_j of s^{-1}. These latter functions can be computed using the τ-functions defined by Segal & Wilson [25]. Later this will be explained in the context of the θ-function solutions for (1.2).

4.2 The finite lattice.

The condition for a finite lattice is $v_j = 0$ for all j except $1 \leq j \leq n$. This condition requires that $W_j(t) = H_+$, except for $1 \leq j \leq n$, for all t. Clearly this condition cannot be satisfied by the construction in section 3, but we can modify the x, t evolution in Fl to get this. All that is required is that we replace ζ in the one-parameter subgroups $\{\exp(x\zeta)\}$ and $\{\exp(t\zeta^*)\}$ by the finite rank operator which equals ζ on the subspace generated by $\{1, \ldots, \zeta^{n-1}\}$ and whose kernel is the orthogonal complement of this subspace. Rather than pursue this, it is quicker to realise that, for the elliptic equations (1.2), this amounts to working with the system

$$\begin{aligned}\Phi^{-1}\partial\Phi &= -\partial s.s^{-1} - s^{-1}\lambda s \\ \Phi^{-1}\bar{\partial}\Phi &= s^{-1}\bar{\partial}s + s\lambda^t s^{-1}.\end{aligned} \qquad (4.8)$$

Here everything takes values in GL_{n+1} or its Lie algebra: Φ takes values in U_{n+1}, s^{-1} is still $diag(s_0, \ldots, s_n)$ and λ is the nilpotent matrix characterised by $\lambda\delta_0 = 0$ and $\lambda\delta_{j+1} = \delta_j$ for $1 \leq j \leq n$. [In particular, there is no need for the spectral parameter ζ here.]

Proposition 4 *Let Φ be any global solution of (4.8), then the map $\phi : \mathbf{R}^2 \to U_{n+1}/T$ defined by $\phi(x, \bar{x}) = \Phi(x, \bar{x})T$ is primitive and therefore equi-harmonic. Consequently, for each column ϕ_k of Φ the map $\langle\phi_k\rangle : \mathbf{R}^2 \to \mathbf{CP}^n$ is also equi-harmonic.*

The proof follows at once from the observation that $\lambda \in \mathbf{g}_1^C$ in the periodic grading on $\mathbf{g}^C = gl_{n+1}$ described earlier.

In particular, to obtain the 'trivial solution' from (4.8) we now take $\Phi^{(0)}$ to be the unitary factor in the Iwasawa decomposition, in GL_{n+1}, of $\exp(x\lambda)^*$. As in the infinite lattice case, there is a 'natural action' of the group (in this case GL_{n+1}) on the space of solutions to (4.8). Given any $g \in GL_{n+1}$ we define Φ to be the unitary factor in the Iwasawa decomposition of $g\Phi^{(0)}$, then Φ also satisfies (4.8) for some s.

It is quite easy to compute the harmonic maps $\langle\phi_0\rangle$ into $\mathbf{CP}^n$ which arise from the orbit of $\Phi^{(0)}$ (here ϕ_0 is the first column of Φ). The vector $\Phi\delta_0$ generates the same line in $\mathbf{C}^{n+1}$ as $g.\exp(x\lambda)^*\delta_0$ (since $\Phi B = g.\exp(x\lambda)^*B$, where B is the subgroup of upper triangular matrices, by virtue of the Iwasawa decomposition). Hence $\langle\phi_0\rangle$ is simply the map

$$(x, \bar{x}) \mapsto \langle g.(1, \bar{x}, \ldots, \frac{\bar{x}^n}{n!})^t\rangle.$$

For each g this is a smooth anti-holomorphic map into $\mathbf{CP}^n$ which extends from $\mathbf{R}^2$ to the whole sphere S^2. The image is just a smooth rational curve of degree n (indeed as g varies over GL_{n+1} this gives all such curves of degree n in $\mathbf{CP}^n$). The other maps $\langle\phi_k\rangle$ are obtained by applying the 'Gram-Schmidt' process to the columns of $g.\exp(x\lambda)^*$. All these maps extend to S^2 and using (4.8) it is easy to show that they belong to the same finite harmonic sequence: in particular there is a diagonal g for which this is the *Veronese sequence* explicitly computed in [5].

Remark 3. There is a version of both (4.6) and (4.8) for any choice of simple Lie algebra. In that case ν is the Coxeter-Killing automorphism and Λ is the 'cyclic element' defined to be the sum of simple root vectors together with a root vector for minus the maximal root (the pair ν, Λ play a pivotal role in the study of generalised Toda lattices in [10]). For λ we take the sum of simple root vectors. Because both $L(G, \nu)$ and G have Iwasawa decompositions for any simple G we can repeat the constructions above to obtain primitive harmonic maps into any full flag manifold (*i.e.* the quotient of a compact group by its maximal torus) and subsequently, by homogeneous projection, equi-harmonic maps into the appropriate symmetric spaces.

5 θ-function solutions.

In [25] Segal & Wilson showed how to extract the θ-function solutions of the (generalised) KdV equations from certain 'algebro-geometric' points in Gr. However, to get something non-trivial for the Toda equations we have to adandon working over $|\zeta| = 1$ and proceed as follows. Here I will only present an outline: the details (at least in the case of SL_{n+1}) are given in [20].

The first step is to replace the loop group LG by another infinite dimensional group ΛG. This is a subgroup of the group $\Lambda_C G$ of real-analytic maps $C \to GL_{n+1}$ where C is the union $C_1 \cup C_2$ of circles with radii ϵ, ϵ^{-1} respectively, where $0 < \epsilon < 1$. As in [1, 13, 14] we consider C to be the boundary of two regions on the Riemann sphere: the union of discs I and the annulus E. An element of $\Lambda_C G$ should be thought of as a pair (g_1, g_2) of loops, whose domain is C_1 and C_2 respectively. The subgroup ΛG we want is defined by

$$\Lambda G = \{(g_1, g_2) \in \Lambda_C G : g_2(\zeta) = g_1(\bar{\zeta}^{-1})^{*-1}\}.$$

This group has a global decomposition which plays the role of the Iwasawa decomposition. Let us define the following subgroups of ΛG:

- $\Lambda_E G = \{(g_1, g_2) :$ there is a holomorphic $g : E \to GL_{n+1}$ with boundaries $g_1, g_2\}$,

- $\Lambda_I G = \{(g_1, g_2) :$ there is a holomorphic $g : I \to GL_{n+1}$ with boundaries g_1, g_2 with $g_1(0)$ upper unipotent$\}$,

- $D = \{(s, s^{-1}) : s$ is diagonal, constant and positive definite$\}$.

Theorem 2 *Every $g \in \Lambda G$ has a unique decomposition $g = udn$ where $u \in \Lambda_E G$, $d \in D$ and $n \in \Lambda_I G$.*

The proof is given in the appendix.

To obtain solutions of (1.2) we work with the subgroup $\Lambda(G,\nu)$ of pairs of ν-equivariant loops. The identification between LG and $L(G,\nu)$ (via $GL_{res}^{(n+1)}$) essentially identifies ΛG with $\Lambda(G,\nu)$ (although to be precise we must alter the radius ϵ in passing between these two). In particular, when the previous result is applied to elements of $\Lambda(G,\nu)$ each of the factors lies in $\Lambda(G,\nu)$. As a consequence we have:

Proposition 5 *Let $g \in \Lambda(G,\nu)$ and let $\Phi^{(0)}$ denote the $\Lambda_E(G,\nu)$-valued function corresponding to $\exp(-x\zeta\Lambda + \bar{x}\zeta^{-1}\Lambda^t)$. We write $g\Phi^{(0)} = \Phi dn$ according to the factorisation in the previous theorem. Then Φ satisfies the equations (4.6). Consequently the map $\phi : \mathbf{R}^2 \to U_{n+1}/T$ defined by $\phi = \Phi_\alpha T$, $|\alpha| = 1$, is harmonic.*

Here we have identified Φ with its extension to E. The proof is straightforward, by comparison of Fourier series for Φ on C_1 and C_2. Of course this result is another variation of the 'natural action' of [13, 14] (*cf.* remark 2).

Now I will explain where the θ-function solutions fit in. First observe that in the previous proposition we may as well have taken g to lie in the subgroup generated by D and $\Lambda_I(G,\nu)$, since left multiplication of Φ by an element of $\Lambda_E(G,\nu)$ does not alter (4.6) and makes an insignificant change to ϕ. In that case, $\Phi = I$ at $x, \bar{x} = 0$, so we can compute Φ as follows. On C_2 we have, by definition, $\Phi = g_2 \Phi^{(0)} n_2^{-1} s$. So to compute Φ we need only know $\Psi \equiv \Phi^{(0)} n_2^{-1}$ and s, for then g_2^{-1} is given by evaluating Ψs at $x, \bar{x} = 0$. First I will describe how to compute Ψ using certain algebro-geometric points in the Grassmannian $Gr^{(n+1)}$ corresponding to the Hilbert space H where H is now $L^2(C_2, \mathbf{C})$.

According to [25] we get a point in $Gr^{(n+1)}$ by choosing a Riemann surface X admitting a degree $n+1$ covering $\pi : X \to \mathbf{CP}^1$ together with a line bundle $\mathcal{L}$ over X equipped with a trivialisation φ_∞ over $p_\infty = \pi^{-1}(\infty)$. This means that p_∞ must be a ramification point of X, with ramification index n. The trivialisation φ_∞ is defined over the closure of an open disc D_∞ about p_∞ which must contain no other ramification points. In that case we can use π to define a local parameter ζ^{-1} about p_∞ for which the boundary of D_∞ is C_2. With all this we define a subspace $W^{an} \subset H$ whose elements are given by restricting holomorphic sections of $\mathcal{L}$ over $X - D_\infty$ to C_2 and then applying φ_∞ to get $\mathbf{C}$-valued functions. When $\mathcal{L}$ has degree equal to the genus p of X, the L^2-closure W of W^{an} belongs to $Gr^{(n+1)}$. When W_0 comes from $(\mathcal{L}, \varphi_\infty)$ we obtain a periodic flag $\{W_j\}$ by using $(\mathcal{L} \otimes \mathcal{O}_X(jp_\infty - jp_0), \zeta^j \varphi_\infty)$ to construct W_j, where $p_0 = \pi^{-1}(0)$ is required to be another ramification point (also of index n). To get a 1-parameter family of flags $\{W_j(t)\}$ of the form (3.1) we replace $\mathcal{L}$ by the line bundle $\mathcal{L} \otimes [e^{t\zeta^{-1}}]$, where $[e^{t\zeta^{-1}}]$ is the degree zero line bundle constructed using $e^{t\zeta^{-1}}$ as a transition function over C_1 (considered as the boundary of a disc about p_0).

The Baker functions for this 1-parameter family produce solutions of (1.1) via (3.5). To get real solutions of the elliptic equations (1.2) we have to impose reality conditions on the data $(X, \pi, \mathcal{L})$ which correspond to passing from $\Lambda_C(G,\nu)$ to the subgroup $\Lambda(G,\nu)$. Namely, we require that X possess an anti-holomorphic involution $\rho : X \to X$ which covers the map $z \mapsto \bar{z}^{-1}$ on $\mathbf{CP}^1$. We also require $\mathcal{L}$ to satisfy the reality condition

$$\overline{\rho_* \mathcal{L}} \simeq \mathcal{O}_X(R) \otimes \mathcal{L}^{-1}, \tag{5.1}$$

where R is the ramification divisor of π in $X - \{p_0, p_\infty\}$ (it has degree $2p$).

The $n+1$ Baker functions we obtain from $\{W_j(t)\}$ provide the columns for a matrix function whose Fourier series on C_2 is of the form $(I + O(\zeta^{-1})) \exp(x\zeta\Lambda)$. This matrix is constructed in the same way that we construct an element of $L(G, \nu)$ from $n+1$ columns of an operator from $GL_{res}^{(n+1)}$ (see Subsection 4.1). Then with $t = -\bar{x}$ it is the inverse of this which equals Ψ: we will call this the matrix Baker function (in [20] it is the inverse of this Ψ which is called the matrix Baker function).

Finally, we use the τ-functions for each $W_j(\dot{z})$, together with the formula

$$\tau_j/\tau_{j+1} = s_j^2, \tag{5.2}$$

to compute $s^{-1} = diag(s_0, \ldots, s_n)$ and thus complete the computation of Φ. These functions τ_j are all defined in terms of the single function called $\tau_W(x)$ in [25]. Let us rename this $\tau(x; W)$, then we define

$$\tau_j(x, t) = \tau(x; W_j(t)).$$

When the flag comes from the data $(X, \pi, \mathcal{L})$, in the manner described above, each of these functions may be computed in terms of the Riemann θ-function for X. Their computation is detailed in [20].

Remark 4. Let me briefly the explain the origin of the formula (5.2). The proof lies in the *Birkhoff factorisation* of elements of GL_{res}. In particular, suppose we have a flag of the form (2.2). When each W_j lies in the big cell we can factorise A into $A_- A_0 A_+$ where A_- (resp. A_+) is 'lower' (resp. 'upper') unipotent and A_0 is diagonal. Clearly the flag only depends upon the factor A_-. Now let $A(t)_-$ correspond to the flag $\{W_j(t)\}$. It is essentially by definition of the τ-functions that τ_j/τ_{j+1} gives the j-th diagonal entry of the middle factor in the Birkhoff factorisation of $e^{-x\zeta} A(t)_-$. If we compare this with the equation (3.9) in the proof of Corollary 1, with $t = -\bar{x}$, we find the formula (5.2).

5.1 Harmonic tori

We will have harmonic tori of the form $\phi = \Phi_\alpha T$ whenever Φ_α is doubly periodic. To find these we look first for doubly periodic solutions of (1.2) and then investigate when the monodromy of (4.6) is trivial for some value of ζ on the unit circle. One expects to be able to show, using elliptic operator arguments similar to those in [3, 15], that doubly periodic solutions of (1.2) can only come from θ-function solutions. In that case it is not too difficult to characterise them. Geometrically, a solution of (1.2) corresponds to a 2-parameter family of line bundles $\{\mathcal{L} \otimes [e^{x\zeta}] \otimes [e^{-\bar{x}\zeta^{-1}}]\}$ where $\mathcal{L}$ satisfies the reality condition (5.1). Here $[e^{x\zeta}]$ is the degree zero line bundle whose transition function over C_2 is $e^{x\zeta}$: it follows that every bundle in the 2-parameter family satisfies (5.1). The solution is doubly periodic precisely when the real subgroup $\{[e^{x\zeta}] \otimes [e^{-\bar{x}\zeta^{-1}}]\}$ of the Jacobi variety $J(X)$ is topologically a 2-torus. This can be phrased as a number of conditions on the moduli of X. In particular, this is independent of $\mathcal{L}$. It follows that these solutions always come in families with (at least) p real dimensions: this is the dimension of the real subgroup J_R of $J(X)$ whose action preserves the reality condition (5.1). The effect of this action is to translate the 2-torus around inside J_R: two of these degrees of freedom account for conformal automorphisms of the 2-torus while the other $p-2$ yield genuine deformations of the map.

To satisfy the monodromy conditions is a harder problem. It can be shown that the monodromy of Φ is conjugate to that of Ψ and therefore its eigenvalues (which are all we are interested in) can be computed from the Baker functions. In the case $n = 1$ (the case of the elliptic sinh-Gordon equation) we can use the results of [3, 11] to produce harmonic tori in U_2/T each of which is the Gauss map of a surface of constant mean curvature in $\mathbf{R}^3$. Although these maps come in p-dimensional families (for the reasons given above) there is only a discrete collection of Riemann surfaces X which admit them, hence the problem of locating them was non-trivial. In the case $n \geq 2$ the monodromy conditions are at least as difficult to satisfy and the problem has not, to my knowledge, been tackled.

5.2 Appendix: the proof of Theorem 2.

Only one tool is required for the proof of theorem 2, namely, the Birkhoff factorisation [24] for LG in the following form. Every $g \in LG$ can be factorised into $g = g_- g_+$, where:

- $g_+ = g_+(0)(I + O(\zeta))$ with $g_+(0)$ upper unipotent,

- g_- extends holomorphically to the punctured disc about infinity but may have poles at infinity. For readers familiar with the Birkhoff factorisation it is this factor which includes a representative of the affine Weyl group.

This factorisation holds regardless of the radius of the circle over which LG is defined.

So let $(g, \bar{g}) \in \Lambda G$, where $\bar{g}$ is shorthand for $g(\bar{\zeta}^{-1})^{*-1}$. Over C_1 the loop g has factorisation $g_- g_+$ and g_- extends holomorphically to E. Define $h = \bar{g}_-^{-1} g_-$, then over $|\zeta| = 1$ we simply have $h = g_-^* g_-$. Hence the Birkhoff factorisation for h over E comes from the Iwasawa decomposition for g_- over the unit circle. For, if $g_- = a_u s a_+$ is the Iwasawa decomposition (i.e. a_u is unitary, s is diagonal, constant, positive definite and a_+ is as g_+ above), then $h = a_+^* s^2 a_+$ over the unit circle. But in fact this holds throughout E, since Birkhoff factors have the same domain of definition as their product. Hence we have

$$(g, \bar{g}) = (a_u, \bar{a}_u)(s, s^{-1})(a_+ g_+, \bar{a}_+ \bar{g}_+),$$

where the first factor belongs to $\Lambda_E(G,\nu)$ since $a_u, \bar{a}_u$ both extend to E and are equal on the unit circle (hence everywhere).

The uniqueness of this factorisation is clear. For if $udn = u'd'n'$ then $u^{-1}u'$ must be globally holomorphic and therefore a constant unitary matrix. But at $\zeta = 0$ it equals an upper triangular matrix with positive definite diagonal, hence it is the identity.

Bibliography

[1] M.J. Bergvelt and M.A. Guest, *Actions of loop groups on harmonic maps*, Trans. Amer. Math. Soc. **326** (1991), 861-886.

[2] M. Black, *Harmonic maps into homogeneous spaces*, Pitman Research Notes in Math. 255, Longman, Harlow, 1991.

[3] A.I. Bobenko, *All constant mean curvature tori in R^3, S^3 and H^3 in terms of theta-functions*, Math. Ann. **290** (1991), 209-245.

[4] A.I. Bobenko, *Constant mean curvature surfaces and integrable equations*, Uspekhi Mat. Nauk **46:4** (1991), 3-42; English Transl.: Russ. Math. Surveys **46:4** (1991), 1-45.

[5] J. Bolton, G.R. Jensen, M. Rigoli and L.M. Woodward, *On conformal minimal immersions of S^2 into $\mathbf{CP}^n$*, Math. Ann. **279** (1988), 599-620.

[6] J. Bolton, F.Pedit and L.M. Woodward, *Minimal surfaces and the Toda field model*, preprint, University of Durham (1993).

[7] J. Bolton and L.M. Woodward, *The affine Toda equations and minimal surfaces*, this volume.

[8] F.E. Burstall, *Harmonic tori in spheres and complex projective spaces*, in preparation.

[9] F.E. Burstall and J.H. Rawnsley, *Twistor theory for Riemannian symmetric spaces with applications to harmonic maps of Riemann surfaces*, Lect. Notes in Math. 1424, Springer, Berlin, 1990.

[10] V.G. Drinfel'd and V.V. Sokolov, *Lie algebras and equations of Korteweg-de Vries type*, Itogi Nauki i Tekhnike ser Sov. Prob. Mat. **24** (1984), 81-180; English transl.: J. Soviet Math. **30** (1985), 1975-2036.

[11] N. Ercolani, H. Knörrer and E. Trubowitz, *Hyperelliptic curves that generate constant mean curvature tori in $\mathbf{R}^3$*, preprint, E.T.H. Zürich (1991).

[12] R.S. Farwell and M. Minami, *Derivation and solution of the two-dimensional Toda lattice equations by use of the Iwasawa decomposition*, J. Phys. A: Math. Gen. **15** (1982), 25-46.

[13] M.A. Guest and Y Ohnita, *Group actions and deformations for harmonic maps*, J. Math. Soc. Japan (to appear).

[14] M.A. Guest and Y Ohnita, *Loop group actions on harmonic maps and their applications*, this volume.

[15] N.J. Hitchin, *Harmonic maps from a 2-torus to the 3-sphere*, J. Diff. Geom. **31** (1990), 627-710.

[16] I.M. Krichever, appendix in: B.A. Dubrovin, *Theta functions and non-linear equations*, Uspekhi Mat. Nauk **36:2** (1981), 11-80; English transl.: Russ. Math. Surveys **36:2** (1981), 11-92.

[17] B.A. Kupershmidt and G. Wilson, *Conservation laws and symmetries of generalized sine-Gordon equations*, Comm. Math. Phys. 81 (1981), 189-202.

[18] A.N. Leznov and M.V. Saveliev, *Representation of zero curvature for the system of non-linear partial differential equations $x_{\alpha,z\bar{z}} = \exp(kx)_\alpha$ and its integrability*, Lett. Math. Phys. **3** (1979), 489-494.

[19] A.N. Leznov and M.V. Saveliev, *Representation theory and integration of nonlinear spherically symmetric equations to gauge theories*, Comm. Math. Phys. **74** (1980), 111-118.

[20] I. McIntosh, *Global solutions of the elliptic 2D periodic Toda lattice*, preprint, University of Newcastle upon Tyne (1993).

[21] M. Mañas, *The principal chiral model as an integrable system*, this volume.

[22] A.V. Mikhailov, *Integrability of a two-dimensional generalisation of the Toda chain*, Pis'ma Zh. Eksp. Teor. Fiz. **30** (1979), 443-448; English transl.: JETP Letters **30** (1979), 414-418.

[23] A.V. Mikhailov, M.A. Olshanetsky and A.M. Perelomov, *Two-dimensional generalized Toda lattice*, Comm. Math. Phys. **79** (1981), 473-488.

[24] A. Pressley and G. Segal, *Loop groups*, Oxford Math. Monographs, Clarendon Press, Oxford, 1985.

[25] G.Segal and G. Wilson, *Loops groups and equations of KdV type*, Publ. Math. IHES **61** (1985), 5-65.

[26] G. Wilson, *The modified Lax and two-dimensional Toda equations associated with simple Lie algebras*, Ergod. Th. and Dyn. Sys. **1** (1981), 361-380.

[27] G. Wilson, *Infinite dimensional Lie groups and algebraic geometry in soliton theory*, Phil. Trans. R. Soc. London **A315** (1985), 393-404.

[28] J.C. Wood, *Harmonic maps into symmetric spaces and integrable systems*, this volume.

[29] V.E. Zakharov and A.V. Shabat, *Integration of the non-linear equations of mathematical physics by the inverse scattering method*, Funks. Anal. Appl. **13** (1979), 13-22; English transl.: Funct. Anal. Appl. **13** (1979), 166-174.

Harmonic maps via Adler–Kostant–Symes theory

F.E. Burstall and F. Pedit

Introduction

Over the past few years significant progress has been made in the understanding of various completely integrable nonlinear partial differential equations (soliton equations) and their relationship to classical problems in differential geometry. It has been shown in a series of recent papers [42, 32, 21, 25, 15, 8, 12] that constant mean and Gauss curvature surfaces, Willmore surfaces, minimal surfaces in spheres and projective spaces and generally harmonic maps from a Riemann surface M into various homogeneous spaces may be described as solutions to various soliton equations (see [7]). Moreover, these solutions are *algebraic* in the sense that they are obtained by integrating *ordinary* differential equations of Lax type which linearise on the Jacobian of an appropriate algebraic curve.

Links between harmonic maps and integrable systems have been known to exist for some time: for instance, Uhlenbeck [55] showed that S^1-equivariant harmonic maps $\mathbb{R} \times S^1 \to S^n$ amount to solutions of the Neumann system describing motion on S^n in a quadratic potential—a classical completely integrable system (see [60] and for related results on S^1-equivariant harmonic maps, see [33, 24]). However, a significant interaction between differential geometry and soliton theory did not emerge until after Wente's resolution [59] of the Hopf conjecture on the existence of a constant mean curvature torus in $\mathbb{R}^3$.

Let us consider this problem in more detail as it contains the seeds of all subsequent developments in this area. Recall that a surface M in $\mathbb{R}^3$ has constant mean curvature if and only if its Gauss map $\phi : M \to S^2$ is harmonic. We are therefore led to pose the following problem: given a compact Riemann surface M of genus g, find all harmonic maps ϕ from M into the 2-sphere. The harmonicity of ϕ implies that the $(2,0)$-part η of the quadratic differential $(d\phi, d\phi)$ is holomorphic. If M has genus 0 then this differential vanishes identically, i.e. ϕ is conformal and thus ±holomorphic. In this sense the harmonic map equation reduces to the linear Cauchy–Riemann equation. A similar phenomenon occurs when the target space is replaced by a compact symmetric space where the reduction to the Cauchy–Riemann equation is achieved by various twistorial constructions (see [35, 60] in this volume for further details and references).

The situation changes drastically when M has higher genus. There is no systematic theory when M has genus larger then 1 so we shall concentrate on the case where M is a 2-torus. Here, the differential $\eta = (d\phi, d\phi)^{(2,0)}$ either vanishes identically (in which case we again have ±holomorphic maps) or is nowhere zero. In the latter case one can choose a global complex coordinate z so that $(d\phi, d\phi)^{(2,0)} = dz^2$ and $(d\phi, d\phi)^{(1,1)} = \cosh(\omega) |dz|^2$. Then the harmonicity of ϕ amounts to the sinh-Gordon equation for ω, a well known soliton equation. Starting from this observation, Pinkall–Sterling [42] show that all doubly-periodic solutions to the sinh-Gordon equation (and thus all constant mean curvature tori) are obtained by integration of a family of completely integrable finite dimensional systems of ODE. These ODE linearise on the Jacobian of an algebraic curve the study of which enabled Bobenko [6] to show that the solutions can be expressed rather explicitly in terms of theta functions. That this behaviour extends to more general target spaces was indicated in [32] for S^3, [25] for S^4 and culminated in [15] where a rather comprehensive theory of harmonic 2-tori in compact Riemannian symmetric spaces is developed.

One may summarise this theory as follows: the starting point is the basic fact [43, 56, 65, 66] that the harmonic map equation can be reformulated as a zero-curvature equation involving an auxiliary parameter or, in other words, as the Maurer–Cartan equation for a certain loop algebra valued 1-form. One can obtain solutions to these equations by integrating a pair of commuting Hamiltonian vector fields on certain finite-dimensional subspaces of loop algebras. In this way, we get harmonic maps of $\mathbb{R}^2$ which we call *harmonic maps of finite type*. Finally, under suitable nondegeneracy assumptions, one shows that these procedures account for all doubly periodic harmonic maps, that is, for all harmonic 2-tori.

It is the purpose of this article to provide a unified account of these ideas and, in particular, the results of [8, 12, 15, 25, 42]. A framework for this is given by the theory of Adler–Kostant–Symes [1, 37, 54]. This theory provides a scheme for producing and integrating non-trivial commuting Hamiltonian flows on Lie algebras. The basic setting for the scheme is this: one has a Lie algebra $\mathfrak{g}$ which admits a (vector space) direct sum decomposition into subalgebras

$$\mathfrak{g} = \mathfrak{k} \oplus \mathfrak{b}. \tag{0.1}$$

In the presence of a suitable inner product, $\mathfrak{k}$ inherits a Poisson structure from the Lie–Poisson structure on $\mathfrak{b}^*$. With this Poisson structure, $\mathfrak{k}$ has a large supply of Poisson commuting functions provided by the restriction to $\mathfrak{k}$ of the invariant functions on $\mathfrak{g}$. To integrate the Hamiltonian flows so obtained, we use the method of Symes: let G be a Lie group with Lie algebra $\mathfrak{g}$ and let K, B be the subgroups corresponding to $\mathfrak{k}, \mathfrak{b}$. Suppose that there is a decomposition

$$G = K \times B \tag{0.2}$$

corresponding to (0.1). Then the Hamiltonian flow through an initial condition $\xi_o \in \mathfrak{k}$ is given by

$$\operatorname{Ad} k^{-1} \xi_o,$$

where k is the projection onto K via (0.2) of a suitable geodesic in G.

We apply this theory to a family of twisted loop algebras which admit a decomposition (0.1) of Iwasawa type. A key point is that the subspaces of Laurent polynomial loops of fixed degree are Poisson submanifolds so that we obtain commuting ODE on *finite-dimensional* Poisson manifolds. From the Hamiltonian flows on these subspaces we get solutions to zero-curvature equations which give rise to (framings of) harmonic maps of $\mathbb{R}^2$ into various homogeneous spaces.

In fact, we get in this way both harmonic maps into Riemannian symmetric spaces and primitive maps into k-symmetric spaces. These last are rather special harmonic maps into a class of reductive homogeneous spaces that generalise Riemannian symmetric spaces (the involutions of a Riemannian symmetric space are replaced by automorphisms of order k) and include all flag manifolds. A map into such a space is primitive if it satisfies a first order condition not unlike a Cauchy-Riemann equation defined by the geometry of the k-symmetric space. The motivation for studying primitive maps comes from the fact that they include twistor lifts of minimal maps in spheres and complex projective spaces and also, in case the target is a flag manifold, are in bijective correspondence with (periodic) Toda fields.

The method of Symes also works in this infinite-dimensional loop algebra setting. There are Iwasawa decompositions of the loop groups that correspond to our loop algebras and projection of (complex) geodesics provides maps into loop groups which are essentially the *extended solutions* in the sense of Uhlenbeck [56] for the harmonic maps we have produced.

Finally, we obtain sufficient conditions on a harmonic 2-torus to arise from our constructions. The basic result here is due to Burstall-Ferus-Pedit-Pinkall [15] and the method of Symes is applied to translate that result to our present setting.

The second author was partially supported by NSF grant DMS-9205293 during the preparation of this article.

1 The Adler-Kostant-Symes scheme

One of the main themes of this volume is the construction of harmonic maps from commuting Hamiltonian flows on loop algebras. A setting for such results is provided by the celebrated analysis of the (open) Toda lattice by Kostant and Symes [37, 54], where a general scheme is described for producing and integrating commuting Hamiltonian flows on Lie algebras (see also [1]). Since this beautiful circle of ideas may be unfamiliar to Riemannian geometers, we begin by rehearsing the main points of the theory in a form suitable for our applications.

1.1 Poisson structures and Lie algebra decompositions

Let M be a manifold. A *Poisson structure* on M is a Lie algebra structure on $C^\infty(M)$, with bracket denoted $(f,g) \mapsto \{f,g\}$, for which ad f is a derivation over multiplication:

$$\{f, gh\} = \{f, g\}h + g\{f, h\},$$

for all $f, g, h \in C^\infty(M)$. As a consequence, each $f \in C^\infty(M)$ gives rise to a *Hamiltonian vector field* X_f by

$$X_f g = \{f, g\}.$$

It follows from the Jacobi identity for $\{\,,\,\}$ that $f \mapsto X_f$ is a Lie algebra homomorphism $C^\infty(M) \to C^\infty(TM)$ (see [60] for more details).

For our basic example of a Poisson structure, let $\mathfrak{b}$ be a (real) Lie algebra with dual $\mathfrak{b}^*$. When $f \in C^\infty(\mathfrak{b}^*)$ and $x \in \mathfrak{b}^*$, we have $df_x \in \mathfrak{b}^{**} \cong \mathfrak{b}$ and, using this identification, we define a Poisson structure on $\mathfrak{b}^*$ by

$$\{f, g\}(x) = \langle x, [df_x, dg_x]\rangle.$$

The Ad^*-invariant functions on $\mathfrak{b}^*$ commute with respect to this Poisson structure but, unfortunately, they also commute with all other functions on $\mathfrak{b}^*$ and so have trivial Hamiltonian flows. To get non-trivial commuting flows, we must introduce some extra structure.

Suppose then that we have a Lie algebra $\mathfrak{g}$ which admits a (vector space) direct sum decomposition into subalgebras:

$$\mathfrak{g} = \mathfrak{k} \oplus \mathfrak{b}.$$

Suppose further that there is a (real) non-degenerate symmetric bilinear form on $\mathfrak{g}$, denoted $(\,,\,)$, which is invariant:

$$([\xi, \eta], \zeta) = -(\eta, [\xi, \zeta]).$$

The bilinear form induces a musical isomorphism $\mathfrak{g} \cong \mathfrak{g}^*$ and so, by restriction, an isomorphism $\mathfrak{k}^\circ \cong \mathfrak{b}^*$. Here $\mathfrak{k}^\circ$ is the polar of $\mathfrak{k}$ with respect to $(\,,\,)$:

$$\mathfrak{k}^\circ = \{\eta \in \mathfrak{g} : (\eta, \xi) = 0 \text{ for all } \xi \in \mathfrak{k}\}.$$

Thus $\mathfrak{k}^\circ$ acquires a Poisson structure from that of $\mathfrak{b}^*$ and this is the Poisson structure that will be important for us.

To describe the Poisson structure on $\mathfrak{k}^\circ$ explicitly, we introduce some notation. Let $\pi_\mathfrak{k}, \pi_\mathfrak{b}$ be the projections onto $\mathfrak{k}, \mathfrak{b}$ along $\mathfrak{b}, \mathfrak{k}$ respectively. Further, for $f \in C^\infty(\mathfrak{k}^\circ)$, let $\tilde{f}$ be some extension of f to $\mathfrak{g}$ and let $\nabla \tilde{f}$ denote its gradient with respect to $(\,,\,)$. It is easy to check that the projected gradient

$$\pi_\mathfrak{b} \nabla \tilde{f}$$

depends only on f. A straightforward calculation now shows that the Poisson bracket on $\mathfrak{k}^\circ$ is given by

$$\{f, g\}(\xi) = (\xi, [\pi_{\mathfrak{b}} \nabla \tilde{f}_\xi, \pi_{\mathfrak{b}} \nabla \tilde{g}_\xi]). \tag{1.1}$$

To see the Hamiltonian vector field of f, let $\pi_{\mathfrak{k}^\circ}$ denote projection onto $\mathfrak{k}^\circ$ along $\mathfrak{b}^\circ$. Then, using the invariance of $(\,,\,)$, we have

$$\begin{aligned}\{f, g\}(\xi) &= ([\xi, \pi_{\mathfrak{b}} \nabla \tilde{f}_\xi], \pi_{\mathfrak{b}} \nabla \tilde{g}_\xi) \\ &= (\pi_{\mathfrak{k}^\circ}[\xi, \pi_{\mathfrak{b}} \nabla \tilde{f}_\xi], \nabla \tilde{g}_\xi) \\ &= d\tilde{g}(\pi_{\mathfrak{k}^\circ}[\xi, \pi_{\mathfrak{b}} \nabla \tilde{f}_\xi]) = dg(\pi_{\mathfrak{k}^\circ}[\xi, \pi_{\mathfrak{b}} \nabla \tilde{f}_\xi]).\end{aligned}$$

Thus, for $\xi \in \mathfrak{k}^\circ$,

$$X_f(\xi) = \pi_{\mathfrak{k}^\circ}[\xi, \pi_{\mathfrak{b}} \nabla \tilde{f}_\xi]. \tag{1.2}$$

1.2 Poisson submanifolds

Let M be a Poisson manifold with bracket $\{\,,\,\}_M$. A submanifold V of M is a *Poisson submanifold* if

1. V has a Poisson structure $\{\,,\,\}_V$;

2. the inclusion $i : V \hookrightarrow M$ is a *Poisson morphism*:

$$\{f \circ i, g \circ i\}_V = \{f, g\}_M \circ i, \tag{1.3}$$

for $f, g \in C^\infty(M)$.

Since $i^* : C^\infty(M) \to C^\infty(V)$ is surjective, it is clear that (1.3) completely determines the Poisson structure on V. Thus a submanifold V is Poisson as soon as the prescription (1.3) gives a well-defined bracket $\{\,,\,\}_V$ on $C^\infty(V)$.

There is a simple condition for this to be the case: a submanifold V is Poisson if and only if the restriction of any Hamiltonian vector field to V is tangent to V (see [58, Lemma 1.1]). In the setting of Section 1.1, we use this and the form of the Hamiltonian vector fields (1.2) to obtain the following

Proposition 1.1 *Let $V \subset \mathfrak{k}^\circ$ be a linear subspace satisfying*

$$\pi_{\mathfrak{k}^\circ}[V, \mathfrak{b}] \subset V. \tag{1.4}$$

Then V is a Poisson submanifold of $\mathfrak{k}^\circ$.

Such considerations will be important when we come to discuss finite-dimensional subspaces of infinite-dimensional loop algebras.

Remark The above discussion is merely the translation into our setting of the fact that ad^* $\mathfrak{b}$-invariant subspaces of $\mathfrak{b}^*$ are Poisson submanifolds of $\mathfrak{b}^*$ which in turn follows from the fact that the symplectic leaves of $\mathfrak{b}^*$ are precisely the co-adjoint orbits.

1.3 Commuting Hamiltonians

We now come to the main point of our constructions: ad-invariant functions on $\mathfrak{g}$ restrict to Poisson commuting functions on $\mathfrak{k}^\circ$. Recall (cf. [60]) that $f \in C^\infty(\mathfrak{g})$ is ad-invariant if

$$df_\xi([\xi, \mathfrak{g}]) = 0,$$

or, equivalently,

$$[\nabla f_\xi, \xi] = 0, \qquad (1.5)$$

for all $\xi \in \mathfrak{g}$. If $f, g \in C^\infty(\mathfrak{g})$ are two ad-invariant functions, then their restrictions to $\mathfrak{k}^\circ$ have Poisson bracket

$$\begin{aligned}\{f,g\}(\xi) &= (\xi, [\pi_\mathfrak{b} \nabla f_\xi, \pi_\mathfrak{b} \nabla g_\xi]) \\ &= ([\xi, \pi_\mathfrak{b} \nabla f_\xi], \pi_\mathfrak{b} \nabla g_\xi).\end{aligned}$$

We now use the invariance of f which implies $[\xi, \pi_\mathfrak{k} \nabla f_\xi] = -[\xi, \pi_\mathfrak{b} \nabla f_\xi]$ to get

$$\begin{aligned}\{f,g\}(\xi) &= -([\xi, \pi_\mathfrak{k} \nabla f_\xi], \pi_\mathfrak{b} \nabla g_\xi) \\ &= -(\xi, [\pi_\mathfrak{k} \nabla f_\xi, \pi_\mathfrak{b} \nabla g_\xi]) \\ &= ([\xi, \pi_\mathfrak{b} \nabla g_\xi], \pi_\mathfrak{k} \nabla f_\xi) \\ &= -([\xi, \pi_\mathfrak{k} \nabla g_\xi], \pi_\mathfrak{k} \nabla f_\xi) = (\xi, [\pi_\mathfrak{k} \nabla f_\xi, \pi_\mathfrak{k} \nabla g_\xi])\end{aligned}$$

which last vanishes since $[\pi_\mathfrak{k} \nabla f_\xi, \pi_\mathfrak{k} \nabla g_\xi] \in \mathfrak{k}$ and $\xi \in \mathfrak{k}^\circ$. As for the corresponding Hamiltonian vector fields, from (1.2) and (1.5), we see that the restriction to $\mathfrak{k}^\circ$ of an ad-invariant f has Hamiltonian vector field

$$X_f(\xi) = \pi_{\mathfrak{k}^\circ}[\pi_\mathfrak{k} \nabla f_\xi, \xi] = [\pi_\mathfrak{k} \nabla f_\xi, \xi], \qquad (1.6)$$

where the last equality comes from $[\mathfrak{k}, \mathfrak{k}^\circ] \subset \mathfrak{k}^\circ$.

1.4 Group decompositions and the method of Symes

We have therefore found a family of commuting flows on $\mathfrak{k}^\circ$ and our next task is to integrate them. For this, we globalise the situation following Symes [54]: let G be a Lie group with Lie algebra $\mathfrak{g}$, let K, B be subgroups with Lie algebras $\mathfrak{k}, \mathfrak{b}$ and suppose that multiplication $K \times B \to G$ is a diffeomorphism onto. Thus any $g \in G$ may be uniquely written as a product

$$g = kb,$$

with $k \in K$, $b \in B$. We further demand that the adjoint action of G on $\mathfrak{g}$ preserve $(,)$ (which follows from the invariance of $(,)$ if G is connected) and we consider Ad-invariant functions on $\mathfrak{g}$: that is, functions $f \in C^\infty(\mathfrak{g})$ satisfying

$$f(\operatorname{Ad} g\, \xi) = f(\xi), \qquad (1.7)$$

for all $\xi \in \mathfrak{g}$, $g \in G$ (again, when G is connected, this is equivalent to (1.5)). Differentiating equation (1.7) with respect to ξ, we deduce that ∇f is equivariant in the sense that

$$\nabla f_{\operatorname{Ad} g \xi} = \operatorname{Ad} g \, \nabla f_\xi, \tag{1.8}$$

for all $\xi \in \mathfrak{g}$, $g \in G$.

So let $f \in C^\infty(\mathfrak{g})$ be Ad-invariant and fix an initial condition $\xi_o \in \mathfrak{k}^o$. Consider the geodesic $g : \mathbb{R} \to G$ given by $g(t) = \exp -t\nabla f_{\xi_o}$ and factorise to get curves $k : \mathbb{R} \to K$, $b : \mathbb{R} \to B$ with

$$g(t) = k(t)b(t),$$

for all $t \in \mathbb{R}$. Define $\xi : \mathbb{R} \to \mathfrak{k}^o$ by $\xi = \operatorname{Ad} k^{-1}\xi_c$. Then ξ is the integral curve of X_f with $\xi(0) = \xi_o$. Indeed,

$$\dot{\xi} = -[k^{-1}\dot{k}, \xi]$$

while

$$-\nabla f_{\xi_o} = \dot{g}g^{-1} = \dot{k}k^{-1} + \operatorname{Ad} k \, \dot{b}b^{-1}$$

so that taking $\operatorname{Ad} k^{-1}$ of both sides and projecting onto $\mathfrak{k}$ gives

$$k^{-1}\dot{k} = -\pi_{\mathfrak{k}} \operatorname{Ad} k^{-1}\nabla f_{\xi_o} = -\pi_{\mathfrak{k}}\nabla f_\xi,$$

where the last inequality follows from the equivariance (1.8) of ∇f. Thus

$$\dot{\xi} = [\pi_{\mathfrak{k}}\nabla f_\xi, \xi].$$

To summarise: we have seen how to equip $\mathfrak{k}^o$ with a Poisson structure for which the restriction of Ad-invariant functions on $\mathfrak{g}$ commute. Moreover, the corresponding Hamiltonian vector fields are of Lax form (1.6) and are integrated using the projection onto K of geodesics in G.

Remark In fact, the invariant inner product $(\,,\,)$, although present in all our applications, is not strictly necessary in the above development. If one replaces $\mathfrak{k}^o$ with $\mathfrak{k}^\perp \subset \mathfrak{g}^*$, the annihilator of $\mathfrak{k}$, then, as above, $\mathfrak{k}^\perp \cong \mathfrak{b}^*$ and so acquires a Poisson structure with respect to which the restrictions to $\mathfrak{k}^\perp$ of Ad^*-invariant functions on $\mathfrak{g}^*$ commute. Moreover, the Symes method of integration also works in this setting *mutatis mutandis*.

1.5 The Toda lattice

We conclude this discussion with a brief exposition of the original application of the theory to the open Toda lattice. Besides completing the circle of ideas, this will also give us an opportunity to develop concepts we shall need later.

Let $\mathfrak{g}$ be a complex semisimple Lie algebra and $\mathfrak{k}$ a compact real form. Fix a maximal toral subalgebra $\mathfrak{t} \subset \mathfrak{k}$ and set $\mathfrak{a} = i\mathfrak{t}$ where $i = \sqrt{-1}$. Then $\mathfrak{h} = \mathfrak{t} \oplus \mathfrak{a}$ is a Cartan subalgebra of $\mathfrak{g}$. Let $\Delta(\mathfrak{g}, \mathfrak{h}) \subset \mathfrak{h}^*$ be the root system of $\mathfrak{g}$ with respect to $\mathfrak{h}$ and fix a choice of positive roots $\Delta^+ \subset \Delta(\mathfrak{g}, \mathfrak{h})$ with corresponding simple roots $\alpha_1, \ldots, \alpha_l$.

The Toda lattice is a Hamiltonian system on $T^*\mathfrak{a} = \mathfrak{a} \times \mathfrak{a}^*$, equipped with its canonical symplectic structure. The Hamiltonian is given by

$$H(q,p) = |p|^2/2 + \sum_{i=1}^{l} e^{2\alpha_i(q)}$$

where the metric is given by the real part of the Killing form κ of $\mathfrak{g}$. One may think of this system as describing l particles moving on a line with exponential interactions governed by the Dynkin diagram of $\mathfrak{g}$, see [37]. (Note that the roots are *real* linear functionals on $\mathfrak{a}$.)

To put this Hamiltonian system into our frame-work, we follow Bloch–Flaschka–Ratiu [5] and use the Iwasawa decomposition of $\mathfrak{g}$: for each root $\alpha \in \Delta(\mathfrak{g}, \mathfrak{h})$, let $\mathfrak{g}^\alpha$ be the corresponding root space and define a nilpotent subalgebra $\mathfrak{n} \subset \mathfrak{g}$ by

$$\mathfrak{n} = \sum_{\alpha \in \Delta^+} \mathfrak{g}^\alpha.$$

We then have a direct sum decomposition

$$\mathfrak{g} = \mathfrak{k} \oplus \mathfrak{a} \oplus \mathfrak{n},$$

the *Iwasawa decomposition* of $\mathfrak{g}$ (cf. [30, page 275]). Moreover, there is a global analogue of this decomposition: let G be a Lie group with Lie algebra $\mathfrak{g}$, K the maximal compact subgroup with Lie algebra $\mathfrak{k}$ and A, N the analytic subgroups with Lie algebras $\mathfrak{a}, \mathfrak{n}$. Then multiplication $K \times A \times N \to G$ is a diffeomorphism onto.

Now $[\mathfrak{a}, \mathfrak{n}] \subset \mathfrak{n}$ so that $\mathfrak{b} = \mathfrak{a} \oplus \mathfrak{n}$ is a solvable subalgebra of $\mathfrak{g}$ and we may apply our preceding discussion to $\mathfrak{g} = \mathfrak{k} \oplus \mathfrak{b}$. As our invariant symmetric form, we take the imaginary part of the Killing form: $(\,,\,) = \text{Im}\,\kappa$. The Killing form is positive definite on $\mathfrak{a}$, negative definite on $\mathfrak{k}$ and vanishes identically on $\mathfrak{n}$ so that both $\mathfrak{k}$ and $\mathfrak{b}$ are isotropic for $(\,,\,)$ giving

$$\mathfrak{k}^\circ = \mathfrak{k}, \qquad \mathfrak{b}^\circ = \mathfrak{b}.$$

Thus we have a Poisson structure on $\mathfrak{k} \cong \mathfrak{b}^*$.

For $1 \leq i \leq l$, choose non-zero vectors $X_i \in \mathfrak{g}^{\alpha_i}$ and consider

$$\mathcal{O} = \{H + \sum_{i=1}^{l} a_i(X_i + \overline{X}_i) : H \in \mathfrak{t},\, a_i > 0\}.$$

It is not too difficult to see that $\mathcal{O}$ is a symplectic leaf of the Poisson structure on $\mathfrak{k}$. Indeed, the co-adjoint action of $B = AN$ on $\mathfrak{b}^*$ induces an action of B on $\mathfrak{k}$, the orbits of which are the symplectic leaves on $\mathfrak{k}$ and one can show that $\mathcal{O}$ is the B-orbit of $\sum_i (X_i + \overline{X}_i)$.

In fact, $\mathcal{O}$ is symplectomorphic with the phase space $T^*\mathfrak{a}$ of the Toda lattice: use $(\,,\,)$ to identify $\mathfrak{a}^*$ with $\mathfrak{t}$ and define $\phi: T^*\mathfrak{a} \to \mathcal{O}$ by

$$\phi(q,p) = p + \sum_{i=1}^{l} c_i e^{\alpha_i(q)}(X_i + \overline{X}_i),$$

where the constants c_i are positive constants to be chosen later. By checking the Poisson brackets of suitable co-ordinate functions, it is straight-forward to prove that ϕ is a Poisson morphism and hence a symplectomorphism for any choice of the constants c_i.

Now take as an Ad-invariant function $f : \mathfrak{g} \to \mathbb{R}$ half the negative of the real part of the Killing form:

$$f(\eta) = -\operatorname{Re} \kappa(\eta, \eta)/2.$$

On $\mathcal{O}$, we have

$$f(H + \sum_i a_i(X_i + \overline{X}_i)) = -\kappa(H,H)/2 - \sum_i a_i^2 \kappa(X_i, \overline{X}_i)$$

so that

$$\begin{aligned} f \circ \phi(q,p) &= |p|^2/2 - \sum_i c_i^2 \kappa(X_i, \overline{X}_i) e^{2\alpha_i(q)} \\ &= |p|^2/2 + \sum_i c_i^2 |X_i|^2 e^{2\alpha_i(q)}. \end{aligned}$$

Thus choosing $c_i = 1/|X_i|$, we see that $f \circ \phi$ is the Toda Hamiltonian.

The preceding theory may now be applied in a number of ways: the invariant polynomials on $\mathfrak{g}$ restrict to give l functionally independent Poisson commuting conserved quantities on $\mathcal{O}$ so that the Toda lattice is completely integrable in the sense of Liouville. Moreover, the method of Symes can be used to integrate the Toda flow leading eventually to explicit formulae for the flows in terms of matrix coefficients of the fundamental representations of G. For this and for further details on the ideas we have been discussing, the Reader is referred to the original papers and the books [29, 41].

2 Twisted loop algebras and zero-curvature equations

2.1 What the Maurer–Cartan equations are for

Let G be a Lie group with Lie algebra $\mathfrak{g}$ and let θ be the (left) Maurer–Cartan form of G. Thus θ is the $\mathfrak{g}$-valued 1-form on G given by

$$\theta_g(X) = (g^{-1})_* X \in T_e G = \mathfrak{g},$$

for $X \in T_g G$. A simple calculation using the left-invariance of θ shows that

$$d\theta + \tfrac{1}{2}[\theta \wedge \theta] = 0. \tag{2.1}$$

These equations are the *Maurer–Cartan equations*.

Now let $\phi : M \to G$ be a map of a manifold M and set $\alpha = \phi^*\theta$. Then ϕ pulls back (2.1) to give

$$d\alpha + \tfrac{1}{2}[\alpha \wedge \alpha] = 0 \tag{2.2}$$

so that α also satisfies the Maurer–Cartan equations.

Fundamental to what follows is that a partial converse of this is true. We have the following classical theorem (which is proved, for example, in [53]):

Theorem 2.1 *Let α be a $\mathfrak{g}$-valued 1-form on a simply-connected manifold M. Then $\alpha = \phi^*\theta$ for some map $\phi : M \to G$ if and only if α satisfies the Maurer–Cartan equations (2.2).*

In this case, ϕ is unique up to left translation by a constant element of G.

Otherwise said, (2.2) amounts to the assertion that the connection $d + \alpha$ on the trivial principal G-bundle $M \times G \to M$ has vanishing curvature (i.e., is *flat*) while Theorem 2.1 says that there is a gauge transformation ϕ which gauges this flat connection to the trivial connection. It is for this reason that the Maurer–Cartan equations are often called the *zero-curvature equations*.

Notation When G is a matrix group, $\phi^*\theta = \phi^{-1}d\phi$. Throughout this article we shall use this notation even when G is not a matrix group.

The application of the above constructions to the theory of harmonic maps begins with two observations. Firstly, it is well known [43, 56, 65, 66] that, for maps of a Riemann surface into suitable homogeneous spaces, the harmonic map equations are equivalent to the flatness of a certain loop of connections, that is, to certain solutions of the Maurer–Cartan equations (2.2) where α is a 1-form with values in an appropriate loop algebra.

On the other hand, solutions of (2.2) arise in the setting of the previous sections from the simultaneous integration of several commuting Hamiltonian flows on $\mathfrak{k}^\circ$. Indeed, let $f^1, \ldots, f^m$ be Ad-invariant functions on $\mathfrak{g}$ with corresponding Hamiltonian vector fields X_i on $\mathfrak{k}^\circ$. Fixing an initial condition $\xi_o \in \mathfrak{k}^\circ$, we may integrate these vector fields (when they are complete) to get $\xi : \mathbb{R}^m \to \mathfrak{k}^\circ$ satisfying

$$d\xi = \sum_i X_i(\xi)\, dt^i$$
$$\xi(0) = \xi_o.$$

Moreover, the Symes method shows that $\xi = \operatorname{Ad} k^{-1}\xi_o$ where $k : \mathbb{R}^m \to K$ satisfies

$$k^{-1}dk = -\pi_\mathfrak{k} \operatorname{Ad} k^{-1} \sum_i \nabla f^i_{\xi_o}\, dt^i = -\pi_\mathfrak{k} \sum_i \nabla f^i_\xi\, dt^i.$$

In particular, $\alpha = -\pi_\mathfrak{k} \sum_i \nabla f^i_\xi\, dt^i$ is the pull-back by k of the Maurer–Cartan form on K and so satisfies the Maurer–Cartan equations.

Harmonic maps via Adler–Kostant–Symes theory

The point now is that when $\mathfrak{k}$ is an appropriate loop algebra and the f^i are suitably chosen, the 1-form α is of precisely the right form to produce a harmonic map. Moreover, the map k will be an "extended solution" as described in the contribution of Guest–Ohnita to this volume [28].

2.2 Passage to infinite dimensions

Our intention is to carry out this programme for a family of twisted loop algebras. We therefore begin by considering the adjustments that must be made when carrying the Adler–Kostant–Symes scheme over to an infinite-dimensional setting. So let $\mathcal{G}$ be a reflexive Banach Lie algebra and suppose we have

(i) a decomposition $\mathcal{G} = \mathcal{K} \oplus \mathcal{B}$ into *closed* subalgebras;

(ii) a continuous ad-invariant symmetric bilinear form $(\,,\,)$ which is non-degenerate in the *weak* sense that the induced map $i : \mathcal{G} \to \mathcal{G}^*$ is injective;

(iii) $\mathcal{K}, \mathcal{B}$ are both isotropic for $(\,,\,)$.

It follows that $\mathcal{B}$ is reflexive so that we may canonically identify $\mathcal{B}$ with $\mathcal{B}^{**}$ and so obtain a Poisson structure on $\mathcal{B}^*$ as before. By virtue of the isotropy of $\mathcal{K}$, the injection $\mathcal{G} \to \mathcal{G}^*$ restricts to give an injection $\mathcal{K} \to \mathcal{B}^*$ but this need no longer be an isomorphism and so we cannot transfer the Poisson structure of $\mathcal{B}^*$ to $\mathcal{K}$.

However, suppose that $V \subset \mathcal{K}$ is a linear subspace satisfying

1. $\pi_\mathcal{K}[V, \mathcal{B}] \subset V$;

2. $i(V)$ is closed and has a closed complement in $\mathcal{B}^*$ (trivially true if V is finite-dimensional).

In this case, V has a Poisson structure for which $i : V \to \mathcal{B}^*$ is a Poisson morphism. Indeed, by virtue of the second hypothesis, $i^* : C^\infty(\mathcal{B}^*) \to C^\infty(V)$ is a surjection so it suffices to show that the prescription

$$\{f \circ i, g \circ i\}_V = \{f, g\}_{\mathcal{B}^*} \circ i$$

is well-defined. This, in turn, follows from the first hypothesis as in Proposition 1.1.

We can now proceed as before with the *caveat* that not all functions on $\mathcal{G}$ have gradients with respect to $(\,,\,)$. This prompts the following terminology:

Definition $F \in C^\infty(\mathcal{G})$ is *admissible* if it has a gradient with respect to $(\,,\,)$.

For admissible functions $F, G \in C^\infty(\mathcal{G})$, we can compute the Poisson bracket of their restrictions to V as before:

$$\{F, G\}(\xi) = (\xi, [\pi_\mathcal{B} \nabla F_\xi, \pi_\mathcal{B} \nabla G_\xi]),$$

for $\xi \in V$.

Finally, an admissible function $F \in C^\infty(\mathcal{G})$ is *invariant* if

$$[\xi, \nabla F_\xi] = 0,$$

for all $\xi \in \mathcal{G}$. For such functions, the analysis of the previous section goes through as before and is summarised in the following theorem.

Theorem 2.2 *Let F and G be invariant, admissible functions on $\mathcal{G}$. Then their restrictions to V Poisson commute.*

Moreover, the Hamiltonian vector field corresponding to F is given by

$$X_F(\xi) = [\pi_\mathcal{K} \nabla F_\xi, \xi],$$

for $\xi \in V$.

2.3 Twisted loop algebras

To introduce the loop algebras to which these ideas will be applied, we begin by fixing the following ingredients:

1. A compact semisimple Lie algebra $\mathfrak{g}$.

2. An automorphism $\tau : \mathfrak{g} \to \mathfrak{g}$ of finite order k with fixed set $\mathfrak{k}$.

3. The primitive k-th root of unity $\omega = e^{2\pi i/k}$.

In addition, we fix an Iwasawa decomposition of $\mathfrak{k}^\mathbb{C}$:

$$\mathfrak{k}^\mathbb{C} = \mathfrak{k} \oplus \mathfrak{b},$$

where $\mathfrak{b}$ is a solvable subalgebra of $\mathfrak{k}^\mathbb{C}$. Such a decomposition exists since $\mathfrak{k}$ is compact so that $\mathfrak{k}^\mathbb{C}$ is reductive.

Define a loop algebra by

$$\Lambda \mathfrak{g}_\tau^\mathbb{C} = \{\xi : S^1 \to \mathfrak{g}^\mathbb{C} : \xi(\omega\lambda) = \tau\,\xi(\lambda) \text{ for all } \lambda \in S^1\}$$

and equip it with the Sobolev H^r-topology for some $r > \frac{1}{2}$. Thus $\Lambda \mathfrak{g}_\tau^\mathbb{C}$ is a Banach (indeed, Hilbert) Lie algebra under point-wise bracket. There is an analogue of the Iwasawa decomposition for $\Lambda \mathfrak{g}_\tau^\mathbb{C}$: let $\Lambda \mathfrak{g}_\tau$ be the real form

$$\Lambda \mathfrak{g}_\tau = \{\xi \in \Lambda \mathfrak{g}_\tau^\mathbb{C} : \xi : S^1 \to \mathfrak{g}\}$$

and define a complementary subalgebra by

$$\Lambda_+ \mathfrak{g}_\tau^\mathbb{C} = \{\xi \in \Lambda \mathfrak{g}_\tau^\mathbb{C} : \xi \text{ extends holomorphically to } \xi : D \to \mathfrak{g}^\mathbb{C} \text{ and } \xi(0) \in \mathfrak{b}\},$$

where D is the disc $\{|\lambda| < 1\}$.

A loop $\xi \in \Lambda \mathfrak{g}_\tau^\mathbb{C}$ has a Fourier decomposition:

$$\xi = \sum_{n \in \mathbb{Z}} \xi_{-n} \lambda^n$$

Harmonic maps via Adler–Kostant–Symes theory

with each $\xi_n \in \mathfrak{g}^{\mathbb{C}}$ satisfying $\tau \xi_n = \omega^{-n} \xi_n$. (Our convention of labelling the coefficient of λ^n by $-n$ is to prevent a welter of negative indices in subsequent formulae.) In terms of this decomposition, the conjugation on $\Lambda \mathfrak{g}^{\mathbb{C}}_\tau$ across the real form $\Lambda \mathfrak{g}_\tau$ is given by

$$\overline{\xi} = \sum_{n \in \mathbb{Z}} \overline{\xi_{-n}} \lambda^{-n}$$

where the conjugation in $\mathfrak{g}^{\mathbb{C}}$ is with respect to the real form $\mathfrak{g}$. Moreover, our subalgebras are given by

$$\Lambda \mathfrak{g}_\tau = \{\xi \in \Lambda \mathfrak{g}^{\mathbb{C}}_\tau : \overline{\xi_n} = \xi_{-n}\};$$
$$\Lambda_+ \mathfrak{g}^{\mathbb{C}}_\tau = \{\xi \in \Lambda \mathfrak{g}^{\mathbb{C}}_\tau : \xi_n = 0 \text{ for } n > 0; \xi_0 \in \mathfrak{b}\}.$$

From this it is clear that

$$\Lambda \mathfrak{g}^{\mathbb{C}}_\tau = \Lambda \mathfrak{g}_\tau \oplus \Lambda_+ \mathfrak{g}^{\mathbb{C}}_\tau$$

is a vector space decomposition into closed subalgebras.

Now let κ be the Killing form of $\mathfrak{g}^{\mathbb{C}}$ and introduce a (weakly) non-degenerate invariant symmetric bilinear form $(\,,\,)$ on $\Lambda \mathfrak{g}^{\mathbb{C}}_\tau$ by taking the imaginary part of the L^2 inner product:

$$(\xi, \eta) = \mathrm{Im} \int_{S^1} \kappa(\xi, \eta).$$

Observe that both $\Lambda \mathfrak{g}_\tau$ and $\Lambda_+ \mathfrak{g}^{\mathbb{C}}_\tau$ are isotropic for this form so that we are in the situation of the previous discussion.

Let $\pi_{\mathcal{K}}, \pi_{\mathcal{B}}$ be the projections onto $\Lambda \mathfrak{g}_\tau$, $\Lambda_+ \mathfrak{g}^{\mathbb{C}}_\tau$ along $\Lambda_+ \mathfrak{g}^{\mathbb{C}}_\tau$, $\Lambda \mathfrak{g}_\tau$ respectively. To calculate these, write $\xi \in \Lambda \mathfrak{g}^{\mathbb{C}}_\tau$ as

$$\xi = \xi^+ + \xi^0 + \xi^-,$$

where

$$\xi^+ = \sum_{n>0} \xi_{-n} \lambda^n; \qquad \xi^0 = \xi_0 \in \mathfrak{k}^{\mathbb{C}}; \qquad \xi^- = \sum_{n<0} \xi_{-n} \lambda^n.$$

Further, write $\xi^0 = \xi^0_{\mathfrak{k}} + \xi^0_{\mathfrak{b}}$ for the Iwasawa decomposition of ξ^0 into $\mathfrak{k}$ and $\mathfrak{b}$ parts. Then

$$\xi = (\xi^- + \xi^0_{\mathfrak{k}} + \overline{\xi^-}) + (\xi^+ - \overline{\xi^-} + \xi^0_{\mathfrak{b}})$$

and the first summand is in $\Lambda \mathfrak{g}_\tau$ while the second is in $\Lambda_+ \mathfrak{g}^{\mathbb{C}}_\tau$. In particular, we have

$$\pi_{\mathcal{K}} \xi = \xi^- + \xi^0_{\mathfrak{k}} + \overline{\xi^-}. \qquad (2.3)$$

Recall that finite-dimensional subspaces $V \subset \Lambda \mathfrak{g}_\tau$ satisfying

$$\pi_{\mathcal{K}}[V, \Lambda_+ \mathfrak{g}^{\mathbb{C}}_\tau] \subset V$$

inherit a Poisson structure from $(\Lambda_+ \mathfrak{g}_\tau^{\mathbb{C}})^*$. Our main examples of such subspaces are the spaces of polynomial loops given by

$$\Lambda_d = \{\xi \in \Lambda\mathfrak{g}_\tau : \xi_n = 0 \text{ for all } |n| > d\},$$

where $d \in \mathbb{N}$. Indeed, if $\xi \in \Lambda_d$ and $\eta \in \Lambda_+\mathfrak{g}_\tau^{\mathbb{C}}$ then $\lambda^d \xi$ extends holomorphically to D so that $[\lambda^d \xi, \eta]$ does also. This means that $[\xi, \eta] = \lambda^{-d}[\lambda^d \xi, \eta]$ has a pole of order at most d at $0 \in D$ so that

$$\pi_\mathcal{K}[\xi,\eta] = [\xi,\eta]^- + [\xi,\eta]_{\mathfrak{k}}^0 + \overline{[\xi,\eta]^-} \in \Lambda_d,$$

as required. We have therefore proved:

Proposition 2.3 *For $d \in \mathbb{N}$, Λ_d is a Poisson manifold on which the restrictions of invariant admissible functions on $\Lambda\mathfrak{g}_\tau^{\mathbb{C}}$ Poisson commute.*

2.4 Zero-curvature equations

We now come to our main construction: for $d \in \mathbb{N}$, we will integrate a pair of commuting Hamiltonian flows on the finite-dimensional Poisson manifold Λ_d and, from that data, construct a $\Lambda\mathfrak{g}_\tau$-valued 1-form which solves the Maurer–Cartan equations. In Section 3, we shall see how such 1-forms are related to harmonic maps.

Fix $d \in \mathbb{N}$ with $d \equiv 1 \bmod k$. Define $f^1, f^2 \in C^\infty(\Lambda\mathfrak{g}_\tau^{\mathbb{C}})$ by

$$f^1(\xi) = -\tfrac{1}{2}\operatorname{Im}\int_{S^1} \lambda^{d-1}\kappa(\xi,\xi), \qquad f^2(\xi) = -\tfrac{1}{2}\operatorname{Re}\int_{S^1} \lambda^{d-1}\kappa(\xi,\xi).$$

It is easy to see that the f^i are invariant admissible functions with gradients given by

$$\nabla f_\xi^1 = -\lambda^{d-1}\xi, \qquad \nabla f_\xi^2 = -i\lambda^{d-1}\xi,$$

(here we have used $d - 1 \equiv 0 \bmod k$ to ensure that the ∇f^i so defined take values in $\Lambda\mathfrak{g}_\tau^{\mathbb{C}}$).

We restrict f^1, f^2 to Λ_d and let X_1, X_2 denote the corresponding Hamiltonian vector fields. Thus

$$X_1(\xi) = [\xi, \pi_\mathcal{K}\lambda^{d-1}\xi]; \qquad X_2(\xi) = [\xi, \pi_\mathcal{K}i\lambda^{d-1}\xi],$$

for $\xi \in \Lambda_d$. That these vector fields are complete is a consequence of two observations: firstly, the L^2 inner product

$$\langle \xi, \eta \rangle = \int_{S^1} \kappa(\xi, \eta).$$

is negative definite on $\Lambda\mathfrak{g}_\tau$ (and thus on Λ_d). Secondly, the X_i are of Lax form, i.e., of the form

$$X(\xi) = [\xi, A(\xi)].$$

For any integral curve ξ of such an X we have

$$\begin{aligned} d/dt|\xi|^2 &= 2\langle \dot\xi, \xi\rangle \\ &= 2\langle [\xi, A(\xi)], \xi\rangle = -2\langle A(\xi), [\xi, \xi]\rangle = 0\end{aligned}$$

so that ξ takes values in a sphere in Λ_d whence X is complete.

Let X_1^s, X_2^t be the flows corresponding to the X_i. Since f^1 and f^2 Poisson commute, the commutator $[X_1, X_2]$ vanishes so that

$$X_1^s \circ X_2^t = X_2^t \circ X_1^s,$$

for all $s, t \in \mathbb{R}$. Now fix an initial condition $\xi_o \in \Lambda_d$ and define $\xi : \mathbb{R}^2 \to \Lambda_d$ by

$$\xi(s,t) = X_1^s \circ X_2^t(\xi_o).$$

Then ξ satisfies

$$d\xi = X_1(\xi)\,ds + X_2(\xi)\,dt; \qquad \xi(0) = \xi_o. \tag{2.4}$$

Our interest in such maps comes from the possibility of constructing loops of flat connections from them:

Theorem 2.4 *Let $\xi : \mathbb{R}^2 \to \Lambda_d$ be the solution of (2.4) and define a $\Lambda\mathfrak{g}_\tau$-valued 1-form α on $\mathbb{R}^2$ by*

$$\alpha = -\pi_{\mathcal{K}}(\nabla f_\xi^1 \,ds + \nabla f_\xi^2 \,dt).$$

Then α solves the Maurer–Cartan equations:

$$d\alpha + \tfrac{1}{2}[\alpha \wedge \alpha] = 0.$$

This theorem can be proved by direct calculation or by recourse to the method of Symes. The first approach is an exercise for the Reader while the second will be carried out in Section 4.

We have thus obtained solutions to the Maurer–Cartan equations by integrating Hamiltonian ordinary differential equations on some Λ_d. Our application to harmonic maps will depend on the precise algebraic form of the solutions so obtained and it is to this that we now turn. We have

$$\alpha = -\pi_{\mathcal{K}}(\nabla f_\xi^1\,ds + \nabla f_\xi^2\,dt) = \pi_{\mathcal{K}}(\lambda^{d-1}\xi\,ds + i\lambda^{d-1}\xi\,dt).$$

This simplifies if we introduce the complex co-ordinate $z = s + it$ on $\mathbb{R}^2$:

$$\alpha = \pi_{\mathcal{K}}(\lambda^{d-1}\xi\,dz).$$

We now use (2.3), together with the fact that ξ takes values in Λ_d, to get

$$\begin{aligned}\alpha &= (\lambda^{d-1}\xi)^-\,dz + (\lambda^{d-1}\xi\,dz)_t^0 + \overline{(\lambda^{d-1}\xi)^-}\,d\bar z \\ &= \lambda^{-1}\xi_d\,dz + (\xi_{d-1}\,dz)_t + \lambda\xi_{-d}\,d\bar z.\end{aligned}$$

To compute the λ-independent term, we must understand the Iwasawa decomposition of $\mathfrak{k}^{\mathbb{C}}$: let $\mathfrak{t}$ be the given maximal torus in $\mathfrak{k}$, $\mathfrak{n}$ the nilpotent subalgebra given by the positive root spaces and set $\mathfrak{h} = \mathfrak{t}^{\mathbb{C}}$. Then we have

$$\mathfrak{k}^{\mathbb{C}} = \mathfrak{n} \oplus \mathfrak{h} \oplus \bar{\mathfrak{n}}, \qquad \mathfrak{k} = (i\mathfrak{t}) \oplus \mathfrak{n}.$$

Corresponding to this decomposition of $\mathfrak{k}^{\mathbb{C}}$, write $\eta \in \mathfrak{k}^{\mathbb{C}}$ as

$$\eta = \eta_{\mathfrak{n}} + \eta_{\mathfrak{h}} + \eta_{\bar{\mathfrak{n}}}.$$

It is then easy to check that

$$(\eta\, dz)_{\mathfrak{k}} = (\eta_{\bar{\mathfrak{n}}} + \tfrac{1}{2}\eta_{\mathfrak{h}})\, dz + \overline{(\eta_{\bar{\mathfrak{n}}} + \tfrac{1}{2}\eta_{\mathfrak{h}})}\, d\bar{z}. \tag{2.5}$$

This suggests that we define $r : \mathfrak{k}^{\mathbb{C}} \to \mathfrak{k}^{\mathbb{C}}$ by

$$r(\eta) = \eta_{\bar{\mathfrak{n}}} + \tfrac{1}{2}\eta_{\mathfrak{h}}.$$

Then, denoting the $(1,0)$-part of α by α', we conclude that

$$\alpha' = (\lambda^{-1}\xi_d + r(\xi_{d-1}))\, dz. \tag{2.6}$$

We can now summarise our discussion in the following theorem:

Theorem 2.5 *For each $d \equiv 1 \bmod k$ and $\xi_o \in \Lambda_d$, there is a unique solution $\xi : \mathbb{R}^2 \to \Lambda_d$ to*

$$\frac{\partial \xi}{\partial z} = [\xi, \lambda^{-1}\xi_d + r(\xi_{d-1})]; \qquad \xi(0) = \xi_o \tag{2.7}$$

and then the $\Lambda \mathfrak{g}_\tau$-valued 1-form α given by

$$\alpha = (\lambda^{-1}\xi_d + r(\xi_{d-1}))\, dz + (\lambda \xi_{-d} + \overline{r(\xi_{d-1})})\, d\bar{z} \tag{2.8}$$

satisfies the Maurer–Cartan equations.

Remark When $\mathfrak{g}$ has rank greater then one, we may generalise these methods to produce Maurer–Cartan solutions on $\mathbb{C}^n$ by simultaneous integration of $2n$ commuting flows. Briefly, the idea is to consider ad-invariant polynomials $P : \mathfrak{g}^{\mathbb{C}} \to \mathbb{C}$ and define functions f^1, f^2 on $\Lambda \mathfrak{g}_\tau^{\mathbb{C}}$ by

$$f^1(\xi) = \operatorname{Im} \int_{S^1} \lambda^m P(\xi); \qquad f^2(\xi) = \operatorname{Re} \int_{S^1} \lambda^m P(\xi)$$

which are easily seen to be admissible and invariant. Any collection of such functions, with P ranging over ad-invariant polynomials, will Poisson commute and so give rise to a solution to the Maurer–Cartan equations. By choosing the exponents m appropriately, one obtains solutions α with the same kind of λ-dependence as in (2.6). From such α one can construct *pluriharmonic* maps of $\mathbb{C}^n$.

We shall not pursue this topic further in this chapter but, instead, refer the Reader to the discussion in [15].

2.5 The r-matrix approach

Elsewhere in this volume (and in the papers [25, 15]), a different approach to the construction of commuting flows on loop algebras is employed, to wit: the r-matrix formalism of Reyman–Semenov-Tian-Shansky [49]. In this section, we take a break from our main development to briefly review this method and show how it gives the same Poisson structures on $\Lambda \mathfrak{g}_\tau$ as the Adler–Kostant–Symes scheme. For brevity of exposition, we shall ignore questions of completeness of metrics, admissibility of functions and so on.

We begin with a (possibly infinite-dimensional) Lie algebra $\mathcal{G}$. A linear map $R : \mathcal{G} \to \mathcal{G}$ is called an r-matrix if the bracket defined by

$$[\xi, \eta]_R = [R\xi, \eta] + [\xi, R\eta]$$

satisfies the Jacobi identity. This is easily seen to be the case if R solves the (modified) classical Yang–Baxter equations:

$$R[\xi, \eta]_R - [R\xi, R\eta] = a[\xi, \eta], \tag{2.9}$$

for some fixed $a \in \mathbb{C}$. Given an r-matrix, $\mathcal{G}$ acquires a second Lie algebra structure so that the construction which opens Section 1.1 equips $\mathcal{G}^*$ with a second Poisson structure.

Suppose also that $\mathcal{G}$ is equipped with an ad-invariant inner product $\langle \, , \, \rangle$. We use this to transfer the new Poisson structure from $\mathcal{G}^*$ to $\mathcal{G}$ and thus arrive at a Poisson structure on $\mathcal{G}$ with bracket:

$$\{f, g\}_R(\xi) = \langle \xi, [\operatorname{grad} f_\xi, \operatorname{grad} g_\xi]_R \rangle, \tag{2.10}$$

where grad denotes the gradient with respect to $\langle \, , \, \rangle$.

Such Poisson structures enjoy many of the properties of those constructed by the Adler–Kostant–Symes scheme. In particular, ad-invariant functions on $\mathcal{G}$ Poisson commute and give rise to solutions of the Maurer–Cartan equations.

Let us show that both these constructions coincide in our setting: define a linear map $R : \Lambda \mathfrak{g}_\tau \to \Lambda \mathfrak{g}_\tau$ by

$$R\xi = i\xi^+ + R_0 \xi^0 - i\xi^-.$$

Here $R_0 : \mathfrak{k} \to \mathfrak{k}$ is given by

$$R_0 \eta = i\eta_\mathfrak{n} - i\eta_{\bar{\mathfrak{n}}}.$$

One observes that the $\pm i$-eigenspaces of R are subalgebras of $\Lambda \mathfrak{g}_\tau^{\mathbb{C}}$ while the 0-eigenspace is the abelian subalgebra $\mathfrak{h}$. From this, it is easy to check that R solves the modified classical Yang–Baxter equations (2.9) with $a = -1$ and so is an r-matrix.

Recall that $\Lambda \mathfrak{g}_\tau$ carries the ad-invariant, negative definite inner product,

$$\langle \xi, \eta \rangle = \int_{S^1} \kappa(\xi, \eta)$$

so that R induces a Poisson structure on $\Lambda\mathfrak{g}_\tau$.

We claim that this Poisson structure is the same as the one described in Section 2.3. To see this, it suffices to compare Hamiltonian vector fields. So let f be some function on $\Lambda\mathfrak{g}_\tau$ and recall that its Hamiltonian vector field with respect to the Adler–Kostant–Symes Poisson structure is given by

$$\begin{aligned}\pi_\mathcal{K}[\xi,\pi_B\nabla f_\xi] &= [\xi,\pi_B\nabla f_\xi] - \pi_B[\xi,\pi_B\nabla f_\xi]\\ &= [\xi,\pi_B\nabla f_\xi] - \pi_B[\xi,\nabla f_\xi],\end{aligned}$$

for $\xi \in \Lambda\mathfrak{g}_\tau$. Here the last equality is due to the fact that $\pi_B[\xi,\pi_\mathcal{K}\nabla f_\xi]$ vanishes since $\Lambda\mathfrak{g}_\tau$ is a subalgebra.

On the other hand, we note that R is skew-symmetric for $\langle,\rangle$ and use this, together with (2.10), to see that the Hamiltonian vector field of f with respect to the R-Poisson structure is given by

$$[\xi, R\operatorname{grad} f_\xi] - R[\xi,\operatorname{grad} f_\xi] = [\xi,(R+i)\operatorname{grad} f_\xi] - (R+i)[\xi,\operatorname{grad} f_\xi].$$

That these vector fields are the same is an immediate consequence of two easily verified facts. Firstly, the relation between the two gradients is given by

$$\operatorname{grad} f_\xi = \tfrac{1}{2i}(\nabla f_\xi - \overline{\nabla f_\xi}),$$

for $\xi \in \Lambda\mathfrak{g}_\tau$. Secondly, one deduces from (2.3) and (2.5) that

$$\pi_B\xi = \tfrac{1}{2i}(R+i)(\xi - \bar\xi),$$

for $\xi \in \Lambda\mathfrak{g}_\tau^\mathbf{C}$. From this it is clear that, for $\xi \in \Lambda\mathfrak{g}_\tau$,

$$[\xi,\pi_B\nabla f_\xi] - \pi_B[\xi,\nabla f_\xi] = [\xi,(R+i)\operatorname{grad} f_\xi] - (R+i)[\xi,\operatorname{grad} f_\xi]$$

so that the Poisson structures coincide.

Example Let $\mathfrak{g} = \mathfrak{so}(5)$ and let $\tau : \mathfrak{g} \to \mathfrak{g}$ be conjugation by

$$\begin{pmatrix} 1 & & & & \\ & 1 & & & \\ & & -1 & & \\ & & & -1 & \\ & & & & -1 \end{pmatrix}.$$

Then τ is an automorphism of order 4 with fixed set a maximal torus $\mathfrak{t}$ of $\mathfrak{so}(5)$. The Iwasawa decomposition of $\mathfrak{t}^\mathbf{C}$ is $\mathfrak{t}^\mathbf{C} = \mathfrak{t} \oplus i\mathfrak{t}$ with no n-part so that the r-matrix in this case is given by

$$R\xi = i\xi^+ - i\xi^-.$$

This is (up to a factor of $\tfrac{1}{2}$) the r-matrix used in [25] to construct commuting flows from which all minimal non-superminimal 2-tori are obtained.

Remark In fact the r-matrix formalism constitutes a strict generalisation of the Adler–Kostant–Symes scheme. In the notation of Section 1.1, one may define an r-matrix on $\mathfrak{g}$ by setting $R = \frac{1}{2}(\pi_\mathfrak{b} - \pi_\mathfrak{k})$ and then check that $\mathfrak{k}^\circ$ is a Poisson submanifold of $\mathfrak{g}$ with respect to the Poisson structure obtained from R. Moreover, the induced Poisson structure on $\mathfrak{k}^\circ$ coincides with that provided by the Adler–Kostant–Symes scheme (cf. [41]).

However, when it can be applied, the Adler–Kostant–Symes scheme has advantages. In particular, the symplectic leaves are known to be the B-orbits on $\mathfrak{k}^\circ$ and so Poisson submanifolds are more readily identified.

3 Harmonic and primitive maps

We now turn from symplectic geometry to Riemannian geometry and apply the theory of the preceding sections to the construction of harmonic maps from $\mathbb{R}^2$ into certain homogeneous spaces. We begin by describing the class of such spaces with which we shall be concerned.

3.1 Symmetric and k-symmetric spaces

Recall the data of Section 2.3: a compact semisimple Lie algebra $\mathfrak{g}$; an automorphism $\tau : \mathfrak{g} \to \mathfrak{g}$ of order k with fixed set $\mathfrak{k}$; the primitive k-th root of unity $\omega = e^{2\pi i/k}$. We have an eigenspace decomposition of $\mathfrak{g}^{\mathbb{C}}$:

$$\mathfrak{g}^{\mathbb{C}} = \sum_{i \in \mathbb{Z}_k} \mathfrak{g}_i$$

where $\mathfrak{g}_i$ is the ω^i-eigenspace of τ. Clearly, $\mathfrak{g}_0 = \mathfrak{k}^{\mathbb{C}}$, $\overline{\mathfrak{g}_i} = \mathfrak{g}_{-i}$ and

$$[\mathfrak{g}_i, \mathfrak{g}_j] \subset \mathfrak{g}_{i+j},$$

for all $i, j \in \mathbb{Z}_k$ (here, of course, all arithmetic is modulo k). In particular, define $\mathfrak{m} \subset \mathfrak{g}$ by

$$\mathfrak{m}^{\mathbb{C}} = \sum_{i \in \mathbb{Z}_k \setminus \{0\}} \mathfrak{g}_i.$$

Then

$$\mathfrak{g} = \mathfrak{k} \oplus \mathfrak{m} \tag{3.1}$$

and $[\mathfrak{k}, \mathfrak{m}] \subset \mathfrak{m}$ so that (3.1) is a reductive decomposition. Moreover, when $k = 2$, $\mathfrak{m}^{\mathbb{C}} = \mathfrak{g}_1$ so that

$$[\mathfrak{m}, \mathfrak{m}] \subset \mathfrak{k}$$

and, in this case, (3.1) is a symmetric decomposition of $\mathfrak{g}$.

Let G be a compact semisimple Lie group with Lie algebra $\mathfrak{g}$ and suppose that τ exponentiates to give an order k automorphism, also called τ, of G. Further let $(G^\tau)_0 \subset K \subset G^\tau$ so that K has Lie algebra $\mathfrak{k}$.

Consider the coset space $N = G/K$ with base-point $o = eK$. Define $\hat{\tau}: N \to N$ by
$$\hat{\tau}(g \cdot o) = \tau(g) \cdot o,$$
for $g \in G$. Similarly, for $x = g \cdot o \in N$, define $\hat{\tau}_x : N \to N$ by
$$\hat{\tau}_x = g \circ \hat{\tau} \circ g^{-1}.$$
Then each $\hat{\tau}_x$ is a diffeomorphism or order k having x as an isolated fixed point. Moreover, we may use the Killing form of $\mathfrak{g}$ to equip N with a metric for which each of the $\hat{\tau}_x$ is an isometry so that N has the structure of a (regular) k-symmetric space in the sense of [38]. Of course, the 2-symmetric spaces are just the familiar Riemannian symmetric spaces of compact type.

The algebraic situation in $\mathfrak{g}$ transfers to the reductive homogeneous space N as follows: for $x = g \cdot o \in N$, the map $\mathfrak{g} \to T_x N$ given by
$$\xi \mapsto \left.\frac{d}{dt}\right|_{t=0} \exp t\xi \cdot x$$
restricts to an isomorphism $\operatorname{Ad} g \, \mathfrak{m} \to T_x N$. The inverse map $\beta_x : T_x N \to \operatorname{Ad} g \, \mathfrak{m} \subset \mathfrak{g}$ may be viewed as a $\mathfrak{g}$-valued 1-form β on N which, after [16], we call the *Maurer–Cartan form* of N.

This prompts in part the following notation: if $\mathfrak{l} \subset \mathfrak{g}^{\mathbb{C}}$ is an $\operatorname{Ad} K$-invariant subspace, let $[\mathfrak{l}]$ denote the subbundle of the trivial bundle $\underline{\mathfrak{g}^{\mathbb{C}}} = N \times \mathfrak{g}^{\mathbb{C}}$ defined by
$$[\mathfrak{l}]_{g \cdot o} = \operatorname{Ad} g \, \mathfrak{l}.$$
In particular, $\beta: TN \to [\mathfrak{m}]$ is a bundle isomorphism while $[\mathfrak{k}]_x$ is the Lie algebra of the stabiliser of $x \in N$. Moreover, the ω^i-eigenspace of $d\hat{\tau}_x$ at $x \in N$ corresponds to $[\mathfrak{g}_i]_x$ under the isomorphism β_x.

3.2 Harmonic maps

A map $\phi: M \to N$ of Riemannian manifolds is *harmonic* if it extremizes the energy functional
$$E(\phi) = \tfrac{1}{2} \int_D |d\phi|^2 \, dvol \tag{3.2}$$
on all compact sub-domains $D \subset M$ (for an introduction to harmonic maps in this volume, see [60]). In case that the target is a reductive homogeneous space $N = G/K$, the corresponding Euler–Lagrange equations have a particularly simple form. Indeed, if such an N has metric induced from an invariant metric on $\mathfrak{g}$, we may use these metrics to identify TN with T^*N and $\mathfrak{g}^*$ with $\mathfrak{g}$. In this way, TN becomes a symplectic manifold on which G acts symplectically and the moment map for this action is precisely the Maurer–Cartan form $\beta: TN \to \mathfrak{g} \cong \mathfrak{g}^*$. A Noether analysis of the energy functional (3.2) now gives [45] (cf. also [60]):

Harmonic maps via Adler–Kostant–Symes theory

Lemma 3.1 ϕ *is harmonic if and only if the pull-back of the Maurer–Cartan form is co-closed:*

$$d^* \phi^* \beta = 0.$$

We can gain a different perspective on the harmonic map equations by lifting everything to the frame bundle. Let $\pi : G \to N = G/K$ be the coset projection and $\phi : M \to N$ be a map. A *framing of* ϕ is a map $\Phi : M \to G$ such that $\pi \circ \Phi = \phi$ so that the following diagram commutes:

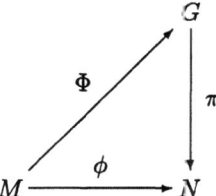

Such lifts always exist locally.

So let $\phi : M \to N$ be a map with framing Φ and set $\alpha = \Phi^{-1} d\Phi$. Corresponding to the reductive decomposition $\mathfrak{g} = \mathfrak{k} \oplus \mathfrak{m}$ is a decomposition of α,

$$\alpha = \alpha_\mathfrak{k} + \alpha_\mathfrak{m}$$

and one may easily verify that

$$\phi^* \beta = \operatorname{Ad} \Phi \, \alpha_\mathfrak{m}.$$

We can now express the harmonic map equations in terms of α:

$$\begin{aligned} d * \phi^* \beta &= d(\operatorname{Ad} \Phi * \alpha_\mathfrak{m}) \\ &= \operatorname{Ad} \Phi \{d * \alpha_\mathfrak{m} + [\alpha \wedge *\alpha_\mathfrak{m}]\}, \end{aligned}$$

where we have used $d(\operatorname{Ad} \Phi) = \operatorname{Ad} \Phi \circ \operatorname{ad} \alpha$. Thus ϕ is harmonic if and only if

$$d * \alpha_\mathfrak{m} + [\alpha \wedge *\alpha_\mathfrak{m}] = 0. \tag{3.3}$$

Henceforth, we will restrict attention to the case where M is two-dimensional. Then the energy is conformally invariant so that, if M is oriented, we may take M to be a Riemann surface. We now have a type decomposition $\alpha_\mathfrak{m} = \alpha'_\mathfrak{m} + \alpha''_\mathfrak{m}$ where $\alpha'_\mathfrak{m}$ is an $\mathfrak{m}^\mathbb{C}$-valued $(1,0)$-form with complex conjugate $\alpha''_\mathfrak{m}$. Then

$$*\alpha_\mathfrak{m} = -i\alpha'_\mathfrak{m} + i\alpha''_\mathfrak{m}$$

so that (3.3) becomes

$$\begin{aligned} 0 &= -(d\alpha'_\mathfrak{m} + [\alpha \wedge \alpha'_\mathfrak{m}]) + (d\alpha''_\mathfrak{m} + [\alpha \wedge \alpha''_\mathfrak{m}]) \\ &= -(d\alpha'_\mathfrak{m} + [\alpha_\mathfrak{k} \wedge \alpha'_\mathfrak{m}] + [\alpha''_\mathfrak{m} \wedge \alpha'_\mathfrak{m}]) + (d\alpha''_\mathfrak{m} + [\alpha_\mathfrak{k} \wedge \alpha''_\mathfrak{m}] + [\alpha'_\mathfrak{m} \wedge \alpha''_\mathfrak{m}]) \\ &= -(d\alpha'_\mathfrak{m} + [\alpha_\mathfrak{k} \wedge \alpha'_\mathfrak{m}]) + (d\alpha''_\mathfrak{m} + [\alpha_\mathfrak{k} \wedge \alpha''_\mathfrak{m}]). \end{aligned} \tag{3.4}$$

On the other hand, we have the Maurer–Cartan equations for α whose m- and ℓ-parts read

$$d\alpha'_m + [\alpha_\ell \wedge \alpha'_m] + d\alpha''_m + [\alpha_\ell \wedge \alpha''_m] + [\alpha'_m \wedge \alpha''_m]_m = 0 \qquad (3.5)$$

$$d\alpha_\ell + \tfrac{1}{2}[\alpha_\ell \wedge \alpha_\ell] + [\alpha'_m \wedge \alpha''_m]_\ell = 0. \qquad (3.6)$$

In particular, (3.4) and (3.5) are equivalent to

$$d\alpha'_m + [\alpha_\ell \wedge \alpha'_m] = -\tfrac{1}{2}[\alpha'_m \wedge \alpha''_m]_m.$$

Suppose now that $[\alpha'_m \wedge \alpha''_m]_m$ vanishes (certainly true when N is a Riemannian symmetric space since then $[m, m] \subset \ell$). The harmonic map equations combine with the Maurer–Cartan equations to give

$$d\alpha'_m + [\alpha_\ell \wedge \alpha'_m] = d\alpha''_m + [\alpha_\ell \wedge \alpha''_m] = 0 \qquad (3.7)$$

$$d\alpha_\ell + \tfrac{1}{2}[\alpha_\ell \wedge \alpha_\ell] + [\alpha'_m \wedge \alpha''_m] = 0. \qquad (3.8)$$

Remark These last equations have a gauge-theoretic formulation: let A denote the connection $d + \alpha_\ell$ with curvature F_A and (temporarily) denote α'_m by Ψ. Then our equations read

$$d^A \Psi = 0$$
$$F_A = -[\Psi \wedge \overline{\Psi}]$$

which are the Yang–Mills–Higgs equations for the connection A and the Higgs field Ψ. (See Hitchin [31] for a far-reaching study of these equations when N is the *non-compact* symmetric space $G^{\mathbb{C}}/G$.)

The equations (3.7) and (3.8) are invariant under an S^1-action: for $\lambda \in S^1$ and α a $\mathfrak{g}$-valued 1-form, set

$$\lambda \cdot \alpha = \alpha_\lambda = \lambda^{-1} \alpha'_m + \alpha_\ell + \lambda \alpha''_m. \qquad (3.9)$$

This is clearly an action of S^1 on $\mathfrak{g}$-valued 1-forms which preserves the solution set of (3.7) and (3.8). In fact, more is true: compare coefficients of λ to see that (3.7) and (3.8) hold for α precisely when

$$d\alpha_\lambda + \tfrac{1}{2}[\alpha_\lambda \wedge \alpha_\lambda] = 0$$

for all $\lambda \in S^1$. Otherwise said, we have written the harmonic map equations as a loop of zero-curvature equations (cf. [43, 56, 65, 66]).

Conversely, suppose that M is simply connected and let α_λ be a loop of 1-forms of the form (3.9) such that

1. $[\alpha'_m \wedge \alpha''_m]_m = 0$;

2. $d\alpha_\lambda + \tfrac{1}{2}[\alpha_\lambda \wedge \alpha_\lambda] = 0$, for all $\lambda \in S^1$.

Then, by Theorem 2.1, for each $\lambda \in S^1$ we can find a map $\Phi_\lambda : M \to G$ such that
$$\Phi_\lambda^{-1} d\Phi_\lambda = \alpha_\lambda$$
and, since α_λ satisfies (3.7) and (3.8), we conclude that each $\phi_\lambda = \pi \circ \Phi_\lambda : M \to N$ is harmonic. In particular, harmonic maps of this kind come in S^1-families.

This analysis applies to harmonic maps into any reductive homogeneous space N (with metric induced by one on $\mathfrak{g}$) which satisfy the auxiliary condition
$$[\alpha'_\mathfrak{m} \wedge \alpha''_\mathfrak{m}]_\mathfrak{m} = 0$$
(which is the same as demanding that $[\phi^*\beta' \wedge \phi^*\beta'']$ be $[\mathfrak{k}]$-valued). This last condition is trivially satisfied when N is a 2-symmetric space and then our discussion is valid for all harmonic maps.

So let us consider the case when N is a Riemannian symmetric space and show how the results of Section 2 provide harmonic maps of $\mathbb{R}^2 \to N$. For this, let τ be the involution determining N so that
$$\mathfrak{g}_0 = \mathfrak{k}^\mathbb{C}; \qquad \mathfrak{g}_1 = \mathfrak{m}^\mathbb{C}$$
and recall the loop algebra $\Lambda \mathfrak{g}_\tau$. Consider the algebraic structure of α_λ in (3.9): the coefficient of λ^{-1} is a $\mathfrak{g}_1$-valued $(1,0)$-form, the constant (in λ) term is $\mathfrak{k}$-valued while the λ-coefficient is a $\mathfrak{g}_1$-valued $(0,1)$-form. In particular, α_λ may be viewed as a $\Lambda \mathfrak{g}_\tau$-valued 1-form. Moreover, we see from (2.8) that this is precisely the type of $\Lambda \mathfrak{g}_\tau$-valued 1-form that arises as the solution to the zero-curvature equations in Theorem 2.5 so that combining that theorem with our present discussion gives

Theorem 3.2 *For each $d \in 2\mathbb{N}+1$ and $\xi_o \in \Lambda_d$, there is a unique solution to*
$$\frac{\partial \xi}{\partial z} = [\xi, \lambda^{-1}\xi_d + r(\xi_{d-1})]; \qquad \xi(0) = \xi_o$$
and then there is a harmonic map $\phi : \mathbb{R}^2 \to N$ with framing $\Phi : \mathbb{R}^2 \to G$ satisfying $\Phi^{-1} \partial \Phi / \partial z = \xi_d + r(\xi_{d-1})$.

We have therefore succeeded in constructing harmonic maps of $\mathbb{R}^2$ into Riemannian symmetric spaces N from our commuting Hamiltonian flows on the Poisson manifolds Λ_d. We call the maps so obtained *harmonic maps of finite type*.

When is a harmonic map $\mathbb{R}^2 \to N$ of finite type? The framing Φ of such a map has $\alpha'_\mathfrak{m} = \xi_d\, dz$ and this leads to necessary conditions. Indeed, comparing coefficients in the differential equation for ξ gives
$$d\xi_d = [\xi_d, (r(\xi_{d-1}) - \xi_{d-1})\, dz + \overline{r(\xi_{d-1})}\, d\bar{z}]$$
which takes values in $[\xi_d, \mathfrak{k}^\mathbb{C}]$. Otherwise said, $d\xi$ takes values in the tangent space at ξ_d to the $\operatorname{Ad} K^\mathbb{C}$-orbit through ξ_d. We therefore deduce from the uniqueness of solutions to ODE:

Lemma 3.3 $\xi_d : \mathbb{R}^2 \to \mathfrak{m}^\mathbb{C}$ *(and hence $\alpha'_\mathfrak{m}(\partial/\partial z)$) takes values in a single $\operatorname{Ad} K^\mathbb{C}$-orbit in $\mathfrak{m}^\mathbb{C}$.*

This has the effect of ensuring that certain polynomial invariants of a finite type harmonic map are constant: if $P : \mathfrak{m}^{\mathbf{C}} \to \mathbb{C}$ is an $\operatorname{Ad} K^{\mathbf{C}}$-invariant polynomial of degree r then clearly $P(\alpha'_{\mathfrak{m}}(\partial/\partial z))$ is constant. On the other hand, P gives rise to a G-invariant, and hence parallel, section $\widehat{P}$ of $S^r T^* N^{\mathbf{C}} \cong S^r[\mathfrak{m}^{\mathbf{C}}]$ by

$$\widehat{P}_{g \cdot o} = P \circ \operatorname{Ad} g^{-1}$$

so that we have

$$\phi^* \widehat{P}^{(r,0)} = P(\alpha'_{\mathfrak{m}}(\partial/\partial z)) \, \mathrm{d}z^r = c \, \mathrm{d}z^r$$

for some constant c.

One important class of harmonic maps which satisfy such constraints is that of doubly periodic harmonic maps or, equivalently, lifts to the universal cover of harmonic 2-tori. Indeed, for any harmonic map, it is known that these polynomial invariants are holomorphic differentials: one may write the harmonic map equations as

$$\nabla^\phi_{\partial/\partial \bar{z}} \partial \phi/\partial z = 0,$$

where ∇^ϕ is the pull-back of the Levi-Civita connection on N (see [60]) and then, for any parallel field of polynomials $\widehat{P}$ we have

$$\frac{\partial}{\partial \bar{z}} \widehat{P}(\partial \phi/\partial z, \ldots, \partial \phi/\partial z) = r \widehat{P}(\nabla^\phi_{\partial/\partial \bar{z}} \partial \phi/\partial z, \partial \phi/\partial z, \ldots, \partial \phi/\partial z) = 0.$$

In particular, when ϕ is doubly periodic, each $\phi^* \widehat{P}^{(r,0)}$ is constant by Liouville's theorem.

Each common level set of the invariant polynomials on $\mathfrak{m}^{\mathbf{C}}$ comprises but a finite number of $\operatorname{Ad} K^{\mathbf{C}}$-orbits so that harmonic 2-tori come close to satisfying the necessary condition to be of finite type provided by Lemma 3.3. In fact, in this case, that necessary condition is almost sufficient: we shall see in Section 5 that the following theorem is a consequence of the results of [15].

Theorem 3.4 *Let $\phi : \mathbb{R}^2 \to N$ be a doubly periodic harmonic map into a Riemannian symmetric space. Suppose that some (and hence every) framing of ϕ has $\alpha'_{\mathfrak{m}}(\partial/\partial z)$ taking values in a single semisimple $\operatorname{Ad} K^{\mathbf{C}}$-orbit in $\mathfrak{m}^{\mathbf{C}}$. Then ϕ is of finite type.*

Here an orbit is semisimple if it consists of elements $\xi \in \mathfrak{m}^{\mathbf{C}}$ which are semisimple in the sense that $\operatorname{ad} \xi$ is diagonalisable.

Let us conclude this section with an example which gives geometric meaning to the "semisimple orbit" hypothesis. For a rank 1 symmetric space, the semisimple orbits are just the non-zero level sets of the Killing form. The corresponding quadratic form on $TN^{\mathbf{C}}$ is the complexified metric h so that the constant holomorphic differential for a doubly periodic map ϕ is just the obstruction to conformality $\phi^* h^{(2,0)}$. We therefore conclude

Theorem 3.5 *A doubly periodic non-conformal harmonic map $\phi : \mathbb{R}^2 \to N$ into a rank 1 symmetric space is of finite type.*

While this result explicitly excludes the geometrically interesting case of minimal 2-tori, it is already strong enough to be very useful. For instance, the non-conformal harmonic 2-tori in S^2 are precisely the Gauss maps of constant mean curvature tori in $\mathbb{R}^3$ and, in this case, the theorem reproduces the basic result of Pinkall–Sterling [42].

For spheres and complex projective spaces, this result will be improved in Section 3.5.

3.3 Primitive maps

We now extend our theory to treat certain harmonic maps into k-symmetric spaces N where $k > 2$. Let us begin by introducing the class of harmonic maps that our constructions will produce.

So let $N = G/K$ be a k-symmetric space, $k > 2$, with automorphism τ and associated eigenspace decomposition

$$\mathfrak{g}^\mathbb{C} = \sum_{i \in \mathbb{Z}_k} \mathfrak{g}_i.$$

In particular, $\mathfrak{g}_{-1} = \overline{\mathfrak{g}_1}$ and, since $k > 2$, $\mathfrak{g}_1 \cap \mathfrak{g}_{-1} = \{0\}$.

Definition A map $\phi : M \to N$ of a Riemann surface is *primitive* if $\phi^* \beta'$ takes values in $[\mathfrak{g}_{-1}]$. Equivalently, ϕ is primitive if and only if any framing Φ has $\alpha'_{\mathfrak{m}}$ taking values in $\mathfrak{g}_{-1}$.

Example Consider the case $k = 3$: here we have

$$\mathfrak{m}^\mathbb{C} = \mathfrak{g}_1 \oplus \mathfrak{g}_{-1}$$

so that N acquires an invariant almost complex structure with $T^{(1,0)}N \cong [\mathfrak{g}_{-1}]$. This is the non-integrable almost complex structure discussed by Salamon [50] who viewed 3-symmetric spaces as twistor spaces for Riemannian symmetric spaces (in this regard, see also [11, 16]). In this setting, primitive maps are just (almost) holomorphic maps.

In general, primitive maps are examples of maps which are f-holomorphic with respect to a horizontal f-structure in the sense of Black [4]. Under rather general conditions, such maps enjoy a number of interesting and useful properties which, in our setting, stem ultimately from the relation

$$[\mathfrak{g}_1, \mathfrak{g}_{-1}] \subset \mathfrak{g}_0 = \mathfrak{k}^\mathbb{C}. \tag{3.10}$$

To be more precise, under the additional assumption that all irreducible subrepresentations of the adjoint representation of K on $\mathfrak{m}^\mathbb{C}$ occur with multiplicity one, Black proves:

1. All primitive maps are *equiharmonic*, that is, harmonic with respect to any invariant metric on N;

2. If $p : N = G/K \to G/H$ is any homogeneous projection and $\phi : M \to N$ is equiharmonic then so is $p \circ \phi$.

Even without such extra assumptions, restricted versions of these results are still available. For this, consider a framing of a primitive map with associated 1-form α. We know that α'_m takes values in $\mathfrak{g}_{-1}$ so that, in view of (3.10), we have

$$[\alpha'_m \wedge \alpha''_m]_m = 0.$$

The projections of the Maurer–Cartan equations for α onto $\mathfrak{g}_{-1}$, $\mathfrak{g}_1$ and $\mathfrak{g}_0$ therefore read

$$d\alpha'_m + [\alpha_{\mathfrak{k}} \wedge \alpha'_m] = 0 \tag{3.11}$$

$$d\alpha''_m + [\alpha_{\mathfrak{k}} \wedge \alpha''_m] = 0 \tag{3.12}$$

$$d\alpha_{\mathfrak{k}} + \tfrac{1}{2}[\alpha_{\mathfrak{k}} \wedge \alpha_{\mathfrak{k}}] + [\alpha'_m \wedge \alpha''_m] = 0. \tag{3.13}$$

However, these are just the harmonic maps equations (3.7) and (3.8) and we conclude:

Theorem 3.6 *A primitive map $\phi : M \to N$ is harmonic with respect to the metric on N induced by that on $\mathfrak{g}$.*

Primitive maps are also well-behaved with respect to projections: let G/H be a reductive homogeneous space with $K \subset H$ and reductive decomposition

$$\mathfrak{g} = \mathfrak{h} \oplus \mathfrak{p}$$

which we assume to be orthogonal and stable under τ. As usual, equip G/H with the metric induced by that on $\mathfrak{g}$ and let $p : N = G/K \to G/H$ be the homogeneous projection. We have

Theorem 3.7 *If $\phi : M \to N$ is primitive then $p \circ \phi : M \to G/H$ is harmonic.*

Proof. Let $\mathfrak{l}$ be the orthogonal complement of $\mathfrak{k}$ in $\mathfrak{h}$ so that we have a τ-stable decomposition

$$\mathfrak{g} = \mathfrak{k} \oplus \mathfrak{l} \oplus \mathfrak{p}$$

with

$$\mathfrak{h} = \mathfrak{k} \oplus \mathfrak{l}; \qquad \mathfrak{m} = \mathfrak{l} \oplus \mathfrak{p}.$$

Both $\mathfrak{l}^{\mathbb{C}}$ and $\mathfrak{p}^{\mathbb{C}}$ decompose into eigenspaces of τ and we have various relations between these eigenspaces: $[\mathfrak{l}_1, \mathfrak{p}_{-1}] \subset \mathfrak{g}_0 = \mathfrak{k}^{\mathbb{C}}$ but $[\mathfrak{l}^{\mathbb{C}}, \mathfrak{p}^{\mathbb{C}}] \subset [\mathfrak{h}^{\mathbb{C}}, \mathfrak{p}^{\mathbb{C}}] \subset \mathfrak{p}^{\mathbb{C}}$ so that

$$[\mathfrak{l}_1, \mathfrak{p}_{-1}] = 0. \tag{3.14}$$

Harmonic maps via Adler–Kostant–Symes theory

We also have $[\mathfrak{p}_1, \mathfrak{p}_{-1}] \subset \mathfrak{k}^{\mathbb{C}}$ so that

$$[\mathfrak{p}_1, \mathfrak{p}_{-1}]_{\mathfrak{p}} = 0. \tag{3.15}$$

Any framing for ϕ is one for $p \circ \phi$ also so it suffices to prove that, for such a framing,

$$[\alpha'_{\mathfrak{p}} \wedge \alpha''_{\mathfrak{p}}]_{\mathfrak{p}} = 0; \tag{3.16}$$

$$d\alpha'_{\mathfrak{p}} + [\alpha_{\mathfrak{h}} \wedge \alpha'_{\mathfrak{p}}] = 0.$$

However, $\alpha'_{\mathfrak{m}} = \alpha'_{\mathfrak{f}} + \alpha'_{\mathfrak{p}}$ takes values in $\mathfrak{g}_{-1}$ so that $\alpha'_{\mathfrak{p}}$ is $\mathfrak{p}_{-1}$-valued and (3.16) follows from (3.15). Finally,

$$\begin{aligned} d\alpha'_{\mathfrak{p}} + [\alpha_{\mathfrak{h}} \wedge \alpha'_{\mathfrak{p}}] &= d\alpha'_{\mathfrak{p}} + [\alpha_{\mathfrak{k}} \wedge \alpha'_{\mathfrak{p}}] + [\alpha''_{\mathfrak{f}} \wedge \alpha'_{\mathfrak{p}}] \\ &= d\alpha'_{\mathfrak{p}} + [\alpha_{\mathfrak{k}} \wedge \alpha'_{\mathfrak{p}}] \end{aligned}$$

in view of (3.14) and this last is just the $\mathfrak{p}$-part of $d\alpha'_{\mathfrak{m}} + [\alpha_{\mathfrak{k}} \wedge \alpha'_{\mathfrak{m}}]$ which vanishes by (3.11). □

In particular, primitive maps may give rise, by projection, to harmonic maps into Riemannian symmetric spaces and it is from this possibility that our principal interest in them derives.

Consider now the loop (3.9) of 1-forms

$$\alpha_\lambda = \lambda^{-1} \alpha'_{\mathfrak{m}} + \alpha_{\mathfrak{k}} + \lambda \alpha''_{\mathfrak{m}},$$

where $\alpha = \Phi^{-1} d\Phi$ for a framing Φ of a primitive map ϕ. Then $\alpha'_{\mathfrak{m}}$ is $\mathfrak{g}_{-1}$-valued so that we may view α_λ as a $\Lambda \mathfrak{g}_\tau$-valued 1-form.

Since ϕ is harmonic and $[\alpha'_{\mathfrak{m}} \wedge \alpha''_{\mathfrak{m}}]_{\mathfrak{m}}$ vanishes, the discussion in Section 3.2 applies so that, for each $\lambda \in S^1$,

$$d\alpha_\lambda + \tfrac{1}{2}[\alpha_\lambda \wedge \alpha_\lambda] = 0.$$

Thus primitive maps also give rise to zero-curvature equations.

Conversely, suppose that M is simply connected and α_λ is a $\Lambda \mathfrak{g}_\tau$-valued 1-form of the form (3.9) such that

$$d\alpha_\lambda + \tfrac{1}{2}[\alpha_\lambda \wedge \alpha_\lambda] = 0.$$

Then, by Theorem 2.1, for each $\lambda \in S^1$, we can find a map $\Phi_\lambda : M \to G$ such that

$$\Phi_\lambda^{-1} d\Phi_\lambda = \alpha_\lambda$$

and then $\phi_\lambda = \pi \circ \Phi_\lambda$ will be harmonic. In fact, more is true: since

$$(\alpha_\lambda)'_{\mathfrak{m}} = \lambda^{-1} \alpha'_{\mathfrak{m}}$$

takes values in $\mathfrak{g}_{-1}$, we conclude that each ϕ_λ is primitive. In particular, primitive maps of simply connected surfaces come in S^1-families.

Finally let us show how the results of Section 2 provide primitive maps $\mathbb{R}^2 \to N$. For this, observe that $\Lambda \mathfrak{g}_\tau$-valued 1-forms of the form (3.9) are precisely the kind of 1-form that arises as the solution to the zero-curvature equations in Theorem 2.5 so that we get

Theorem 3.8 *For each $d \equiv 1 \bmod k$ and $\xi_o \in \Lambda_d$, there is a unique solution to*

$$\frac{\partial \xi}{\partial z} = [\xi, \lambda^{-1}\xi_d + r(\xi_{d-1})]; \qquad \xi(0) = \xi_o$$

and then there is a primitive map $\phi : \mathbb{R}^2 \to N$ with framing $\Phi : \mathbb{R}^2 \to G$ satisfying $\Phi^{-1}d\Phi(\partial/\partial z) = \xi_d + r(\xi_{d-1})$.

We call the primitive maps so obtained *primitive maps of finite type*.

A similar analysis to that in Section 3.2 can be carried out for primitive maps to give sufficient conditions for a primitive map to be of finite type. Just as before, such a map must have $\xi_d = \alpha'_m(\partial/\partial z)$ taking values in a single Ad $K^\mathbb{C}$-orbit in $\mathfrak{g}_{-1}$ and, for doubly periodic maps, this condition is almost sufficient. Indeed, as we shall see in Section 5, the results of [12, 15] imply:

Theorem 3.9 *Let $\phi : \mathbb{R}^2 \to N$ be a doubly periodic primitive map into a k-symmetric space, $k > 2$. Suppose that some (and hence every) framing of ϕ has $\alpha'_m(\partial/\partial z)$ taking values in a single semisimple Ad $K^\mathbb{C}$-orbit in $\mathfrak{g}_{-1}$. Then ϕ is of finite type.*

We conclude this section with an alternative formulation of the "semisimple orbit" condition which will be useful below. First, the holomorphic differentials argument of Section 3.2 can be extended to primitive maps so that one can conclude that a doubly periodic primitive map of $\mathbb{R}^2$ has $\alpha'_m(\partial/\partial z)$ taking values in a single common level set of the Ad $K^\mathbb{C}$-invariant polynomials on $\mathfrak{g}_{-1}$. On the other hand, Vinberg [57] proves that each such level set contains a single semisimple orbit and that this is the unique closed orbit in the level set. We therefore deduce:

Theorem 3.10 *Let $\phi : \mathbb{R}^2 \to N$ be a doubly periodic primitive map into a k-symmetric space, $k > 2$. Suppose that, for some (and hence every) framing of ϕ, $\alpha'_m(\partial/\partial z)$ is semisimple on a dense subset of M. Then ϕ is of finite type.*

3.4 Example: Toda fields

As an illustration of our ideas, we describe some results of Bolton–Pedit–Woodward [8] which relate certain primitive maps into flag manifolds with (periodic) Toda fields (for further details, see [9] in this volume). We begin by introducing the basic ingredients of Toda field theory.

Let $\mathfrak{g}$ be a compact, *simple* Lie algebra with Killing form κ and, as usual, fix a maximal toral subalgebra $\mathfrak{t}$ and set $\mathfrak{a} = i\mathfrak{t}$ so that $\mathfrak{h} = \mathfrak{t}^\mathbb{C} = \mathfrak{t} \oplus \mathfrak{a}$ is a Cartan subalgebra of $\mathfrak{g}^\mathbb{C}$. We also fix a choice of positive roots $\Delta^+ \subset \Delta(\mathfrak{g}^\mathbb{C}, \mathfrak{h})$ with corresponding simple roots $\alpha_1, \ldots, \alpha_l$. Since $\mathfrak{g}$ is simple, we have another distinguished root: the *highest root* μ and we set

$$\Pi = \{\alpha_1, \ldots, \alpha_l, -\mu\}.$$

Thus Π is the set of roots labelling the nodes of the extended Dynkin diagram of $\mathfrak{g}^\mathbb{C}$. From the elementary properties of root systems (see, for example, [34]), we have that if $\alpha, \beta \in \Pi$ and $X_\alpha \in \mathfrak{g}^\alpha$, $X_{-\beta} \in \mathfrak{g}^{-\beta}$ then

Harmonic maps via Adler–Kostant–Symes theory

$$[X_\alpha, X_{-\beta}] = \begin{cases} 0 & \text{if } \alpha \neq \beta \\ \kappa(X_\alpha, X_{-\beta}) H_\alpha & \text{if } \alpha = \beta \end{cases} \tag{3.17}$$

where $H_\alpha \in \mathfrak{a}$ is the Killing dual of α.

With this understood, we can describe Toda field theory. This is a nonlinear Lagrangian field theory where the fields are maps $u : \mathbb{R}^2 \to \mathfrak{a}$ and the Lagrangian is given by

$$\int_{\mathbb{R}^2} \tfrac{1}{2}\kappa(du, du) - \sum_{\alpha \in \Pi} e^{\alpha(u)} \, dvol.$$

The Euler–Lagrange equations for this functional are

$$d^* du - \sum_{\alpha \in \Pi} e^{\alpha(u)} H_\alpha = 0 \tag{3.18}$$

and solutions of this are *Toda fields*.

Remark The alert Reader will notice that our Toda field equations differ from those in [8, 9]. The difference is a matter of normalisation with that in [8, 9] chosen to ensure that zero is a solution. We have given a normalisation which emphasises the relationship with the Toda lattice of Section 1.5.

We can find a zero-curvature formulation of (3.18) via the following *ansatz*: fix, once and for all, root vectors $X_\alpha \in \mathfrak{g}^\alpha$, $\alpha \in \Pi$ with

$$\kappa(X_\alpha, \overline{X_\alpha}) = -\tfrac{1}{4}$$

and set $X = \sum_{\alpha \in \Pi} X_\alpha$. Now for $u : \mathbb{R}^2 \to \mathfrak{a}$, define a loop of $\mathfrak{g}$-valued 1-forms

$$\alpha_\lambda = \lambda^{-1} \alpha'_m + \alpha_t + \lambda \alpha''_m$$

by

$$\alpha'_m = \operatorname{Ad} \exp(u/2) X \, dz = \sum_{\alpha \in \Pi} e^{\alpha(u)/2} X_\alpha \, dz$$

$$\alpha_t = \tfrac{i}{2} * du = \tfrac{1}{2}\left(\frac{\partial u}{\partial z} dz - \frac{\partial u}{\partial \bar{z}} d\bar{z}\right)$$

$$\alpha''_m = \overline{\alpha'_m} = \operatorname{Ad} \exp(-u/2) \overline{X} \, d\bar{z}.$$

In particular, α_t is $\mathfrak{t}$-valued so that $[\alpha_t \wedge \alpha_t]$ vanishes. For any field u, it is easy to verify that

$$d\alpha'_m + [\alpha_t \wedge \alpha'_m] = 0 = d\alpha''_m + [\alpha_t \wedge \alpha''_m]$$

while, in view of (3.17),

$$\begin{aligned} d\alpha_t + \tfrac{1}{2}[\alpha_t \wedge \alpha_t] + [\alpha'_m \wedge \alpha''_m] &= \tfrac{i}{2} d * du + \sum_{\alpha \in \Pi} e^{\alpha(u)} \kappa(X_\alpha, \overline{X_\alpha}) H_\alpha \, dz \wedge d\bar{z} \\ &= \tfrac{i}{2}(d * du + \sum_{\alpha \in \Pi} e^{\alpha(u)} H_\alpha * 1) \\ &= -\tfrac{i}{2} * (d^* du - \sum_{\alpha \in \Pi} e^{\alpha(u)} H_\alpha). \end{aligned}$$

Thus we conclude:

Proposition 3.11 $u : \mathbb{R}^2 \to \mathfrak{a}$ *is a Toda field if and only if*
$$d\alpha_\lambda + [\alpha_\lambda \wedge \alpha_\lambda] = 0,$$
for all $\lambda \in S^1$.

Remark There are numerous other (gauge-equivalent) zero-curvature formulations of the Toda field equations (see, for example, [39]). We have chosen this one because it gives real-valued α_λ.

We now turn to a geometric interpretation of the Toda field equations. Let G be the adjoint group of $\mathfrak{g}$ (so that, in particular, G has trivial centre) and let T, A be the analytic subgroups with Lie algebras $\mathfrak{t}$, $\mathfrak{a}$. Consider the flag manifold G/T—we equip this with the structure of a k-symmetric space as follows. Let $\xi_j \in \mathfrak{t}$, $1 \leq j \leq l$ be defined by
$$\alpha_i(\xi_j) = \sqrt{-1}\delta_{ij}$$
and set $\xi = \sum \xi_j$. In particular, $\alpha(\xi) \in \sqrt{-1}\mathbb{Z}$, for all $\alpha \in \Delta(\mathfrak{g}^\mathbb{C}, \mathfrak{h})$, and we define $k \in \mathbb{N}$ by
$$\mu(\xi) = \sqrt{-1}(k-1).$$
We take τ to be conjugation by $\exp(-2\pi\xi/k)$: this is an order k automorphism of G. Then T is the identity component of the fixed set of τ so that G/T is a k-symmetric space.

Remark This is an example of a rather general construction: any generalised flag manifold $G^\mathbb{C}/P$, where P is a parabolic subgroup, can be given a canonical k-symmetric structure [16, p. 52].

In the case at hand, observe that in the decomposition of $\mathfrak{g}^\mathbb{C}$ into eigenspaces of τ, we have
$$\mathfrak{g}_0 = \mathfrak{h}; \quad \mathfrak{g}_{-1} = \sum_{\alpha \in \Pi} \mathfrak{g}^\alpha$$
so that we can immediately deduce the following result from Theorem 2.1 and Proposition 3.11:

Proposition 3.12 *Let* $u : \mathbb{R}^2 \to \mathfrak{a}$ *be a Toda field with loop of 1-forms* α_λ. *Then each* $\alpha_\lambda = \Phi_\lambda^{-1} d\Phi_\lambda$ *for a framing* $\Phi_\lambda : \mathbb{R}^2 \to G$ *of a primitive map* $\phi_\lambda : \mathbb{R}^2 \to G/T$.

We call such a framing of a primitive map a *Toda frame*.

When does a primitive map $\mathbb{R}^2 \to G/T$ have a Toda frame? Observe that, by construction, a Toda frame has $\alpha'_m(\partial/\partial z)$ taking values in the Ad A-orbit of X. In fact, the converse is true: it is easy to see that A acts simply transitively on $\text{Ad}(A)X$ while $\exp : \mathfrak{a} \to A$ is a diffeomorphism so that, if a framing has $\alpha'_m(\partial/\partial z)$ taking values in $\text{Ad}(A)X$ then we may take logarithms to find $u : \mathbb{R}^2 \to \mathfrak{a}$ satisfying

$$\alpha'_m = \operatorname{Ad}\exp(u/2) X\, dz.$$

Moreover, one can show that α_t is determined completely by α'_m and the Maurer–Cartan equations so that

$$\alpha_t = \tfrac{i}{2} * du.$$

Thus

Lemma 3.13 [8] *A frame is Toda if and only if $\alpha'_m(\partial/\partial z)$ takes values in $\operatorname{Ad}(A)X$.*

To go further, we must introduce some ideas of Kostant [36]: define the set of cyclic elements $\mathcal{C} \subset \mathfrak{g}_{-1}$ by

$$\mathcal{C} = \{\sum_{\alpha\in\Pi} c_\alpha X_\alpha :\quad c_\alpha \in \mathbb{C}\setminus\{0\}\}.$$

Concerning these, Kostant proves

1. any element of $\mathcal{C}$ is (regular) semisimple;

2. if $P_1,\ldots,P_l$ are algebraically independent homogeneous generators of the invariant polynomials on $\mathfrak{g}^{\mathbb{C}}$ with $\deg P_1 \leq \cdots \leq \deg P_l$ then

$$P_i|_{\mathfrak{g}_{-1}} \equiv 0 \quad \text{for } 1 \leq i \leq l-1$$

and, for $\xi \in \mathfrak{g}_{-1}$, $\xi \in \mathcal{C}$ if and only if $P_l(\xi) \neq 0$.

3. For $c \neq 0$, T acts simply transitively on the set of $\operatorname{Ad} A$-orbits in $P_l^{-1}\{c\}$.

From this, we deduce that if a framing Φ satisfies $P_l(\alpha'_m(\partial/\partial z)) \equiv P_l(X)$ then there is a *unique* gauge transformation taking Φ to a Toda frame. More generally, when $P_l(\alpha'_m(\partial/\partial z))$ is a non-zero constant, we can make a linear change of co-ordinate $cz = w$ so that $P_l(\alpha'_m(\partial/\partial w)) \equiv P_l(X)$ and so obtain a Toda frame.

To summarise:

Theorem 3.14 [8] *A primitive map $\phi : \mathbb{R}^2 \to G/T$ has a Toda frame (after a linear change of co-ordinate) if and only if, for some (and hence, every) framing we have that $P_l(\alpha'_m(\partial/\partial z))$ is a non-zero constant. Moreover, in this case, the Toda frame is unique.*

From the uniqueness assertion, we also get

Corollary 3.15 [8] *Let $\phi : \mathbb{R}^2 \to G/T$ be a primitive map with Toda frame Φ and corresponding Toda field u. Then ϕ is doubly periodic if and only if Φ is doubly periodic and then u is doubly periodic.*

We remark that the usual holomorphic differentials argument shows that a doubly periodic primitive map has a Toda frame if and only if $\alpha'_m(\partial/\partial z)$ is cyclic at some (and hence, every) point of $\mathbb{R}^2$.

We can now apply Theorem 3.9 to obtain such doubly periodic primitive maps from commuting flows. Indeed, a primitive map with Toda frame necessarily satisfies the "semisimple orbit" hypothesis since X is semisimple so that we have

Theorem 3.16 [8] *A doubly periodic primitive map* $\phi : \mathbb{R}^2 \to G/T$ *with non-zero* $P_l(\alpha'_m(\partial/\partial z))$ *is of finite type.*

In particular, we obtain doubly periodic Toda fields this way from Corollary 3.15. However, more is true: while it is not the case that doubly-periodic Toda fields necessarily produce doubly periodic primitive maps (there may be holonomy), Bolton–Pedit–Woodward prove that, for doubly periodic Toda fields, the $\Lambda\mathfrak{g}_\tau$-valued 1-form α_λ *always* arises as in Theorem 2.5 so that all doubly periodic Toda fields are of finite type.

Examples Let us take $G = \mathrm{SO}(5)$ with flag manifold $F = \mathrm{SO}(5)/\mathrm{SO}(2) \times \mathrm{SO}(2)$. We have the homogeneous fibration $p : F \to S^4$ and a homogeneous diffeomorphism between F and the Grassmannian bundle $\widetilde{G}_2(TS^4)$ of oriented 2-planes in TS^4. Under this indentification, a map $\phi : M \to F$ is primitive if and only if it is the Gauss map of a minimal (i.e., conformal and harmonic) map $p \circ \phi : M \to S^4$ (a result originally due to Eells–Salamon [22]). Moreover, ϕ has a Toda frame precisely when the corresponding minimal surface is not superminimal. In this case, the above results reproduce the analysis of Ferus–Pedit–Pinkall–Sterling [25].

Again, let us take $F = \mathrm{SU}(n)/T^{n-1}$ as our flag manifold. A map $\phi : M \to F$ may be viewed as a collection of mutually orthogonal maps $\phi_1, \ldots, \phi_n : M \to \mathbb{CP}^{n-1}$ and then ϕ is primitive with Toda frame if and only if the ϕ_i comprise the Frenet frame of a *superconformal* harmonic map $M \to \mathbb{CP}^{n-1}$ in the sense of [8, 9]. It was for the study of such maps that the methods of [8] were developed.

We shall have more to say about results of this kind in the next section.

We conclude this section by briefly contemplating the geometric significance of open Toda fields as this has received a lot of recent attention in the Physics literature [19, 26, 27, 48, 51]. Open Toda fields are extremals of the modified Lagrangian

$$\int_{\mathbb{R}^2} \tfrac{1}{2}\kappa(du, du) - \sum_{i=1}^{l} e^{\alpha_i(u)} \, dvol$$

where there is no longer an exponential interaction involving $-\mu$. Again there is a zero-curvature representation but this time the primitive maps so obtained are *holomorphic* maps into G/T. In fact, they are superhorizontal holomorphic maps in the sense of [16]. As a consequence, open Toda fields are much simpler objects for which Weierstrass representation formulae are available. See [39] in this volume for a brief discussion of such fields.

3.5 Example: primitive maps and twistor lifts

We have seen how all non-conformal harmonic 2-tori in a rank-1 symmetric space are of finite type. However, this excludes the geometrically interesting case of minimal 2-tori. On the other hand, special cases of the results of the previous section have been used to account for certain minimal 2-tori in S^4 [25], S^n and $\mathbb{C}P^n$ [8] by showing that these tori have lifts which are primitive maps of finite type.

In this section, we present a generalisation of these results which accounts for *all* harmonic 2-tori in a sphere S^n. We begin by recalling some results from the well-developed twistor theory of harmonic maps of a Riemann surface into a sphere.

So let $\phi : M \to S^n$ be a harmonic map of a Riemann surface. Let $T = \phi^{-1}TS^n$ with connection ∇ given by the pull-back of the Levi-Civita connection of S^n. Let z be a local holomorphic co-ordinate on M, set $\nabla' = \nabla_{\partial/\partial z}$, $\nabla'' = \nabla_{\partial/\partial \bar{z}}$ and inductively define $\nabla^i \phi$ by

$$\nabla^1 \phi = \partial \phi / \partial z; \qquad \nabla^{i+1} \phi = \nabla' \nabla^i \phi.$$

For $x \in M$, define $W_x^j \subset T_x^{\mathbb{C}}$ by

$$W_x^j = \text{span}_{\mathbb{C}} \{ \nabla_x^i \phi : 1 \leq i \leq j \}.$$

Clearly, each W_x^j is defined independently of the choice of holomorphic co-ordinate z.

If ϕ is non-constant and (weakly) conformal then each W_x^1 is isotropic for the complexified metric on $T^{\mathbb{C}}$ and is 1-dimensional off a discrete set of points in M. This motivates, in part, the following definition:

Definition The *isotropy dimension* r of a conformal harmonic map $\phi : M \to S^n$ is given by

$$r = \max\{ j : \max_x \dim_{\mathbb{C}} W_x^j = j \text{ and } W_x^j \text{ is isotropic for all } x \}.$$

We make the convention that a non-conformal map has isotropy dimension zero.

The following facts are well known (cf. [61]): if ϕ has isotropy dimension $r > 0$ then

1. For $1 \leq j \leq r+1$, there is a bundle W^j of $T^{\mathbb{C}}$ whose fibre at x coincides with W_x^j except at a discrete set of points;

2. Each W^j is stable under ∇'';

3. For $1 \leq j \leq r$, $\text{rank}\, W^j = j$ and $\text{rank}\, W^{r+1} \leq r+1$.

We therefore distinguish two possibilities: either $W^r = W^{r+1}$ or not[1].

---
[1] These two possibilities regulate the relationship between the isotropy dimension of ϕ and the *isotropy order* of ϕ [2] described by Wood [60] in this volume: if ϕ has isotropy dimension r then ϕ is superminimal if and only if $W^r = W^{r+1}$ and has isotropy order $2r+1$ otherwise.

In the first case, we see that W^r is isotropic and stable under ∇' so that the (complexified) inner products

$$(\nabla^i \phi, \nabla^j \phi) \equiv 0$$

for all $i, j \in \mathbb{N}$. Harmonic maps of this kind are variously called *pseudo-holomorphic* [17], *real isotropic* [23] or *superminimal* [10] and were completely classified by Calabi [17] (see also [3]) who proved that all such were projections of horizontal holomorphic curves in the twistor space of S^n. Indeed, since W^r is stable under both ∇' and ∇'', it is parallel so that

$$(W^r \oplus \overline{W^r}) \cap T$$

is a parallel sub-bundle of T with (real) rank $2r$. It follows that ϕ factors through an equatorial $2r$-sphere in S^n so that, without loss of generality, we may take $2r = n$. Now recall that the *twistor space* Z on S^{2r} may be viewed as the bundle of isotropic r-planes in the complexification of TS^{2r}. This is a complex manifold: in fact, $SO(2r+1)$ acts transitively on each of the two connected components of Z and each component is so realised as the generalised flag manifold $SO(2r+1)/U(r)$. It is clear that W^r defines a map $\psi : M \to Z$ covering ϕ and the condition that W^r be parallel is equivalent to the demand that ψ be holomorphic and horizontal with respect to the twistor fibration $Z \to S^{2r}$. For more on twistor spaces of symmetric spaces and their applications to harmonic maps, the Reader is referred to [16], the surveys [13, 14, 47, 62] and the article by Kobak [35] in this volume.

Remark It is clear that the isotropy dimension r of a map $\phi : M \to S^n$ must satisfy $2r \leq n$. In case that $2r = n$, it follows from the easily verified fact

$$(\nabla^{r+1}\phi, \nabla^i \phi) \equiv 0,$$

for $1 \leq i \leq r$, that $W^r = W^{r+1}$ so that ϕ is superminimal.

Let us now consider the non-superminimal harmonic maps where $W^r \neq W^{r+1}$ or, equivalently,

$$(\nabla^{r+1}\phi, \nabla^{r+1}\phi) \not\equiv 0.$$

It is our contention that ϕ is covered by a primitive map into a suitable k-symmetric space and, moreover, that when M is a 2-torus this primitive map is of finite type.

First we describe the relevant k-symmetric space: for $2r < n$, let $F^r(S^n)$ be the bundle over S^n with fibre

$$F^r_x(S^n) = \{w_1 \subset \cdots \subset w_r \subset (T_x S^n)^{\mathbb{C}} : \text{each } w_j \text{ is an isotropic } j\text{-plane}\}.$$

Now $SO(n+1)$ acts transitively on $F^r(S^n)$ with stabilisers conjugate to

$$\underbrace{SO(2) \times \cdots \times SO(2)}_{r \text{ times}} \times SO(n-2r)$$

Harmonic maps via Adler–Kostant–Symes theory

so that $F^r(S^n)$ is a homogeneous space.

Fix a base-point $f = (w_1 \subset \cdots \subset w_r) \in F_x^r(S^n)$ and orthogonalise to obtain isotropic lines $L_1, \ldots, L_r$ and real subspaces L_{r+1} and $L_0 = \mathrm{span}_{\mathbb{C}}\{x\}$ so that

$$(\mathbb{R}^{n+1})^{\mathbb{C}} = L_0 \oplus \sum_{i=1}^{r}(L_i \oplus \overline{L_i}) \oplus L_{r+1}.$$

Take $k = 2r + 2$, let ω be the usual primitive k-th root of unity and define $Q \in O(n+1)$ by

$$Q = \omega^{-i} \text{ on } L_i.$$

Then conjugation by Q is an order k automorphism of $SO(n+1)$ and the identity component of its fixed set is the stabiliser of f. Thus $F^r(S^n)$ is a k-symmetric space.

Let $\psi : M \to F^r(S^n)$ be a map with projection $\phi : M \to S^n$. We may view ψ as a flag of isotropic sub-bundles

$$\psi^{(1)} \subset \cdots \subset \psi^{(r)} \subset T^{\mathbb{C}}$$

(here again $T = \phi^{-1}TS^n$) and one can prove

Proposition 3.17 [12] $\psi : M \to F^r(S^n)$ *is primitive if and only if*

(i) $\nabla^1 \phi$ *is a (local) section of* $\psi^{(1)}$;

(ii) *each* $\psi^{(i)}$ *is stable under* ∇'';

(iii) *if σ is a local section of $\psi^{(i)}$ then $\nabla'\sigma$ is a local section of $\psi^{(i+1)}$.*

We know from Theorem 3.7 that if ψ is primitive then ϕ is harmonic but, in the present case, we have a converse. Let $\phi : M \to S^n$ be a non-superminimal harmonic map of isotropy order r. It is clear from Proposition 3.17 that the bundle of flags $W^1 \subset \cdots \subset W^r$ defines a primitive map $\psi : M \to F^r(S^n)$. When is ψ of finite type? We have

Lemma 3.18 [12] *A (local) framing of ψ has $\alpha'_m(\partial/\partial z)$ semisimple at $x \in M$ if and only if*

$$(\nabla^{r+1}\phi, \nabla^{r+1}\phi) \neq 0$$

at $x \in M$.

Since $W^r \neq W^{r+1}$, we deduce that this condition holds on a dense open subset of M and so conclude from Theorem 3.10:

Theorem 3.19 [12] *A non-superminimal harmonic map of a 2-torus into a sphere is covered by a primitive map of finite type.*

Thus harmonic 2-tori in spheres are completely accounted for: either they are superminimal and so arise as projections of holomorphic curves or they are obtained by integrating commuting flows.

Example We have seen that a non-superminimal map into S^n must have isotropy dimension r with $2r < n$. Suppose $n = 2m$ and consider maps of isotropy dimension $m - 1$. Now $F^{m-1}(S^{2m})$ is the full flag manifold of $SO(2m + 1)$ and its k-symmetric structure is precisely that described in Section 3.4. Moreover, when M is a 2-torus, the primitive maps we have constructed are precisely those with Toda frames.

Again, if $n = 2m - 1$, the maximal isotropy dimension of a non-superminimal map is $m - 1$ and, viewing S^{2m-1} as an equator of S^{2m}, we again conclude that such tori are covered by primitive maps with Toda frames.

To summarise: non-superminimal harmonic 2-tori of maximal isotropy order are all obtained from doubly-periodic Toda fields of finite type. This result was first proved by Bolton–Pedit–Woodward [8] (see also [19]) who called such maps *superconformal*.

Remark A similar analysis can be carried out for harmonic maps into a complex projective space. Again one has a notion of an isotropic or superminimal map (these are the legs of the Frenet frame of a holomorphic curve) and all non-superminimal harmonic maps are covered by primitive maps into flag manifolds with their canonical k-symmetric structure. Moreover, when M is a 2-torus, these primitive maps are of finite type. The Reader is referred to [12] for more details on this and the other results of this section.

4 Loop groups and extended framings

The Adler–Kostant–Symes scheme produces commuting Hamiltonian flows on Lie algebras and we have seen how to apply this scheme to the twisted loop algebras $\Lambda\mathfrak{g}_\tau^{\mathbb{C}}$ and so obtain loops of flat connections and hence harmonic and primitive maps. However, in Section 1.4, a method of Symes was described for integrating these flows via projection of geodesics. We now show how this method applies in our setting. In so doing, we will introduce an analogue of Uhlenbeck's theory [56] of "extended solutions" (see, also, [28] in this volume) and, incidentally, obtain a conceptual proof of Theorem 2.4.

Terminology To avoid treating the case of k-symmetric spaces with $k = 2$ separately, henceforth we shall talk of primitive harmonic maps, conscious of the fact that the primitive condition is vacuous when $k = 2$ and that the harmonic condition is implied by the primitive condition when $k > 2$ (Theorem 3.6).

4.1 Iwasawa decomposition of $\Lambda G_\tau^{\mathbb{C}}$

We begin by introducing the infinite-dimensional Lie groups that correspond to the Lie algebras $\Lambda\mathfrak{g}_\tau^{\mathbb{C}}$, $\Lambda\mathfrak{g}_\tau$ and $\Lambda_+\mathfrak{g}_\tau^{\mathbb{C}}$ of Section 2.3. So let G be a compact semisimple Lie group with order k automorphism τ whose fixed set is K. Further, let

$$K^{\mathbf{C}} = KB$$

be the Iwasawa decomposition of $K^{\mathbf{C}}$ corresponding to that of $\mathfrak{k}^{\mathbf{C}}$.

Let $\Lambda G_\tau^{\mathbf{C}}$ be the manifold of loops

$$\Lambda G_\tau^{\mathbf{C}} = \{\gamma : S^1 \to G^{\mathbf{C}} : \gamma(\omega\lambda) = \tau\gamma(\lambda) \text{ for all } \lambda \in S^1\}.$$

Thus $\Lambda G_\tau^{\mathbf{C}}$ is an infinite-dimensional Lie group[2] under point-wise multiplication. Further, define subgroups by

$$\Lambda G_\tau = \{\gamma \in \Lambda G_\tau^{\mathbf{C}} : \gamma : S^1 \to G\};$$

$$\Lambda_+ G_\tau^{\mathbf{C}} = \{\gamma \in \Lambda G_\tau^{\mathbf{C}} : \gamma \text{ extends holomorphically to } \gamma : D \to G^{\mathbf{C}}, \gamma(0) \in B\}.$$

Clearly, $\Lambda G_\tau^{\mathbf{C}}$ has Lie algebra $\Lambda \mathfrak{g}_\tau^{\mathbf{C}}$ and so on.

We now have the following Iwasawa decomposition:

Theorem 4.1 [20] *Multiplication* $\Lambda G_\tau \times \Lambda_+ G_\tau^{\mathbf{C}} \to \Lambda G_\tau^{\mathbf{C}}$ *is a diffeomorphism onto.*

This result was deduced by Dorfmeister–Pedit–Wu [20] from the special case $\tau = $ id which was proved by Pressley–Segal [44]. The main idea of Pressley–Segal was that the homogeneous space $\Lambda G_\tau^{\mathbf{C}}/\Lambda_+ G_\tau^{\mathbf{C}}$ can be realised as an infinite-dimensional Grassmannian on which one can show that ΛG_τ acts transitively (cf. [28] in this volume).

4.2 Extended framings and a Symes formula

Let $\alpha_\lambda = \lambda^{-1}\alpha'_{\mathfrak{m}} + \alpha_0 + \lambda\alpha''_{\mathfrak{m}}$ be a $\Lambda\mathfrak{g}_\tau$-valued 1-form on a simply-connected surface M which satisfies the Maurer–Cartan equations. We have seen that, for each $\lambda \in S^1$, there is a smooth map $\Phi_\lambda : M \to G$ framing a primitive harmonic map $\phi_\lambda : M \to G/K$ and satisfying

$$\Phi_\lambda^{-1} d\Phi_\lambda = \alpha_\lambda.$$

Moreover, Φ_λ is unique up to left translation by a constant. We may choose these constants so that $\Phi_\lambda(x)$ depends smoothly on λ for some (and hence every) $x \in M$ and then we may view the Φ_λ as a single smooth map $\hat\Phi : M \to \Lambda G_\tau$ such that the following diagram commutes:

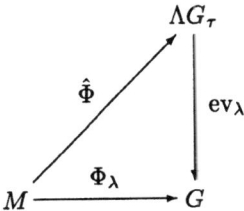

---

[2]We shall not worry about the topology of $\Lambda G_\tau^{\mathbf{C}}$: in fact, it suffices to work with the (Fréchet) C^∞-topology [44].

where $\mathrm{ev}_\lambda : \Lambda G_\tau \to G$ is given by evaluating the loop at $\lambda \in S^1$.

We call such maps $\hat{\Phi}$ *extended framings*. Thus $\hat{\Phi}$ is an extended framing if and only if
$$(\hat{\Phi}^{-1} d\hat{\Phi})_\lambda = \alpha_\lambda = \lambda^{-1}\alpha'_m + \alpha_0 + \lambda \alpha''_m,$$
where α'_m, α_0 and α''_m do not depend on λ. Any extended framing gives rise to a loop of primitive harmonic maps. Moreover, any primitive harmonic map of a simply connected M gives rise to a loop of such maps which are covered in this way by an extended framing. (In the case $k = 2$, these results are due to Rawnsley [46].)

We can find examples of extended framings by recourse to Theorem 4.1 and the method of Symes. For this, let $\Lambda^{(1)} \mathfrak{g}_\tau^{\mathbb{C}} \subset \Lambda \mathfrak{g}_\tau^{\mathbb{C}}$ consist of those elements $\eta \in \Lambda \mathfrak{g}_\tau^{\mathbb{C}}$ for which $\lambda \eta$ extends holomorphically to D. Thus $\eta \in \Lambda^{(1)} \mathfrak{g}_\tau^{\mathbb{C}}$ if and only if
$$\eta = \sum_{n \geq -1} \lambda^n \eta_{-n}.$$

Observe that if $g \in \Lambda_+ G_\tau^{\mathbb{C}}$ and $\eta \in \Lambda^{(1)} \mathfrak{g}_\tau^{\mathbb{C}}$ then $\operatorname{Ad} g \lambda \eta = \lambda \operatorname{Ad} g \eta$ also extends holomorphically to D so that $\Lambda^{(1)} \mathfrak{g}_\tau^{\mathbb{C}}$ is invariant under the adjoint action of $\Lambda_+ G_\tau^{\mathbb{C}}$.

Now choose $\eta_o \in \Lambda^{(1)} \mathfrak{g}_\tau^{\mathbb{C}}$ and define the (complex) geodesic $g : \mathbb{R}^2 \to \Lambda G_\tau^{\mathbb{C}}$ by
$$g(z) = \exp(z \eta_o).$$
Using Theorem 4.1, we find maps $a : \mathbb{R}^2 \to \Lambda G_\tau$, $b : \mathbb{R}^2 \to \Lambda_+ G_\tau^{\mathbb{C}}$ such that
$$g = ab$$
and we now have

Theorem 4.2 *a is an extended framing.*

Proof. We argue as in Section 1.4 to see that
$$\eta_o \, dz = dg \, g^{-1} = da \, a^{-1} + \operatorname{Ad} a (db \, b^{-1})$$
so that
$$a^{-1} da = \pi_{\mathcal{K}}(\operatorname{Ad} a^{-1} \eta_o \, dz).$$
Set $\eta = \operatorname{Ad} a^{-1} \eta_o$ and observe that since
$$\operatorname{Ad} g(z) \, \eta_o = \operatorname{Ad} \exp(z \eta_o) \, \eta_o = \eta_o,$$
we also have
$$\eta = \operatorname{Ad} b \, \eta_o$$
so that η takes values in $\Lambda^{(1)} \mathfrak{g}_\tau^{\mathbb{C}}$. This means that $\eta = \sum_{n \geq -1} \lambda^n \eta_{-n}$ so that, by (2.3) we have
$$a^{-1} da = \pi_{\mathcal{K}}(\eta \, dz) = \lambda^{-1} \eta_1 \, dz + (\eta_0 \, dz)_{\mathfrak{k}} + \lambda \overline{\eta_1} \, d\bar{z} \tag{4.1}$$

and a is an extended framing. □

The extended framings corresponding to finite type primitive harmonic maps all arise in this way: let $d \equiv 1 \bmod k$ and fix an initial condition $\xi_o \in \Lambda_d$. Then $\eta_o = \lambda^{d-1}\xi_o \in \Lambda^{(1)}\mathfrak{g}_\tau^\mathbb{C}$ and we apply Theorem 4.2 to get an extended framing a. Now set $\xi = \operatorname{Ad} a^{-1}\xi_o = \lambda^{1-d}\eta$. Again we have $\xi = \operatorname{Ad} b\,\xi_o$ so that, since $\lambda^d \xi_o$ extends holomorphically to the disc, $\lambda^d \xi$ does also and it follows that $\xi : \mathbb{R}^2 \to \Lambda_d$. Now

$$d\xi = [\xi, a^{-1}da]$$

while, from (4.1) and (2.5), we have

$$a^{-1}da = \pi_\mathcal{K}(\lambda^{d-1}\xi\,dz) = (\lambda^{-1}\xi_d + r(\xi_{d-1}))\,dz + (\lambda\xi_{-d} + \overline{r(\xi_{d-1})})\,d\bar{z}.$$

We therefore conclude that

1. ξ is the solution to the Hamiltonian equations (2.7);

2. the 1-form (2.8) constructed from ξ is precisely $a^{-1}da$ (and so solves the Maurer–Cartan equations so that the proof of Theorem 2.4 is complete);

3. a is precisely the extended framing for the finite type primitive harmonic map corresponding to ξ.

This gives an alternative characterisation of the finite type condition in terms of the extended framing which will be useful below. To be precise, we have proved the following theorem:

Theorem 4.3 *A primitive harmonic map* $\phi : \mathbb{R}^2 \to G/K$ *into a k-symmetric space is of finite type if and only if, for some $d \equiv 1 \bmod k$, there exists $\xi_o \in \Lambda_d$ such that, if $a : \mathbb{R}^2 \to \Lambda G_\tau$, $b : \mathbb{R}^2 \to \Lambda_+ G_\tau^\mathbb{C}$ are defined by*

$$\exp(z\lambda^{d-1}\xi_o) = a(z)b(z),$$

for $z \in \mathbb{R}^2$, then a is an extended framing for ϕ so that

$$\phi = \pi \circ \mathrm{ev}_1 \circ a,$$

where $\pi : G \to G/K$ is the coset projection.

Remark These simple results are the starting point of a loop-group theoretic analysis of the harmonic map equations which, to some extent, parallels that of the KdV equation by Segal–Wilson [52]. We shall return to this elsewhere. For a similar approach to Toda fields, the Reader is referred to the contribution of McIntosh in this volume [39] as well as [21, 40, 64].

5 Another approach

To date, the most comprehensive results concerning the construction of harmonic maps from commuting Hamiltonian flows are to be found in [15]. The approach adopted in that paper is similar to the one we have been describing but differs from it in a number of important respects. Among these are:

1. The Hamiltonian flows arise from the r-matrix formalism described in Section 2.5 and not from the Adler–Kostant–Symes scheme.

2. The main objects of study are harmonic maps into a Lie group G rather than into an arbitrary (k-)symmetric space. Since Lie groups are parallelisable, this means that one can treat such maps directly without recourse to a framing. Maps into (k-)symmetric spaces are then viewed as particular maps into G via a Cartan embedding (see Section 5.3 below).

In the remaining sections of this article, we describe the theory of [15] and its extension to the k-symmetric case in [12]. We will show how it relates to the theory we have been developing above and, in particular, we will see how Theorems 3.4 and 3.9 can be read off a corresponding result in [15].

5.1 Hamiltonian flows in the based loop algebra

We begin by rehearsing Uhlenbeck's zero-curvature reformulation of the harmonic map equations for maps into a Lie group G [56] and the method of Burstall–Ferus–Pedit–Pinkall [15] for producing solutions to those zero-curvature equations. The Reader will find a full account of these results in Wood's contribution to this volume [60] and so we shall be brief here, referring the Reader to [60] or [15] for more details.

So let $\phi : M \to G$ be a map of a Riemann surface into a compact Lie group. As usual, set $\alpha = \phi^{-1} d\phi$. We may view G as the homogeneous space $G/\{e\}$ so that ϕ is its own framing and it then follows from Section 3.2 that ϕ is harmonic if and only if α is co-closed:

$$d^*\alpha = 0. \tag{5.1}$$

Indeed, in this case, $\alpha = \alpha_{\mathfrak{m}}$ in (3.3) while $[\alpha \wedge *\alpha]$ always vanishes.

Combining (5.1) with the Maurer–Cartan equations for α, one concludes with Uhlenbeck that the loop of 1-forms given by

$$A_\lambda = \frac{1 - \lambda^{-1}}{2}\alpha' + \frac{1 - \lambda}{2}\alpha'' \tag{5.2}$$

is flat, i.e. satisfies the Maurer–Cartan equations for each $\lambda \in S^1$. Conversely, given a loop of flat connections (5.2) on a simply connected surface M, we integrate to find a harmonic map $\phi : M \to G$ with $\phi^{-1} d\phi = A_{-1}$.

Observe that the loop A_λ in (5.2) satisfies $A_1 \equiv 0$ and so may be viewed as a 1-form with values in the *based* loop algebra $\Omega \mathfrak{g}$ given by

Harmonic maps via Adler–Kostant–Symes theory

$$\Omega\mathfrak{g} = \{\xi : S^1 \to \mathfrak{g} : \xi(1) = 0\}.$$

View $\Omega\mathfrak{g}$ as a subalgebra of the free loop algebra $\Lambda\mathfrak{g}$ (which is $\Lambda\mathfrak{g}_\tau$ for $\tau = \mathrm{id}$) and set $\Omega_d = \Lambda_d \cap \Omega\mathfrak{g}$. We shall produce flat $\Omega\mathfrak{g}$-valued 1-forms (5.2) on $\mathbb{R}^2$ by integrating commuting flows on Ω_d. For this, fix $d \in \mathbb{N}$ and define vector fields X_1, X_2 on Ω_d by

$$\tfrac{1}{2}(X_1 - iX_2)(\xi) = [\xi, 2i(\lambda^{-1} - 1)\xi_d].$$

As explained in Wood's contribution [60], there is a Poisson structure on $\Omega\mathfrak{g}$ which arises from an r-matrix as in Section 2.5. Moreover, with respect to this Poisson structure, the vector fields X_i are Hamiltonian vector fields of Ad-invariant functions and so commute. Being of Lax type, they are also complete. Thus, if we fix an initial condition $\xi_o \in \Omega_d$, we may simultaneously integrate the X_i to get a unique map $\xi : \mathbb{R}^2 \to \Omega_d$ satisfying

$$\frac{\partial \xi}{\partial z} = [\xi, 2i(\lambda^{-1} - 1)\xi_d]; \qquad \xi(0) = \xi_o.$$

Then, defining an $\Omega\mathfrak{g}$-valued 1-form by

$$A_\lambda = 2i(\lambda^{-1} - 1)\xi_d \, dz - 2i(\lambda - 1)\xi_{-d} \, d\bar{z}$$

produces a solution to the Maurer–Cartan equations and hence a harmonic map $\phi : \mathbb{R}^2 \to G$ with $\phi^{-1}\partial\phi/\partial z = -4i\xi_d$.
Again we shall call the maps so obtained *harmonic maps of finite type*.

Notation Consistency with our previous development compels us to use slightly different notation from that in Wood's contribution [60] and, indeed, [15]. In those papers, the spectral parameter λ is the reciprocal of ours while the coefficients ξ_d differ by a sign.

Remark One should note that the finite-dimensional subspaces Ω_d are not Poisson submanifolds for the Poisson structure on $\Omega\mathfrak{g}$: they are only invariant for the Hamiltonian flows of Ad-invariant functions. Thus, while the ordinary differential equations that we are interested in evolve on a finite-dimensional space, these spaces are not Poisson manifolds in their own right. This is in contrast with the spaces $\Lambda_d \subset \Lambda\mathfrak{g}_\tau$ discussed above.

When is a harmonic map $\phi : \mathbb{R}^2 \to G$ of finite type? Given a harmonic map ϕ with associated $\Omega\mathfrak{g}$-valued 1-form A_λ, we may reformulate the finite type condition as the demand that, for some $d \in \mathbb{N}$, there is a map $\xi : \mathbb{R}^2 \to \Omega_d$ satisfying

$$d\xi = [\xi, A_\lambda]; \tag{5.3}$$
$$\alpha' = -4i\xi_d \, dz. \tag{5.4}$$

As before, a necessary condition for the existence of such a ξ is that $\phi^{-1}\partial\phi/\partial z$ takes values in a single $\mathrm{Ad}\, G^{\mathbb{C}}$-orbit in $\mathfrak{g}^{\mathbb{C}}$. The principal result of [15] is that a partial converse is true for doubly periodic harmonic maps:

Theorem 5.1 [15] *A doubly periodic harmonic map* $\phi : \mathbb{R}^2 \to G$ *is of finite type if* $\phi^{-1} \partial \phi / \partial z$ *takes values in a single semisimple* $\operatorname{Ad} G^{\mathbb{C}}$-*orbit*.

We shall see later that Theorems 3.4 and 3.9 are corollaries of this theorem. Meanwhile, let us see how it is proved. The first step is to find a formal solution to (5.3) and (5.4), that is, a formal Laurent series

$$Y = \sum_{j \geq -1} \lambda^j Y_{-j}$$

with each $Y_j : \mathbb{R}^2 \to \mathfrak{g}^{\mathbb{C}}$ such that $\alpha' = -4iY_1 \, dz$ and

$$dY = [Y, A_\lambda] \tag{5.5}$$

coefficient-wise. Using the semisimplicity hypothesis (and specifically that fact that $\eta \in \mathfrak{g}^{\mathbb{C}}$ is semisimple if and only if $\operatorname{ad} \eta$ is invertible on its image) one can find an explicit recursion formula for the Y_j and so construct Y. It is an immediate consequence of the recursion formula that if A_λ is doubly periodic then so is each Y_j.

The second step is where we make essential use of the double periodicity hypothesis. It follows from (5.5) that each Y_j is a Jacobi field for ϕ and thus is a solution of a linear elliptic equation. When α is doubly periodic, we view each Y_j as being defined on a torus and the compactness of the torus together with standard elliptic theory now implies that the Y_j span a *finite-dimensional* space. Using this, it is easy to construct a solution $\xi : \mathbb{R}^2 \to \Omega_d$, for some d, from some of the Y_j.

5.2 Extended solutions and the Symes method

A variant of the Symes method of Section 4.2 is available for integrating the flows on Ω_d. This is based on the following loop group decomposition: set

$$\Lambda G^{\mathbb{C}} = \{\gamma : S^1 \to G^{\mathbb{C}}\}$$

and define subgroups

$$\Omega G = \{\gamma \in \Lambda G^{\mathbb{C}} : \gamma : S^1 \to G \text{ and } \gamma(1) = e\};$$

$$\Lambda_+ G^{\mathbb{C}} = \{\gamma \in \Lambda G^{\mathbb{C}} : \gamma \text{ extends holomorphically to } \gamma : D \to G^{\mathbb{C}}\}.$$

We have

Theorem 5.2 [44] *Multiplication* $\Omega G \times \Lambda_+ G^{\mathbb{C}} \to \Lambda G^{\mathbb{C}}$ *is a diffeomorphism onto*.

Remark This is very similar to the Iwasawa decomposition in Section 4.1 the main difference being that the uniqueness of the decomposition is ensured by basing the real loops at 1 rather than basing the holomorphic loops at 0.

Harmonic maps via Adler–Kostant–Symes theory

Consider now a loop of 1-forms on a simply connected surface M:

$$A_\lambda = \frac{1-\lambda^{-1}}{2}\alpha' + \frac{1-\lambda}{2}\alpha'$$

which solve the Maurer–Cartan equations. For each $\lambda \in S^1$, we integrate to get maps $\Psi_\lambda : M \to G$ such that

$$\Psi_\lambda^{-1} d\Psi_\lambda = A_\lambda.$$

In particular, since $A_1 \equiv 0$, Ψ_1 is constant and we choose the constants of integration so that $\Psi_1 \equiv e$ and the Ψ_λ form a smooth map $\hat\Psi : M \to \Omega G$. Such $\hat\Psi$ are the *extended solutions* of Uhlenbeck [56] (cf. [28, 60] in this volume). It is clear from Section 5.1 that if $\hat\Psi$ is an extended solution, the map ϕ defined by the commuting diagram

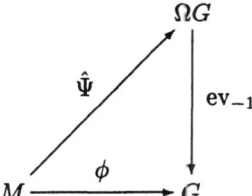

is harmonic and all harmonic maps $M \to G$ of a simply connected M are covered by an extended solution in this way.

The method of Symes used in Section 4.2 can be adapted to this setting. Firstly, if $\eta_o \in \Lambda \mathfrak{g}^{\mathbb{C}}$ is such that $\lambda\eta_o \in \Lambda_+\mathfrak{g}^{\mathbb{C}}$ (i.e., $\lambda\eta_o$ extends holomorphically to D) then we define $a : \mathbb{R}^2 \to \Omega G$, $b : \mathbb{R}^2 \to \Lambda_+ G^{\mathbb{C}}$ by

$$\exp(z\eta_o) = a(z)b(z),$$

for $z \in \mathbb{R}^2$, and we have

Proposition 5.3 *a is an extended solution.*

Secondly, if $\xi_o \in \Omega_d$, put $\eta_o = 2i\lambda^{d-1}\xi_o$ and define a as above. Then the map $\xi : \mathbb{R}^2 \to \Omega\mathfrak{g}$ given by

$$\xi = \operatorname{Ad} a^{-1}\xi_o$$

has the following properties:

1. ξ takes values in Ω_d;
2. $a^{-1}da = 2i(\lambda^{-1} - 1)\xi_d \, dz - 2i(\lambda - 1)\xi_{-d} \, d\bar{z}$;
3. $d\xi = [\xi, a^{-1}da]$.

Thus ξ is the solution of our Hamiltonian flows on Ω_d with initial condition ξ_o and a is the extended solution for the corresponding harmonic map of finite type.

The proofs of these assertions are similar to those in Section 4.2 and consequently left to the Reader.

5.3 Maps into k-symmetric spaces and Cartan embeddings

To make contact with the results of Section 3 and, in particular, to obtain primitive harmonic maps into k-symmetric spaces from the constructions of Section 5.1, we introduce a certain embedding of a k-symmetric space G/K into its group of isometries G. This will enable us to view primitive harmonic maps into G/K as maps into G to which the methods of Section 5.1 can be applied.

So let G/K be a k-symmetric space with involution τ and Maurer–Cartan form $\beta: T(G/K) \to \mathfrak{g}$. Define a map $\iota: G/K \to G$ by

$$\iota(g \cdot o) = \tau(g)g^{-1}.$$

It is clear that ι is well-defined. In fact, it is an immersion, as can be seen from the following

Lemma 5.4 [12] *Let θ be the Maurer–Cartan form of G. Then, for $x \in G/K$,*

$$\iota^*\theta_x = \tau_x \beta_x - \beta_x,$$

where $\tau_{g \cdot o} = \operatorname{Ad} g \circ \tau \circ \operatorname{Ad} g^{-1}$.

Moreover, if K is the fixed set of τ, then ι is an embedding. We call ι the *Cartan embedding* of G/K into G.

When $k = 2$, the Cartan embedding is well-known to be totally geodesic [18] so that if $\phi: M \to G/K$ is harmonic then $\iota \circ \phi: M \to G$ is also. Thus the methods of Section 5.1 apply to $\iota \circ \phi$. This is the approach to harmonic maps into symmetric spaces adopted in [15]. When $k > 2$, ι is no longer totally geodesic but all is not lost: a primitive map $\phi: M \to G/K$ still gives rise to a loop A_λ of the form (5.2) from which ϕ may be recovered.

For this, let $\phi: M \to G/K$ be a primitive harmonic map into a k-symmetric space and set $\delta = \phi^*\beta$. The equations (3.7) for a framing of ϕ can be written in terms of δ as

$$\mathrm{d}\delta' - [\delta' \wedge \delta''] = \mathrm{d}\delta'' - [\delta' \wedge \delta''] = 0.$$

However, these are precisely the coefficients of $(\lambda^{-1} - 1)$ and $(\lambda - 1)$ in the loop of 1-forms

$$A_\lambda = (\lambda^{-1} - 1)\delta' + (\lambda - 1)\delta''. \tag{5.6}$$

Thus a primitive map ϕ gives rise to a loop of flat 1-forms A_λ of the form (5.2). To recover $\psi = \iota \circ \phi$ from A_λ, observe that from Lemma 5.4 we get

$$\psi^*\theta = (\omega^{-1} - 1)\delta' + (\omega - 1)\delta''$$

whence

$$\psi^{-1}\mathrm{d}\psi = A_\omega.$$

Example Let $k = 2$. Then $\omega = -1$ and $\alpha = \psi^{-1}d\psi = -2\delta$ so that the loop A_λ is just
$$\frac{1-\lambda^{-1}}{2}\alpha' + \frac{1-\lambda}{2}\alpha''$$
which is, of course, the loop associated to the harmonic map $\psi = \iota \circ \phi$ by Uhlenbeck.

We have seen in Section 5.1 how to produce loops (5.2) of flat 1-forms from commuting flows on Ω_d. When does this procedure give rise to primitive harmonic maps? This is a question with a pleasantly simple answer: it is just a matter of choosing the right initial condition for the flows as the following theorem shows [12] (see [15] for the case $k = 2$).

Theorem 5.5 *Fix $d \equiv 1 \mod k$ and choose $\xi_o \in \Omega_d \cap \Lambda\mathfrak{g}_\tau$. Let $\xi : \mathbb{R}^2 \to \Omega_d$ satisfy*
$$\frac{\partial \xi}{\partial z} = [\xi, 2i(\lambda^{-1} - 1)\xi_d]; \qquad \xi(0) = \xi_o$$
and let $\psi : \mathbb{R}^2 \to G$ be the map satisfying
$$\psi^{-1}\frac{\partial \psi}{\partial z} = 2i(\omega^{-1} - 1)\xi_d \qquad \psi(0) = 0.$$
Then $\psi = \iota \circ \phi$ for a primitive harmonic map $\phi : \mathbb{R}^2 \to G/K$.

It would be natural to call such maps primitive harmonic maps of finite type. However, this terminology has already been reserved for primitive harmonic maps which possess a framing which arises from commuting flows on $\Lambda_d \subset \Lambda\mathfrak{g}_\tau$. A priori, this is a different condition so, for now, we call the maps provided by Theorem 5.5 *primitive harmonic maps of G-finite type*. Later we shall see that G-finite type maps are indeed of finite type in the sense of Section 3.

Meanwhile, in view of the relation
$$\psi^{-1}\partial\psi/\partial z = (\omega^{-1} - 1)\delta'(\partial/\partial z),$$
a semisimple orbit condition on $\delta'(\partial/\partial z)$ implies one for $\psi^{-1}\partial\psi/\partial z$ and a simple adaptation of the proof of Theorem 5.1 then gives:

Theorem 5.6 [12] *Let $\phi : \mathbb{R}^2 \to G/K$ be a doubly periodic primitive harmonic map such that $\delta'(\partial/\partial z)$ takes values in a single semisimple $\mathrm{Ad}\,G^{\mathbb{C}}$-orbit. Then ϕ is of G-finite type.*

If Φ is a framing for such a map, we have
$$\mathrm{Ad}\,\Phi\alpha_m = \delta$$
so again, an orbit condition on δ' is implied by one on α'_m and we therefore deduce the following analogue of Theorems 3.4 and 3.9:

Theorem 5.7 *Let $\phi : \mathbb{R}^2 \to G/K$ be a doubly periodic primitive harmonic map into a k-symmetric space. Suppose that some (and hence every) framing of ϕ has $\alpha'_m(\partial/\partial z)$ taking values in a single semisimple $\mathrm{Ad}\,K^{\mathbb{C}}$-orbit in $\mathfrak{g}_{-1}$. Then ϕ is of G-finite type.*

5.4 Finite vs. G-finite type

In this section, we shall show that a primitive harmonic map $\phi : \mathbb{R}^2 \to G/K$ of G-finite type is in fact of finite type. Together with Theorem 5.7, this will provide a proof of Theorems 3.4 and 3.9.

Before embarking on this however, let us pause to consider why this rather indirect method is needed to prove Theorems 3.4 and 3.9. The problem is that the approach of Section 3 deals with framings of ϕ and, moreover, framings of a rather special kind: for ϕ to be of finite type, the map $\xi : \mathbb{R}^2 \to \Lambda_d$ produces a framing with several restrictions:

1. $\alpha'_\mathfrak{m}(\partial/\partial z) = \xi_d$ so that, as we have seen, $\alpha'_\mathfrak{m}(\partial/\partial z)$ takes values in a single $\operatorname{Ad} K^\mathbb{C}$-orbit.

2. $\alpha_\mathfrak{k}(\partial/\partial z) = r(\xi_{d-1})$ so that $\alpha_\mathfrak{k}(\partial/\partial z)$ takes values in $\bar{\mathfrak{n}} \oplus \mathfrak{h}$.

3. In view of the equations for ξ we have
$$\frac{\partial \xi_d}{\partial z} = [\xi_d, (r - \operatorname{id})\xi_{d-1}]$$
which amounts to a differential relation between $\alpha'_\mathfrak{m}$ and $\alpha'_\mathfrak{k}$.

These conditions are quite stringent: for example, for the primitive maps into flag manifolds discussed in Section 3.4 they are equivalent to the demand that the framing be a Toda frame.

Now suppose that ϕ does have such a framing with α doubly periodic. The argument of Theorem 5.1 can easily be adapted to produce the necessary $\xi : \mathbb{R}^2 \to \Lambda_d$. Indeed, this is the approach used in [8, 25] to treat primitive maps of flag manifolds since, in this case, by Corollary 3.15, a doubly periodic map with Toda frame has doubly periodic Toda frame. However, in the general case, we know of no direct method to equip a doubly periodic primitive harmonic map with a doubly periodic frame of the right kind. Hence the need to proceed indirectly.

This said, our proof is a simple adaptation of the Symes method we have been developing. So let $\phi : \mathbb{R}^2 \to G/K$ be a primitive harmonic map of G-finite type with $\psi = \iota \circ \phi : \mathbb{R}^2 \to G$. We know from Sections 5.2 and 5.3 that the G-finite type condition means that, for some $d \equiv 1 \bmod k$, there is $\xi_o \in \Omega_d \cap \Lambda\mathfrak{g}_\tau$ so that the map $g : \mathbb{R}^2 \to \Lambda G^\mathbb{C} : z \mapsto \exp(2i\lambda^{d-1}\xi_o)$ has a unique factorisation

$$g = ab, \tag{5.7}$$

$a : \mathbb{R}^2 \to \Omega G$, $b : \mathbb{R}^2 \to \Lambda_+ G^\mathbb{C}$ with

$$a_\omega = \psi.$$

On the other hand, we may view g as a map $\mathbb{R}^2 \to \Lambda G^\mathbb{C}_\tau$ and use the Iwasawa decomposition of Theorem 4.1 to write

$$g = \tilde{a}\tilde{b}$$

with $\tilde{a} : \mathbb{R}^2 \to \Lambda G_\tau$, $\tilde{b} : \mathbb{R}^2 \to \Lambda_+ G_\tau^{\mathbb{C}}$. We know from Theorem 4.3 that $\tilde{a}$ is an extended framing of a primitive harmonic map of finite type given by

$$\pi \circ \mathrm{ev}_1 \circ \tilde{a}$$

and it is our contention that this map is ϕ. For this, it suffices to show that

$$\iota \circ \pi \circ \mathrm{ev}_1 \circ \tilde{a} = \psi.$$

From the definition of the Cartan embedding, we get

$$\iota \circ \pi \circ \mathrm{ev}_1 \circ \tilde{a} = \tau(\tilde{a}_1)\tilde{a}_1^{-1} = \tilde{a}_\omega \tilde{a}_1^{-1}$$

where we have also used the symmetry $\tilde{a}_{\omega\lambda} = \tau \circ \tilde{a}_\lambda$. However, by the uniqueness of the decomposition (5.7), we have

$$a = \tilde{a}\tilde{a}_1^{-1}; \qquad b = \tilde{a}_1 \tilde{b}$$

so that

$$\tilde{a}_\omega \tilde{a}_1^{-1} = a_\omega = \psi.$$

We have therefore proved:

Theorem 5.8 *A primitive harmonic map $\mathbb{R}^2 \to G/K$ of G-finite type is of finite type.*

5.5 Coda: finite type vs. finite uniton number

There is another interesting class of harmonic maps from a Riemann surface to a Lie group G: these are the harmonic maps with finite uniton number discovered by Uhlenbeck [56]. The theory of these maps is fundamental to the twistorial approach to harmonic 2-spheres in symmetric spaces (see [16, 56, 63] and [28, 60] in this volume).

We conclude this article with a simple example which demonstrates that the class of finite type harmonic maps is essentially disjoint from that of maps with finite uniton number.

First let us recall Uhlenbeck's theory: let $\phi : M \to G$ be a harmonic map of a Riemann surface and suppose that ϕ admits an extended solution $\Phi : M \to \Omega G$ (which is certainly true if M is simply connected). The extended solution is unique up to left multiplication by a constant element of ΩG and we say that ϕ has *finite uniton number* if this constant can be chosen so that Φ takes values in the subspace of (Laurent) polynomial loops in ΩG of some fixed degree:

$$\Phi_\lambda(z) = \sum_{|m| \leq d} \lambda^m T_m(z), \tag{5.8}$$

for $z \in M$. Of course, to make sense of (5.8) we must view G as a matrix group.

Such polynomial loops admit a decomposition into "linear" factors which gives rise to a decomposition of the harmonic map into flag factors as described by Wood [60] in this volume.

The main result of Uhlenbeck is that any extended solution on a *compact* M can be normalised to take values in such a space of polynomial loops. As a consequence, any harmonic map $S^2 \to G$ has finite uniton number.

The situation for finite type harmonic maps is completely different: one can prove

Theorem 5.9 *A finite type harmonic map $\phi : \mathbb{R}^2 \to G$ with $\phi^{-1}\partial\phi/\partial z$ semisimple does not have finite uniton number.*

Corollary 5.10 *A doubly periodic harmonic map $\phi : \mathbb{R}^2 \to G$ with $\phi^{-1}\partial\phi/\partial z$ taking values in a semisimple orbit does not have finite uniton number.*

Note that a harmonic map $\phi : S^2 \to G$ has $\phi^{-1}\partial\phi/\partial z$ nilpotent by the usual holomorphic differentials argument.

The proof of Theorem 5.9 would take us too far afield so we shall content ourselves with presenting a simple example which turns out to contain the heart of the matter.

Let $A \in \mathfrak{g}^{\mathbb{C}}$ and suppose that $[A, \overline{A}] = 0$ (from which it follows that A is semisimple). Set $\xi_A = (\lambda^{-1} - 1)A + (\lambda - 1)\overline{A} \in \Omega_1$ and consider the geodesic $g : \mathbb{R}^2 \to \Lambda G^{\mathbb{C}}$ given by

$$g(z) = \exp(z\xi_A).$$

It is easy to see that the extended solution obtained by factorising g is given by

$$a(z) = \exp((\lambda^{-1} - 1)zA + (\lambda - 1)\overline{zA})$$

so that the corresponding harmonic map $\phi : \mathbb{R}^2 \to G$ of finite type is given by

$$\phi(z) = \exp(-2zA - 2\overline{zA}).$$

Thus ϕ is a product of geodesics with commuting generators. The corresponding map $\xi : \mathbb{R}^2 \to \Omega_1$ is constant so that we have a stationary point for the Hamiltonian flows.

Suppose now that ϕ has finite uniton number. Then there is a constant $\gamma \in \Omega G$ for which $\gamma a = \Phi$ has polynomial dependence on λ. But fixing $z_1 \neq z_2 \in \mathbb{R}^2$, this implies that

$$\Phi(z_1)^{-1}\Phi(z_2) = a(z_1)^{-1}a(z_2)$$

is a Laurent polynomial in λ. But

$$a(z_1)^{-1}a(z_2) = \exp((\lambda^{-1} - 1)(z_2 - z_1)A + (\lambda - 1)\overline{(z_2 - z_1)A})$$

which has an essential singularity at 0.

Bibliography

[1] M. Adler and P. van Moerbeke, *Completely integrable systems, Euclidean Lie algebras, and curves*, Adv. Math. **38** (1980), 267–317.

[2] A. Bahy-El-Dien and J.C. Wood, *The explicit construction of all harmonic two-spheres in $G_2(\mathbf{R}^n)$*, J. reine angew. Math. **398** (1989), 36–66.

[3] J. Barbosa, *On minimal immersions of S^2 into S^{2m}*, Trans. Amer. Math. Soc. **210** (1975), 75–106.

[4] M. Black, *Harmonic maps into homogeneous spaces*, Pitman Res. Notes in Math., vol. 255, Longman, Harlow, 1991.

[5] A.M. Bloch, H. Flaschka, and T. Ratiu, *A convexity theorem for isospectral manifolds of Jacobi matrices in a compact Lie algebra.*, Duke Math. J. **61** (1990), 41–65.

[6] A.I. Bobenko, *All constant mean curvature tori in $\mathbf{R}^3$, S^3, H^3 in terms of theta-functions*, Math. Ann. **290** (1991), 209–245.

[7] _____, A.I. Bobenko, *Surfaces in terms of 2 by 2 matrices. Old and new integrable cases*, this volume.

[8] J. Bolton, F. Pedit, and L.M. Woodward, *Minimal surfaces and the affine Toda field model*, preprint, 1993.

[9] J. Bolton and L.M. Woodward, *The affine Toda equations and minimal surfaces*, this volume.

[10] R.L. Bryant, *Conformal and minimal immersions of compact surfaces into the 4-sphere*, J. Diff. Geom. **17** (1982), 455–473.

[11] _____, *Lie groups and twistor spaces*, Duke Math. J. **52** (1985), 223–261.

[12] F.E. Burstall, *Harmonic tori in spheres and complex projective spaces*, preprint, 1993.

[13] _____, *Twistor methods for harmonic maps*, Differential Geometry (V.L. Hansen, ed.), Lect. Notes in Math., vol. 1263, Springer, Berlin, Heidelberg, New York, 1987, pp. 55–96.

[14] _____, *Recent developments in twistor methods for harmonic maps*, Harmonic mappings, twistors and σ-models (P. Gauduchon, ed.), World Scientific, Singapore, 1988, pp. 158–176.

[15] F.E. Burstall, D. Ferus, F. Pedit, and U. Pinkall, *Harmonic tori in symmetric spaces and commuting Hamiltonian systems on loop algebras*, Ann. of Math. **138** (1993), 173–212.

[16] F.E. Burstall and J.H. Rawnsley, *Twistor theory for Riemannian symmetric spaces with applications to harmonic maps of Riemann surfaces*, Lect. Notes in Math., vol. 1424, Springer, Berlin, Heidelberg, New York, 1990.

[17] E. Calabi, *Quelques applications de l'analyse complexe aux surfaces d'aire minima*, Topics in Complex Manifolds, Presses de Université de Montréal, Montréal, 1967, pp. 59–81.

[18] J. Cheeger and D. Ebin, *Comparison Theorems in Riemannian Geometry*, North Holland, Amsterdam, 1975.

[19] A. Doliwa and A. Sym, *The non-linear σ-model on spheres*, preprint, Warsaw University, 1992.

[20] J. Dorfmeister, F. Pedit, and H. Wu, *Weierstrass representation of harmonic maps into symmetric spaces*, in preparation.

[21] J. Dorfmeister and H. Wu, *Constant mean curvature surfaces and loop groups*, J. reine angew. Math., to appear.

[22] J. Eells and S.M. Salamon, *Twistorial construction of harmonic maps of surfaces into four manifolds*, Ann. Scuola Norm. Sup. Pisa **12** (1985), 589–640.

[23] J. Eells and J.C. Wood, *Harmonic maps from surfaces into projective spaces*, Adv. in Math. **49** (1983), 217–263.

[24] D. Ferus and F. Pedit, S^1-*equivariant minimal tori in* S^4 *and* S^1-*equivariant Willmore tori in* S^3, Math. Z. **204** (1990), 269–282.

[25] D. Ferus, F. Pedit, U. Pinkall, and I. Sterling, *Minimal tori in* S^4, J. reine angew. Math. **429** (1992), 1–47.

[26] K. Fujii, *A relation between instantons of Grassmannian σ-models and Toda equations I, II, III*, Yokohama preprints, 1992.

[27] J.-L. Gervais and Y. Matsuo, *W-geometries*, Phys. Lett. B **274** (1992), 309–316.

[28] M.A. Guest and Y. Ohnita, *Loop group actions on harmonic maps and their applications*, this volume.

[29] V. Guillemin and S. Sternberg, *Symplectic techniques in physics*, Camb. Univ. Press, Cambridge, 1984.

[30] S. Helgason, *Differential Geometry, Lie Groups, and Symmetric Spaces*, Academic Press, New York, San Francisco, London, 1978.

[31] N.J. Hitchin, *The self-duality equations on a Riemann surface*, Proc. Lond. Math. Soc. **55** (1987), 59–126.

[32] _____, *Harmonic maps from a 2-torus to the 3-sphere*, J. Diff. Geom. **31** (1990), 627–710.

[33] W.Y. Hsiang and H.B. Lawson, *Minimal submanifolds of low cohomogeneity*, J. Diff. Geom. **5** (1971), 1–38.

[34] J.E. Humphreys, *Introduction to Lie algebras and representation theory*, Springer, New York, Heidelberg, London, 1972.

[35] P.Z. Kobak, *Twistors, nilpotent orbits and harmonic maps*, this volume.

[36] B. Kostant, *The principal three dimensional subgroup and the Betti numbers of a complex simple Lie group*, Amer. J. Math. **81** (1959), 973–1032.

[37] _____, *The solution to a generalized Toda lattice and representation theory*, Adv. Math. **34** (1979), 195–338.

[38] O. Kowalski, *Generalized symmetric spaces*, Lect. Notes in Math., vol. 805, Springer, Berlin, Heidelberg, New York, 1980.

[39] I. McIntosh, *Infinite dimensional Lie groups and the two-dimensional Toda lattice*, this volume.

[40] _____, *Global solutions of the elliptic 2D Toda lattice*, preprint, 1992.

[41] A.M. Perelomov, *Integrable systems of classical mechanics and Lie algebras*, Birkhäuser, Basel, 1990.

[42] U. Pinkall and I. Sterling, *On the classification of constant mean curvature tori*, Ann. of Math. **130** (1989), 407–451.

[43] K. Pohlmeyer, *Integrable Hamiltonian systems and interactions through quadratic constraints*, Commun. Math. Phys. **46** (1976), 207–221.

[44] A.N. Pressley and G. Segal, *Loop Groups*, Oxford Math. Monographs, Clarendon Press, Oxford, 1986.

[45] J.H. Rawnsley, *Noether's theorem for harmonic maps*, Diff. Geom. Methods in Math. Phys. (S. Sternberg, ed.), Reidel, Dortrecht, Boston, London, 1984, pp. 197–202.

[46] ———, *Harmonic 2-spheres*, Quantum Theories and Geometry (M. Cahen and M. Flato, eds.), Kluwer Acad. Publ., Dordrecht, 1988, pp. 175–189.

[47] ———, *Twistor theory for Riemannian manifolds*, New Developments in Lie Theory and Their Applications (J. Tirao and N. Wallach, eds.), Progress in Mathematics, vol. 105, Birkhäuser, Boston, Basel, Stuttgart, 1992, pp. 115–128.

[48] G.L. Rcheulishvili and M.V. Saveliev, *Multidimensional nonlinear systems related to the Grassman manifolds BI, DI*, Funkts. Anal. i Prilozh. **21:4** (1987), 83–84; English transl.: Funct. Anal. Appl. **21** (1987), 332–333.

[49] A.G. Reyman and M.A. Semenov-Tian-Shansky, *Compatible Poisson structures for Lax equations: an r-matrix approach*, Phys. Let. A **130** (1988), 456–460.

[50] S.M. Salamon, *Harmonic and holomorphic maps*, Geometry Seminar L. Bianchi., Lect. Notes Math., vol. 1164, Springer, Berlin, Heidelberg, New York, 1986, pp. 161–224.

[51] M.V. Saveliev, *Classification of exactly integrable embeddings of two-dimensional manifolds. The coefficients of the third fundamental forms.*, Teoret. Mat. Fiz. **60:1** (1984), 9–23; English transl.: Theoret. and Math. Phys. **60:1** (1984), 638–647.

[52] G. Segal and G. Wilson, *Loop groups and equations of KdV type*, Publ. Math. I.H.E.S. **61** (1985), 5–65.

[53] S. Sternberg, *Lectures on differential geometry*, Chelsea, New York, 1983.

[54] W. Symes, *Systems of Toda type, inverse spectral problems and representation theory*, Invent. Math. **159** (1980), 13–51.

[55] K. Uhlenbeck, *Equivariant harmonic maps into spheres*, Harmonic Maps (R.J. Knill, M. Kalka, and H.C.J. Sealey, eds.), Lect. Notes in Math., vol. 949, Springer, Berlin, Heidelberg, New York, 1982, pp. 146–158.

[56] ———, *Harmonic maps into Lie groups (classical solutions of the chiral model)*, J. Diff. Geom. **30** (1989), 1–50.

[57] E.B. Vinberg, *The Weyl group of a graded Lie algebra*, Izv. Akad. Nauk SSSR Ser. Mat. **40** (1976), 488–526, 709; English transl.: Math. USSR-Izv. **10** (1976), 463–495.

[58] A. Weinstein, *The local structure of Poisson manifolds*, J. Diff. Geom. **18** (1983), 523–557.

[59] H.C. Wente, *Counterexample to a conjecture of H. Hopf*, Pac. J. Math. **121** (1986), 193–243.

[60] J.C. Wood, *Harmonic maps into symmetric spaces and integrable systems*, this volume.

[61] ———, *Holomorphic differentials and classification theorems for harmonic maps and minimal immersions*, Global Riemannian Geometry (T.J. Willmore and N.J. Hitchin, eds.), E. Horwood Series in Math. and its Appl., E. Horwood, Chichester, 1984, pp. 168–175.

[62] _____, *Twistor constructions for harmonic maps*, Differential Geometry and Differential Equations (ed. C.H. Gu, M. Berger and R.L. Bryant), Lect. Notes in Math., vol. 1255, Springer, Berlin, Heidelberg, New York, 1988, pp. 130–159.

[63] _____, *Explicit construction and parametrisation of harmonic two-spheres in the unitary group*, Proc. Lond. Math. Soc. **58** (1989), 608–624.

[64] H. Wu, *Banach manifolds of minimal surfaces in the 4-sphere*, Differential Geometry (R.E. Greene and S.T. Yau, eds.), Proc. Symp. Pure Math., vol. 54, Amer. Math. Soc., Providence RI, 1993, pp. 530–539.

[65] V.E. Zakharov and A.V. Mikhaïlov, *Relativistically invariant two-dimensional models of field theory which are integrable by means of the inverse scattering problem method*, Zh. Eksp. Teor .Fiz. **74** (1978), 1953-1973; English transl.: Sov. Phys. JETP **47** (1978) 1017-1027.

[66] V.E. Zakharov and A.B. Shabat, *Integration of non-linear equations of mathematical physics by the method of inverse scattering, II*, Funkts. Anal i Prilozhen **13** (1978), 13–22; English transl.: Func. Anal. Appl.**13** (1979), 166-174

Loop group actions on harmonic maps and their applications

M. A. Guest and Y. Ohnita

1 Introduction

The purpose of this article is to give an exposition of finite dimensional and infinite dimensional group actions on harmonic maps of Riemann surfaces into Lie groups (or symmetric spaces), and their applications to the study of deformations of harmonic maps. The classification theory of harmonic maps of Riemann spheres into the n-dimensional standard sphere S^n was given by Calabi ([13], Chern [15], Barbosa [5]), see [45]. It was shown that any harmonic 2-sphere in S^n can be lifted to a horizontal holomorphic curve in the *twistor space* over S^n. A certain (noncompact simple) complex Lie group acts on the twistor space in a natural way, and this action transforms a horizontal holomorphic curve to another one. Thus it induces a group action on harmonic 2-spheres in S^n. This group action has been used to obtain interesting results on the space of harmonic 2-spheres in S^n (cf. [5],[26],[8]).

¿From the point of view of gauge theory, harmonic maps of Riemann surfaces into symmetric spaces can be regarded as classical solutions of a nonlinear σ-model over 2-dimensional Euclidean space. By virtue of the work of Pohlmeyer [34], Zakharov, Mikhailov, Shabat [48, 49] and Uhlenbeck [42], it is well known that harmonic maps of simply connected Riemann surfaces into compact Lie groups G can be lifted to holomorphic maps into the based loop group ΩG satisfying a certain "horizontality" condition, namely so called "extended solutions". This correspondence is the basic result in this area of harmonic map theory. Indeed, the lift to horizontal holomorphic curves in twistor space, which plays a fundamental role in the theory of harmonic 2-spheres in S^n, can be understood as a special case of an extended solution.

On the other hand, infinitesimal actions of infinite dimensional Lie algebras on solution spaces of nonlinear field equations (like those for nonlinear σ-models and anti-self-dual Yang-Mills fields) appear in mathematical physics (in the theory of integrable systems) where they are known as "hidden symmetries". The affine Kac-Moody algebra of infinitesimal deformations for the $SU(n)$-model was first observed by Dolan ([16]) and developed by Wu ([46]). Zakharov, Mikhailov and Shabat ([48, 49]) exhibited the method of the Riemann-Hilbert transform for 2-dimensional models of field theory, in particular for σ-models. Ueno and Nakamura ([40, 41]) clarified the link between Dolan's infinitesimal action and the so-called infinitesimal Riemann-Hilbert transform. Following the work of Takasaki [39] on anti-self-dual Yang-Mills fields, who used a method similar to that of M. Sato ([37]), Jacques and Saint-Aubin ([25]) interpreted the linear system associated with σ-models (i.e. the equations for extended solutions) in terms of the evolution of a point in an infinite dimensional (formal) Grassmann manifold, and gave a pseudo-action of a (formal) Lie group of certain matrices of infinite degree corresponding to Dolan's infinitesimal action.

Uhlenbeck ([42]) introduced the action of an infinite dimensional group of certain matrix-valued rational functions on harmonic maps into the unitary group $U(n)$, which was closely related with the method of the Riemann-Hilbert transform of [48, 49]. Moreover, Uhlenbeck's action induces the same infinitesimal action as (the harmonic map version of) Dolan's infinitesimal action. The first author's joint paper [7] with Bergvelt gave a simplification of the pseudo-action of [48, 49] and discussed Uhlenbeck's action as a special case of a "dressing pseudo-action" on harmonic maps. In [23], the "natural" action of the complex loop group (a Kac-Moody group) on harmonic maps was considered from the viewpoint of the Grassmannian model in loop group theory (cf. [35]). It was shown that Uhlenbeck's action coincides with this natural action. Since the latter is considerably easier to work with, this gave a further simplification of the theory.

The natural action of the complex loop group LG^c (on harmonic maps with values in G) can be used to study the space of harmonic maps. The idea is to use a one-parameter subgroup to deform a given extended solution into a simpler one. In the case at hand, the orbits of suitable one-parameter subgroups are flow lines of Morse-Bott functions and may be described explicitly. As applications of this idea, we shall mention results on the connected components of spaces of harmonic 2-spheres in specific symmetric spaces like $U(n)$, S^n, the projective spaces $\mathbf{R}P^n$, $\mathbf{C}P^n$, $\mathbf{H}P^n$, and also on the fundamental groups of the spaces of harmonic 2-spheres in S^n and $\mathbf{R}P^n$.

2 Extended solutions and harmonic maps

Let us begin with the definition of extended solutions corresponding to harmonic maps into Lie groups. Let G be a compact connected Lie group and let $\mathfrak{g}$ be its Lie algebra. We equip G with a bi-invariant Riemannian metric. Then G becomes a compact Riemannian symmetric space. Denote by μ the Maurer-Cartan form of G, which satisfies the identity

$$d\mu + \frac{1}{2}[\mu \wedge \mu] = 0.$$

(the Maurer-Cartan equation). Now let $\varphi : \Sigma \longrightarrow G$ be a smooth map of a Riemann surface Σ into G. Set $\alpha = \varphi^*\mu = \varphi^{-1}d\varphi$. This is a 1-form with values in $\mathbf{g}$, which satisfies

$$d\alpha + \frac{1}{2}[\alpha \wedge \alpha] = 0. \tag{1.1}$$

We may decompose the 1-form α into $\alpha = \alpha' + \alpha''$, where α' and α'' denote the $(1,0)$-part and $(0,1)$-part of α with respect to the complex structure of Σ. We have (cf. [45]

Proposition 1 *The map φ is harmonic if and only if*

$$\bar{\partial}\alpha' - \partial\alpha'' = 0. \tag{1.2}$$

More generally, if a smooth map $\varphi : \Sigma \longrightarrow G^c$ satisfies the above equation (1.2), where G^c is the complexification of the compact Lie group G, then we shall say that φ is *harmonic*.

Now let $\beta = \beta' + \beta''$ be a $\mathbf{g}^c$-valued 1-form on Σ, where β' and β'' are the $(1,0)$-part and $(0,1)$-part of β, respectively. For each $\lambda \in \mathbf{C}^* = \mathbf{C} \setminus \{0\}$, define

$$\beta_\lambda = \frac{1}{2}(1 - \lambda^{-1})\beta' + \frac{1}{2}(1 - \lambda)\beta''.$$

Consider the linear first order partial differential equation

$$\Phi_\lambda^* \mu_{G^c} = \beta_\lambda, \text{ for all } \lambda \in \mathbf{C}^*, \tag{1.3}$$

or equivalently

$$\begin{cases} \partial \Phi_\lambda = \frac{1}{2}(1 - \lambda^{-1})\Phi_\lambda \beta', \\ \bar{\partial} \Phi_\lambda = \frac{1}{2}(1 - \lambda)\Phi_\lambda \beta''. \end{cases} \tag{1.4}$$

Then the important observation is that the complete integrability conditions of (1.3) or (1.4), namely

$$d\beta_\lambda + \frac{1}{2}[\beta_\lambda \wedge \beta_\lambda] = 0 \quad \text{for all } \lambda \in \mathbf{C}^*, \tag{1.5}$$

are equivalent to (1.1) and (1.2). A solution Φ_λ of (1.3) or (1.4) is called an *extended solution* (of the harmonic map equation) ([42]).

Thus we obtain:

Theorem 1 ([34, 42, 49, 48]) *Assume that Σ is simply connected. Let $\varphi : \Sigma \longrightarrow G$ be a harmonic map. For $z_0 \in \Sigma$ and $\gamma : \mathbf{C}^* \longrightarrow G^c$, there is a unique extended solution $\Phi : \mathbf{C}^* \times \Sigma \longrightarrow G^c$ such that $\Phi_\lambda(z_0) = \gamma(\lambda)$.*

If we choose the above γ with $\gamma(1) = e$ and $\gamma(S^1) \subset G$, then the extended solution Φ satisfies $\Phi_1 \equiv e$ and $\Phi_\lambda(\Sigma) \subset G$ for each $\lambda \in S^1$. Such an extended solution Φ is said to be *real*.

Let $\Omega G = \{\gamma : S^1 \longrightarrow G \mid \gamma \text{ is smooth}, \gamma(1) = e\}$ be the group of smooth based loops in G. (The completion of) ΩG has a standard infinite dimensional complex manifold structure and Kähler structure (cf. [35]). We identify the complexified tangent space $T_e\Omega G^c$ with the based loop algebra $\Omega \mathbf{g}^c = \bigoplus_{\alpha \neq 0}(\lambda^\alpha - 1)\mathbf{g}^c$. The (left-invariant) complex structure is defined by requiring that

$$(T_e\Omega G)_{1,0} = \bigoplus_{\alpha>0}(\lambda^{-\alpha} - 1)\mathbf{g}^c.$$

The (left invariant) Kähler form is defined as

$$\kappa(X, Y) = \int_{S^1} \langle X', Y \rangle dt,$$

where $\lambda = e^{2\pi\sqrt{-1}t}$. Define $\pi : \Omega G \longrightarrow G$ by $\pi(\gamma) = \gamma(-1)$.

We can regard a real extended solution as a smooth map into ΩG. ¿From (1.3) and (1.4), we see that an extended solution $\Phi : \Sigma \longrightarrow \Omega G$ is a holomorphic map. More generally, a smooth map $\Phi : \Sigma \longrightarrow \Omega G$ satisfying the condition

$$\Phi^*\mu_{\Omega G}(T\Sigma^{1,0}) = \Phi^{-1}d\Phi(T\Sigma^{1,0}) \subset (\lambda^{-1} - 1)\mathbf{g}^c$$

will also be called an *extended solution*. By the above argument, we see that if $\Phi : \Sigma \longrightarrow \Omega G$ is an extended solution, then $\varphi = \pi \circ \Phi : \Sigma \longrightarrow G$ is a harmonic map.

Let $\mathcal{E}$ denote the moduli space of extended solutions $\Sigma \longrightarrow \Omega G$ modulo the left action of elements of ΩG, and let $\mathcal{H}$ denote the moduli space of harmonic maps $\Sigma \longrightarrow G$ modulo the left action of elements of G. Assume that Σ is simply connected. Then we have a bijective correspondence between $\mathcal{E}$ and $\mathcal{H}$ given by $[\Phi] \longleftrightarrow [\varphi = \pi \circ \Phi]$.

Assume that $G = U(n)$. If a harmonic map $\varphi : \Sigma \longrightarrow G$ admits an extended solution $\Phi : \Sigma \longrightarrow \Omega G$ which has finite Laurent expansion

$$\Phi = \sum_{\alpha=0}^{m} \lambda^\alpha T_\alpha,$$

then we say that φ or Φ has *finite uniton number*, (or that $[\varphi] \in \mathcal{H}$ or $[\Phi] \in \mathcal{E}$ has finite uniton number). The minimal such number m is called the *minimal uniton number* of φ (or of $[\varphi]$). This concept was introduced in [42], where it was shown that harmonic maps of finite uniton number can be factorised into so called *unitons*. This factorisation theorem shows how such harmonic maps may be constructed from holomorphic maps into complex Grassmann manifolds. Explicit constructions in various special cases were developed independently by Chern, Wolfson, Burstall, Wood and others (see [18] and [45]). The class of harmonic maps of finite uniton number includes all harmonic maps of the Riemann sphere $\Sigma = S^2$, and more generally all harmonic maps of Riemann surfaces known variously as *superminimal surfaces* ([10]), *pseudo-holomorphic curves* ([13]) or *isotropic harmonic maps* ([18]). Conversely, it is known ([33]) that any harmonic map $\varphi : \Sigma \longrightarrow U(n)$ of finite uniton number is weakly conformal.

Next we recall the notions of Gauss bundles and isotropy (cf. [17, p450] and [45]). Let $\varphi : \Sigma \longrightarrow Gr(\mathbf{C}^n)$ be a harmonic map, where $Gr(\mathbf{C}^n)$ denotes the Grassmann manifold of all complex subspaces of $\mathbf{C}^n$. Let $\underline{\varphi}$ denote the subbundle of the trivial bundle $\underline{\mathbf{C}}^n = \Sigma \times \mathbf{C}^n$ defined by the map φ. Then we can define the ∂'-Gauss bundle $G'(\varphi)$ as the subbundle generated by the image of the ∂'-second fundamental form $A'_\varphi = \pi_\varphi^\perp \partial : \underline{\varphi} \longrightarrow \underline{\varphi}^\perp$, and it is known that this corresponds to a harmonic map $\Sigma \longrightarrow Gr(\mathbf{C}^n)$. Define the i-th ∂'-Gauss bundle $G^{(i)}(\varphi)$ of φ by $G^{(i)}(\varphi) = G'(G^{(i-1)}(\varphi))$, inductively. We say that the harmonic map φ is *strongly isotropic* or *of infinite isotropy order*, or simply *isotropic* if all $G^{(i)}(\varphi)$ ($i \geq 1$) are orthogonal to $\underline{\varphi}$ with respect to the Hermitian inner product of $\mathbf{C}^n$, and otherwise we say that φ is of *finite isotropy order*. Any holomorphic map of a Riemann surface into $Gr(\mathbf{C}^n)$ is isotropic.

Example. The simplest example of a harmonic map is a holomorphic map of a Riemann surface into a Kähler manifold. The harmonic maps $\varphi : \Sigma \longrightarrow U(n)$ of uniton number 1 are those maps of the form $\varphi = a \circ f$, for some $a \in U(n)$ and some holomorphic map $f : \Sigma \longrightarrow Gr(\mathbf{C}^n)$ (cf. [42]). (Here, $Gr(\mathbf{C}^n)$ is embedded in $U(n)$ by the map $V \longmapsto \pi_V - \pi_V^\perp$.) The group $GL(n, \mathbf{C})$ acts on harmonic maps of uniton number 1, by means of the formula $A^\natural \varphi = a(A^\natural_* f)$.

Example. Consider the complex bilinear form $(z, w) = \sum_{i=0}^{2n} z_i w_i$, where $z = (z_1, \ldots, z_{2n})$, $w = (w_1, \ldots, w_{2n}) \in \mathbf{C}^{2n}$. Let Z_n be the space of n-dimensional isotropic subspaces W in $\mathbf{C}^{2n+1}$, that is, n-dimensional subspaces for which $(w, w) = 0$ for each $w \in W$. If $G = SO(2n+1)$, the space Z can be embedded as a complex submanifold of ΩG via the map $j : W \longmapsto \pi_W + \lambda \pi_{W \oplus \bar{W}}^\perp + \lambda^2 \pi_{\bar{W}}$. We have the natural twistor fibration $\pi : Z_n \longrightarrow \mathbf{R}P^{2n}$. It is known ([13]) that any harmonic map $\varphi : S^2 \longrightarrow S^{2n}$ or $\mathbf{R}P^{2n}$ is *isotropic* as a harmonic map $S^2 \longrightarrow (\mathbf{R}P^{2n} \subset) \mathbf{C}P^{2n}$, and that there is a bijective correspondence between full horizontal holomorphic maps $f : \Sigma \longrightarrow Z_n$ and full isotropic harmonic maps $\varphi : \Sigma \longrightarrow \mathbf{R}P^{2n}$, given by $\varphi = \pi \circ f$. Such a horizontal holomorphic map into the twistor space provides a good example of an extended solution: for any horizontal holomorphic map $f : \Sigma \longrightarrow Z_n$, the map $j \circ f : \Sigma \longrightarrow \Omega G$ is an extended solution for φ, of uniton number 2. We observe that the complex Lie group $G^c = SO(2n+1, \mathbf{C})$ acts naturally on the space Z_n, and we denote this action by by $A^\natural W = A(W)$. It may be verified that the action preserves the complex structure and horizontal subspaces. Thus we obtain an action of $G^c = SO(2n+1, \mathbf{C})$ on horizontal holomorphic maps into Z_n, and hence on isotropic harmonic maps into S^{2n}.

Example. Let $F_{r,r+1}(\mathbf{C}^{n+1})$ be the space of flags (E_r, E_{r+1}) in $\mathbf{C}^{n+1}$ with $\dim_\mathbf{C} E_r = r$, $\dim_\mathbf{C} E_{r+1} = r+1$ and $E_r \subset E_{r+1}$. If $G = U(n+1)$, the space $F_{r,r+1}(\mathbf{C}^{n+1})$ can be embedded as a complex submanifold of ΩG via the map $j : (E_r, E_{r+1}) \longmapsto \pi_{E_r} + \lambda \pi_{E_r^\perp \cap E_{r+1}}^\perp + \lambda^2 \pi_{E_{r+1}}^\perp$. We have the natural twistor fibrations $\pi : F_{r,r+1}(\mathbf{C}^{n+1}) \longrightarrow$

CP^n ($0 \leq r \leq n$). It is known (see [17, 18]) that any harmonic map $\varphi : S^2 \longrightarrow CP^n$ is *isotropic*, and that there is a bijective correspondence between pairs (r, f) of integers $0 \leq r \leq n$ and full horizontal holomorphic maps $f : \Sigma \longrightarrow F_{r,r+1}(\mathbf{C}^{n+1})$ on the one hand, and full isotropic harmonic maps $\varphi : \Sigma \longrightarrow CP^n$ on the other hand, given by $\varphi = \pi \circ f$. For any horizontal holomorphic map $f : \Sigma \longrightarrow F_{r,r+1}(\mathbf{C}^{n+1})$, the map $j \circ f : \Sigma \longrightarrow \Omega G$ is an extended solution, which has uniton number 2 if $0 < r < n$, or 1 if $r = 0$ or n. The complex Lie group $G^c = GL(n+1, \mathbf{C})$ acts naturally on the space $F_{r,r+1}(\mathbf{C}^{n+1})$, and we denote this action by $A^\natural W = A(W)$. Again in this case it may be verified that the action preserves the complex structure and horizontal subspaces. Thus we obtain an action of $G^c = GL(n+1, \mathbf{C})$ on horizontal holomorphic maps into $F_{r,r+1}(\mathbf{C}^{n+1})$, and hence on isotropic harmonic maps into CP^n.

3 Dressing, Birkhoff and Uhlenbeck actions

Actions of infinite dimensional groups on harmonic maps may be formulated in terms of the *dressing method* ([49, 48, 7]). Let $\mathcal{G}$ be a group and let $\mathcal{G}_1$, $\mathcal{G}_2$ be two subgroups of $\mathcal{G}$ with $\mathcal{G} = \mathcal{G}_1\mathcal{G}_2$, $\mathcal{G}_1 \cap \mathcal{G}_2 = \{e\}$. Then each $g \in \mathcal{G}$ has a unique decomposition $g = g_1 g_2$, with $g_1 \in \mathcal{G}_1$, $g_2 \in \mathcal{G}_2$. A group action of $\mathcal{G}$ on itself may be defined by $g \cdot h = gh(h^{-1}gh)_2^{-1} = h(h^{-1}gh)_1$, for each $g, h \in \mathcal{G}$. It is easy to check that $g \cdot (g' \cdot h) = (gg') \cdot h$ for $g, g', h \in \mathcal{G}$.

Let $D_0 = \{\lambda \in \mathbf{C} \mid |\lambda| < 1\}$, $D_\infty = \{\lambda \in \mathbf{C} \cup \{\infty\} \mid |\lambda| > 1\}$ and let T be a fixed maximal torus of G. We introduce various fundamental loop groups as follows:

$$\begin{aligned}
LG^c &= \{\gamma : S^1 \longrightarrow G^c \mid \gamma \text{ is smooth}\}, \\
L^+G^c &= \{\gamma \in LG^c \mid \gamma \text{ extends continuously to} \\
&\qquad \text{a holomorphic map } D_0 \longrightarrow G^c\}, \\
L^-G^c &= \{\gamma \in LG^c \mid \gamma \text{ extends continuously to} \\
&\qquad \text{a holomorphic map } D_\infty \longrightarrow G^c\}, \\
L_1^-G^c &= \{\gamma \in L^-G^c \mid \gamma(1) = e\}, \\
\check{T} &= \{\gamma : S^1 \longrightarrow T \mid \gamma \text{ is a homomorphism}\}.
\end{aligned}$$

We recall the Birkhoff decomposition theorem (see [35]).

Theorem 2 *Each $\gamma \in LG^c$ can be decomposed as*

$$\gamma = \gamma_- \delta \gamma_+,$$

for $\gamma_- \in L^-G^c$, $\delta \in \check{T}$ and $\gamma_+ \in L^+G^c$. Moreover, the multiplication

$$L_1^-G^c \times L^+G^c \longrightarrow L^-G^c L^+G^c$$

is a diffeomorphism, and $L^-G^c L^+G^c$ is an open dense subset of the identity component of LG^c.

Loop group actions on harmonic maps and their applications

In the above definition of dressing action, let us take $\mathcal{G} = LG^c$, $\mathcal{G}_1 = L_1^- G^c$ and $\mathcal{G}_2 = L^+ G^c$. As $\mathcal{G}_1 \mathcal{G}_2 \neq \mathcal{G}$ in this case, we have a *pseudo-action*, that is, an action which is defined only for certain group elements. (In fact we have a local group of local transformations, that is, a group germ.)

Definition. The *Birkhoff pseudo-action* of LG^c on itself is defined as follows. For each $\gamma, \delta \in LG^c$ with $\delta^{-1} \gamma \delta \in L_1^- G^c L^+ G^c$,

$$\gamma^\sharp \delta = \gamma \delta (\delta^{-1} \gamma \delta)_+^{-1} = \delta (\delta^{-1} \gamma \delta)_- \in LG^c.$$

The following is the fundamental relation between this pseudo-action and extended solutions (and thus harmonic maps) ([7]).

Proposition 2 *Suppose that $\gamma \in LG^c$ and $\Phi : S^1 \times \Sigma \longrightarrow G^c$ is an extended solution. If $\Phi(z)^{-1} \gamma \Phi(z) \in L_1^- G^c L^+ G^c$ for each $z \in \Sigma$, then $\gamma^\sharp \Phi : S^1 \times \Sigma \longrightarrow G^c$ is also an extended solution.*

We are primarily interested in harmonic maps into G (rather than into G^c), so we must impose a reality condition to obtain a group action on real extended solutions. This may be done by generalizing the Birkhoff pseudo-action as follows. Let $0 < \varepsilon < 1$. We take two circles $C_0 = \{|\lambda| = \varepsilon\}$, $C_\infty = \{|\lambda| = 1/\varepsilon\}$ on the Riemann sphere $\mathbf{C} \cup \{\infty\} = S^2$. Let $I_0 = \{|\lambda| < \varepsilon\}$ and $I_\infty = \{|\lambda| > 1/\varepsilon\}$. Let $C = C_0 \cup C_\infty$, $I = I_0 \cup I_\infty$ and $E = S^2 \setminus (C \cup I)$. We introduce the following loop groups:

$$\begin{aligned}
L^\varepsilon G^c &= \{\gamma : C \longrightarrow G^c \mid \gamma \text{ is smooth }\}, \\
L^{E,\varepsilon} G^c &= \{\gamma \in L^\varepsilon G^c \mid \gamma \text{ extends continuously to} \\
&\qquad \text{a holomorphic map } E \longrightarrow G^c\}, \\
L_1^{E,\varepsilon} G^c &= \{\gamma \in L^{E,\varepsilon} G^c \mid \gamma(1) = e\}, \\
L^{I,\varepsilon} G^c &= \{\gamma \in L^\varepsilon G^c \mid \gamma \text{ extends continuously to} \\
&\qquad \text{a holomorphic map } I \longrightarrow G^c\}.
\end{aligned}$$

With these definitions, we have ([7]):

Proposition 3 *The multiplication*

$$L_1^{E,\varepsilon} G^c \times L^{I,\varepsilon} G^c \longrightarrow L^{E,\varepsilon} G^c L^{I,\varepsilon} G^c$$

is a diffeomorphism, and $L^{E,\varepsilon} G^c L^{I,\varepsilon} G^c$ is an open dense subset of the identity component of $L^\varepsilon G^c$.

In the above definition of dressing action, let us take $\mathcal{G} = L^\varepsilon G^c$, $\mathcal{G}_1 = L_1^{E,\varepsilon} G^c$ and $\mathcal{G}_2 = L^{I,\varepsilon} G^c$. We call the resulting pseudo-action of $\mathcal{G} = L^\varepsilon G^c$ on itself the *Uhlenbeck pseudo-action* ([42],[7], [23]), and denote it by $\sharp$.

If $\gamma \in L^\varepsilon G^c$ satisfies the condition $\gamma(\lambda)^{-1} = \gamma(\bar\lambda^{-1})^*$, then we call γ *real*. Denote by $L_{\mathbf{R}}^\varepsilon G^c$ the subgroup of $L^\varepsilon G^c$ consisting of all real elements. Set $L_{\mathbf{R}}^{E,\varepsilon} G^c = L^{E,\varepsilon} G^c \cap L_{\mathbf{R}}^\varepsilon G^c$ and $L_{\mathbf{R}}^{I,\varepsilon} G^c = L^{I,\varepsilon} G^c \cap L_{\mathbf{R}}^\varepsilon G^c$.

It is easy to check that if $\gamma, \delta \in L_{\mathbf{R}}^\varepsilon G^c$ and $\gamma^\sharp \delta$ is well defined, then $\gamma^\sharp \delta \in L_{\mathbf{R}}^\varepsilon G^c$. Therefore, for $\gamma \in L_{\mathbf{R}}^\varepsilon G^c$ and a real extended solution $\Phi : \Sigma \longrightarrow \Omega G$, we have a real extended solution $\gamma^\sharp \Phi : \Sigma \longrightarrow \Omega G$, provided that $\gamma^\sharp \Phi$ is well defined. Actually, Uhlenbeck ([42]) considered the following smaller groups:

$$\mathcal{A} = \{\text{rational functions on } \mathbf{C} \cup \{\infty\} \text{ with values in } G^c$$
$$\text{which are holomorphic in neighbourhoods of } 0 \text{ and } \infty\},$$
$$\mathcal{A}_{\mathbf{R}} = \{g \in \mathcal{A} \mid g(\bar\lambda^{-1})^* = g(\lambda)^{-1} \text{ for all } \lambda\}.$$

For each integer $k \geq 0$ or $k = \infty$, let

$$\mathcal{X}_k = \{\gamma : \mathbf{C}^* \longrightarrow G^c \mid \gamma \text{ holomorphic}, \gamma(1) = e,$$
$$\text{and } \gamma(\lambda) = \sum_{|\alpha| \leq k} \lambda^\alpha A_\alpha, \gamma^{-1}(\lambda) = \sum_{|\alpha| \leq k} \lambda^\alpha B_\alpha\},$$
$$\mathcal{X}_{k,\mathbf{R}} = \{\gamma \in \mathcal{X}_k \mid \gamma(\bar\lambda^{-1})^* = \gamma(\lambda)^{-1} \text{ for all } \lambda\}.$$

By showing that any element of $\mathcal{A}_{\mathbf{R}}$ decomposes into a product of elements of "simplest type" and that the action is well defined for any element of simplest type, Uhlenbeck concluded that for all $g \in \mathcal{A}_{\mathbf{R}}$ and all $\gamma \in \mathcal{X}_{\infty,\mathbf{R}}$, $g^\sharp \gamma \in \mathcal{X}_{\infty,\mathbf{R}}$ is well defined. See also [6].

For $k < \infty$, the Uhlenbeck pseudo-action on $\mathcal{X}_{k,\mathbf{R}}^+$ (and on extended solutions or harmonic maps of minimal uniton number k) collapses to the action of a certain finite dimensional Lie group $\mathcal{A}_{\mathbf{R}}/\mathcal{A}_{k,\mathbf{R}}$ (see [23], cf. also [1],[2],[3]).

Remark. Recently, I. McIntosh ([28]) showed that

$$L_{\mathbf{R}}^{E,\varepsilon} G^c L_{\mathbf{R}}^{I,\varepsilon} G^c = L_{\mathbf{R}}^\varepsilon G^c.$$

This surprising result shows that the Uhlenbeck pseudo-action of $L_{\mathbf{R}}^\varepsilon G^c$ on itself is actually a *group action*.

4 The Grassmannian model of ΩG

Here we recall the so called Grassmannian model of ΩG (due to Quillen, see [35]). For simplicity, we assume that $G = U(n)$. Let $H = L^2(S^1, \mathbf{C}^n)$ be the Hilbert space of all L^2-functions on S^1 with values in $\mathbf{C}^n$. We have the decomposition $H = H_+ \oplus H_-$, where H_+ is the closed subspace of all functions with Fourier expansion of the form $f = \sum_{k \geq 0} \lambda^k a_k$, and $H_- = (H_+)^\perp$. The Grassmannian model of ΩG is

$$Gr_\infty = \{W \mid W \text{ is a closed subspace of } H, \lambda W \subset W$$
$$pr_+ : W \longrightarrow H_+ \text{ is a Fredholm operator,}$$
$$pr_- : W \longrightarrow H_- \text{ is a Hilbert-Schmidt operator,}$$
$$pr_+(W^\perp), pr_-(W) \text{ consist of smooth functions}\},$$

where pr_+ and pr_- denote the Hermitian projections to H_+ and H_-, respectively. The loop groups LG^c and LG act transitively on Gr_∞, and their isotropy groups at $H_+ \in Gr_\infty$ are L^+G^c and G, respectively. Thus we obtain the identifications

$$Gr_\infty \cong LG^c/L^+G^c \cong LG/G \cong \Omega G$$

and we have the following theorem which may be referred to as the Iwasawa decomposition for loop groups.

Proposition 4 *Each $\gamma \in LG^c$ can be uniquely decomposed as*

$$\gamma = \gamma_u \gamma_+,$$

where $\gamma_u \in \Omega G$ and $\gamma_+ \in L^+G^c$.

Let $W : \Sigma \longrightarrow Gr_\infty$ be the map corresponding to a smooth map $\Phi : \Sigma \longrightarrow \Omega G$, i.e. $W(z) = \Phi(z)H_+$ for all $z \in \Sigma$. Then Φ is holomorphic if and only if W satisfies

$$\frac{\partial}{\partial \bar{z}} C^\infty(W) \subset C^\infty(W).$$

Moreover, a holomorphic map Φ is an extended solution if and only if W satisfies

$$\frac{\partial}{\partial z} C^\infty(W) \subset \lambda^{-1} C^\infty(W).$$

The following result on holomorphic maps into $\Omega U(n)$ is given in [38].

Theorem 3 *Assume that Σ is a compact Riemann surface. Let $\Psi : \Sigma \longrightarrow \Omega U(n)$ be a holomorphic map. Then there exists $\gamma \in \Omega U(n)$ and a complex polynomial $p(\lambda)$ such that*

$$pH_+ \subset \tilde{W}(z) \subset H_+ \quad \text{for } z \in \Sigma,$$
$$\text{Span}\{\tilde{W}(z) \mid z \in \Sigma\} = H_+,$$

where $\tilde{W} = \tilde{\Psi} H_+$, $\tilde{\Psi} = \gamma \Psi$.

We introduce two $\mathbf{C}^*$-actions as follows. For $v \in \mathbf{C}^*$ and $\gamma \in \mathcal{X}_{k,\mathbf{R}} \subset \Omega G$, with $0 \le k \le \infty$, define

$$v^\sharp \gamma = (v \cdot \gamma)\gamma(v^{-1})^{-1},$$

and

$$v^\flat \gamma = (v \cdot \gamma)_u,$$

where $v \cdot \gamma(\lambda) = \gamma(v^{-1}\lambda)$. Note that $u^\sharp \gamma = u^\flat \gamma$ for each $u \in S^1$ and each γ. These $\mathbf{C}^*$-actions have the following properties ([42], [23]).

Proposition 5 *If $u \in \mathbf{C}^*$ and $\Phi : \mathbf{C}^* \times \Sigma \longrightarrow G^c$ is an extended solution, then so are $v^\sharp \Phi : \Sigma \longrightarrow G^c$ and $v^\flat \Phi : \Sigma \longrightarrow G^c$. For the first $\mathbf{C}^*$-action, if $u \in S^1$ and Φ is real, then $u^\sharp \Phi$ is also real. For the second $\mathbf{C}^*$-action, if Φ is real, then $v^\flat \Phi$ is also real for each $v \in \mathbf{C}^*$.*

The first $\mathbf{C}^*$-action on extended solutions was first noticed by Terng (see [42, Section 7]). The second $\mathbf{C}^*$-action on real extended solutions is discussed in [23].

5 The natural action

In the above definition of dressing action, let us take $\mathcal{G} = LG^c$, $\mathcal{G}_1 = \Omega G$ and $\mathcal{G}_2 = L^+G^c$.

Definition. The *natural action* of LG^c on itself is defined as follows. For each $\gamma, \delta \in LG^c$,

$$\gamma^{\natural}\delta = \gamma\delta(\delta^{-1}\gamma\delta)_+^{-1} = \delta(\delta^{-1}\gamma\delta)_u.$$

In particular, if $\gamma \in LG^c$ and $\delta \in \Omega G$, then

$$\gamma^{\natural}\delta = \gamma\delta(\gamma\delta)_+^{-1} = (\gamma\delta)_u$$

and hence

$$\gamma^{\natural}\delta H_+ = \gamma\delta H_+$$

in the Grassmannian model.

For any $\gamma \in LG^c$ and any extended solution $\Phi : \Sigma \longrightarrow \Omega G$, we define a map $\gamma^{\natural}\Phi : \Sigma \longrightarrow \Omega G$ by $(\gamma^{\natural}\Phi)(z) = (\gamma\Phi(z))_u$. Equivalently, in terms of the Grassmannian model, $(\gamma^{\natural}\Phi)(z)H_+ = \gamma\Phi(z)H_+$, for each $z \in \Sigma$. Then we have ([23]):

Proposition 6 *Suppose that $\gamma \in LG^c$ and $\Phi : \Sigma \longrightarrow \Omega G$ is an extended solution. Then $\gamma^{\natural}\Phi : \Sigma \longrightarrow \Omega G$ is also an extended solution.*

Some of the fundamental results of [42] were proved by Segal ([38]), using the Grassmannian model. Here we shall review this approach.

Theorem 4 ([38]) *Let $\varphi : \Sigma \longrightarrow U(n)$ be a harmonic map of finite uniton number. Then there exists an extended solution $\Phi : \Sigma \longrightarrow \Omega U(n)$ of φ and a nonnegative integer m such that the map $W = \Phi H_+ : \Sigma \longrightarrow Gr_\infty$ satisfies (i) $\lambda^m H_+ \subset W(z) \subset H_+$ for all $z \in \Sigma$, and (ii) $\mathrm{Span}\{W(z) \mid z \in \Sigma\} = H_+$.*

An extended solution satisfying conditions (i) and (ii) is said to be *normalized*. There is a *canonical flag* associated to W, namely

$$\lambda^m H_+ \subset W = W_{(m)} \subset W_{(m-1)} \subset \ldots \subset W_{(0)} = H_+,$$

where $W_{(i)} : \Sigma \longrightarrow Gr_\infty$ is the holomorphic map defined by $W_{(i)} = \lambda^{-(m-i)}W \cap H_+$. It follows that $W_{(i)} = \Phi_{(i)}H_+$, for some extended solution $\Phi_{(i)} : \Sigma \longrightarrow \Omega U(n)$. This gives rise to a factorisation of harmonic maps into *unitons*:

$$\varphi = a(\pi_1 - \pi_1^\perp)(\pi_2 - \pi_2^\perp)\ldots(\pi_m - \pi_m^\perp).$$

The factorisation is also closely related to the cell decomposition of the unitary group ([47]). We note the following relation between the factorisation and the natural action of the complex loop group:

Proposition 7 *For any $\gamma \in L^+G^c$, if $\Phi : \Sigma \longrightarrow \Omega G$ is a normalized extended solution, then $\gamma^\natural \Phi : \Sigma \longrightarrow \Omega G$ is also a normalized extended solution. Moreover, the natural action of L^+G^c preserves the canonical flags associated to normalized extended solutions, that is, $\gamma^\natural \Phi_{(i)} = (\gamma^\natural \Phi)_{(i)}$.*

Let $\Omega_{\text{pol}} U(n)$ denote the group of all based polynomial loops in $U(n)$, i.e. $\gamma \in \Omega U(n)$ with $\gamma(\lambda) = \sum_{\alpha=-k}^{k} \lambda^\alpha X_\alpha$ for some nonnegative integer k. Under the diffeomorphism $\Omega U(n) \longrightarrow Gr_\infty$, the image of $\Omega_{\text{pol}} U(n)$ is the subvariety Gr_0 of Gr_∞ consisting of linear subspaces W which satisfy

$$\lambda^k H_+ \subset W \subset \lambda^{-k} H_+$$

for some nonnegative integer k. Let $L_{\text{pol}} GL(n, \mathbf{C})$ denote the group of all loops $\gamma \in LGL(n, \mathbf{C})$ such that $\gamma(\lambda), \gamma(\lambda)^{-1}$ are polynomials in λ, λ^{-1}. Then we obtain the identifications

$$Gr_0 \cong L_{\text{pol}} G^c / L_{\text{pol}}^+ G^c \cong L_{\text{pol}} G / G \cong \Omega_{\text{pol}} G.$$

The following subvariety of Gr_0 was introduced in [29] (see also [38]):

$$F_{n,k} = \{ W \in Gr_0 \mid \lambda^k H_+ \subset W \subset H_+, \dim H_+/W = k \}.$$

If we make the identification $\mathbf{C}^{kn} \cong H_+/\lambda^k H_+ \cong \bigoplus_{0 \le i \le k-1, 1 \le j \le n} \mathbf{C} \lambda^i e_j$, then

$$F_{n,k} \cong \{ E \in Gr_{kn-k}(\mathbf{C}^{kn}) \mid NE \subset E \},$$

where N is the nilpotent operator on $\mathbf{C}^{kn}$ induced by multiplication by λ. The action of $L^+GL(n, \mathbf{C})$ on $F_{n,k}$ collapses to the action of the finite dimensional complex Lie group

$$G_{n,k} = \{ A \in GL(kn, \mathbf{C}) \mid AN = NA \}.$$

Now we have two infinite dimensional group actions on harmonic maps, the Uhlenbeck action and the natural action. We shall see that they are closely related. For any ε with $0 < \varepsilon < 1$, we have an injective homomorphism of real Lie groups

$$L^+ G^c \longrightarrow L_{\mathbf{R}}^{I,\varepsilon} G^c, \quad \gamma \longmapsto \hat{\gamma}$$

defined by

$$\hat{\gamma}(\lambda) = \begin{cases} \gamma(\lambda) & \text{for } |\lambda| \le \varepsilon, \\ \gamma(\bar{\lambda}^{-1})^{-1*} & \text{for } |\lambda| \ge 1/\varepsilon \end{cases}$$

for $\gamma \in L^+G^c$. We then have:

Theorem 5 ([23]) *If $\gamma \in L^+G^c$ and $\delta \in \mathcal{X}_{k,\mathbf{R}} \subseteq \Omega G$ for $0 \le k \le \infty$, then $\hat{\gamma}^\natural \delta \in \mathcal{X}_{k,\mathbf{R}}$ is well defined and*

$$\gamma^\natural \delta = \hat{\gamma}^\natural \delta.$$

Corollary 1 *If $\gamma \in L^+G^c$ and $\Phi : \Sigma \longrightarrow \Omega G$ is an extended solution such that Φ_λ is holomorphic in $\lambda \in \mathbf{C}^*$, then we have*

$$\gamma^\natural \Phi = \hat{\gamma}^\natural \Phi.$$

In Section 3, we mentioned that the Uhlenbeck action on $\mathcal{X}_{k,\mathbf{R}}$ collapses to the action of the finite dimensional Lie group $\mathcal{A}_\mathbf{R}/\mathcal{A}_{k,\mathbf{R}}$, and in this section that the natural action of $L^+GL(n,\mathbf{C})$ on $F_{n,k}$ collapses to the action of the finite dimensional Lie group $G_{n,k}$. It turns out that we have $\mathcal{A}_\mathbf{R}/\mathcal{A}_{k,\mathbf{R}} \cong G_{n,k}$, as real Lie groups. From Theorem 5, we see that the action of $\mathcal{A}_\mathbf{R}$ (or $\mathcal{A}_\mathbf{R}/\mathcal{A}_{k,\mathbf{R}}$) on extended solutions of finite uniton number coincides with the action of $L^+GL(n,\mathbf{C})$ (or $G_{n,k}$). Hence:

Corollary 2 *The Uhlenbeck action of $L_\mathbf{R}^\varepsilon G^c$ on extended solutions (and thus harmonic maps) of finite uniton number coincides with the natural action of L^+G^c.*

6 Harmonic maps into symmetric spaces

In this section we explain how the complex loop group acts on harmonic maps into compact symmetric spaces.

Let G be a compact connected Lie group with trivial centre, and let the Lie algebra of G be $\mathbf{g}$. The Grassmannian model of ΩG is defined as follows ([35]). We consider $H = H^\mathbf{g} = L^2(S^1, \mathbf{g}^c)$, which has the structure of an infinite dimensional complex Lie algebra. The loop groups LG^c and LG act on $H^\mathbf{g}$ via the adjoint representation. Define

$$Gr_\infty^\mathbf{g} = \{W \in Gr_\infty \mid \bar{W}^\perp = \lambda W, \, W^{sm} \text{ is a Lie algebra } \}.$$

Here W^{sm} denotes the subspace of smooth functions in W. Then LG^c acts transitively on $Gr_\infty^\mathbf{g}$ and $\gamma \longmapsto \gamma H_+$ defines a diffeomorphism $\Omega G \longrightarrow Gr_\infty^\mathbf{g}$.

Assume that $M = G/K$ is a compact symmetric space given by a symmetric pair (G, K). This means that there is an involutive automorphism σ of G such that $(G_\sigma)^0 \subset K \subset G_\sigma$, where G_σ and $(G_\sigma)^0$ are the subgroup of all fixed points of σ and its identity component, respectively. Then the map $i : M = G/K \longrightarrow G$ defined by $gK \longmapsto \sigma(g)g^{-1}$ is a totally geodesic immersion, the so called *Cartan immersion*. Set $N^\sigma = \{x \in G \mid \sigma(x)x = e\}$, which is the set of fixed points of the involutive isometry $x \longmapsto \sigma(x^{-1})$ of G. Then each connected component of N^σ is a totally geodesic submanifold of G. The image of i is a symmetric space G/G_σ embedded in G, which is a connected component of N^σ containing the identity element $e \in G$.

There is an infinite dimensional twistor space over M, which is a subspace of ΩG ([17, (9.41)],[36]). To construct this, define subgroups of LG^c and LG by

$$L(G^c, \sigma) = \{\gamma \in LG^c \mid \sigma(\gamma(\lambda)) = \gamma(-\lambda) \text{ for all } \lambda \in S^1\}$$

and $L(G,\sigma) = LG \cap L(G^c,\sigma)$. Then the required twistor space is $L(G,\sigma)/K$. We have a natural embedding $j : L(G,\sigma)/K \longrightarrow LG/G = \Omega G$ defined by $j(\gamma K) = \gamma\gamma(1)^{-1}$, and a surjective map $\hat{\pi} : L(G,\sigma)/K \longrightarrow G/K$ defined by $\hat{\pi}(\gamma K) = \gamma(1)K$. Set

$$\Omega(G,\sigma) = \{\gamma \in \Omega G \mid \sigma(\gamma(\lambda)) = \gamma(-\lambda)\gamma(-1)^{-1} \text{ for all } \lambda \in S^1\},$$

which is the set of fixed points of the involutive holomorphic isometry $(-1)^\natural \circ \sigma$. Thus each connected component of $\Omega(G,\sigma)$ is a totally geodesic complex submanifold of ΩG. We have a surjective map $\pi : \Omega(G,\sigma) \longrightarrow N^\sigma$, which is a restriction of the twistor fibration $\pi : \Omega G \longrightarrow G$. The image of j is contained in $\Omega(G,\sigma)$ and is $L(G,\sigma)/G_\sigma$, which is the preimage of G/G_σ by $\pi : \Omega(G,\sigma) \longrightarrow N^\sigma$. Note that $j : L(G,\sigma)/K \longrightarrow L(G,\sigma)/G_\sigma$ becomes a finite covering map.

Under the diffeomorphism $\Omega G \longrightarrow Gr_\infty^{\mathfrak{g}}$, the space $\Omega(G,\sigma)$ corresponds to the subvariety Gr_∞^σ of $Gr_\infty^{\mathfrak{g}}$ consisting of $W \in Gr_\infty^{\mathfrak{g}}$ satisfying $\tilde{\sigma}(W) = W$, where $\tilde{\sigma}(f)(\lambda) = \sigma(f(-\lambda))$ for $f \in H^{\mathfrak{g}}$. Hence we define the natural action of $L(G^c,\sigma)$ on Gr_∞^σ as a restriction of the natural action of LG^c on ΩG. Define a subgroup of $L(G^c,\sigma)$ by

$$L^+(G^c,\sigma) = L^+G^c \cap L(G^c,\sigma).$$

Then we have $L(G,\sigma)/G_\sigma \cong L(G^c,\sigma)/L^+(G^c,\sigma)$.

Assume that $\varphi : \Sigma \longrightarrow N^\sigma$ is a harmonic map of a simply connected Riemann surface Σ into the symmetric space N^σ. Then there is an extended solution $\Phi : \Sigma \longrightarrow \Omega G$, such that $\varphi = \pi \circ \Phi$ and the image of Φ is contained in $\Omega(G,\sigma)$. Indeed, it follows from $\sigma(\varphi)\varphi = e$ that $d\sigma(\alpha) + \text{Ad}(\varphi)(\alpha) = 0$, that is, $d\sigma(\alpha') + \text{Ad}(\varphi)(\alpha') = 0$ and $d\sigma(\alpha'') + \text{Ad}(\varphi)(\alpha'') = 0$, where $\varphi^*\mu_G = \alpha = \alpha' + \alpha''$. Let $\delta \in \Omega(G,\sigma)$ with $\delta(-1) = \varphi(z_0)$. By Theorem 1, there is a unique extended solution $\Phi : \Sigma \longrightarrow \Omega G$ with $\Phi(z_0) = \delta$ and $\pi \circ \Phi = \varphi$. Then we can check easily that $(\sigma \circ \Phi)^{-1}d(\sigma \circ \Phi) = ((-1)^\natural \Phi)^{-1}d((-1)^\natural \Phi)$. Since $\sigma \circ \Phi(z_0) = \sigma \circ \delta = (-1)^\natural \delta = (-1)^\natural \Phi(z_0)$, we have $\sigma \circ \Phi = (-1)^\natural \Phi$, and hence $\Phi(\Sigma) \subset \Omega(G,\sigma)$.

In particular, if $\varphi : \Sigma \longrightarrow G/K$ is a harmonic map of a simply connected Riemann surface Σ into the symmetric space G/K, then there is a smooth map $\Phi : \Sigma \longrightarrow L(G,\sigma)/K$ such that $j \circ \Phi : \Sigma \longrightarrow \Omega G$ is an extended solution and $\varphi = \hat{\pi} \circ \Phi$.

Proposition 8 *The natural action of $L(G^c,\sigma)$ on Gr_∞^σ induces a group action on extended solutions $\Phi : \Sigma \longrightarrow L(G,\sigma)/K$ and thus on harmonic maps $\varphi : \Sigma \longrightarrow G/K$.*

Next we shall consider in more detail the case of a symmetric space of "inner type", where it is possible to construct a finite dimensional twistor subspace of $L(G,\sigma)/K$. Define $M = \{a \in G \mid a^2 = e\}$, whose connected components are totally geodesic submanifolds of even dimension. These are examples of totally geodesic submanifolds called *polars*, which have been investigated extensively by Nagano ([14, 31, 32]). Thus each connected component of M is congruent to a compact symmetric space embedded (via the Cartan embedding) in G for an involutive inner automorphism σ, after left translation by an element of G. All compact symmetric spaces of inner type can be obtained in this way. Consider

$$\mathcal{F} = \{\omega \in \Omega G \mid \omega \text{ is a homomorphism}\},$$

which is the set of all fixed points of the S^1-action $\sharp$. Thus each connected component of $\mathcal{F}$ is a totally geodesic complex submanifold of ΩG (see [17]). Let $\omega \in \mathcal{F}$. There is an element P in $\mathbf{g}$ with $\exp(tP) = \omega(\lambda)$, where $\lambda = e^{2\pi\sqrt{-1}t}$, and we may construct the generalized flag manifold G/C_P, where

$$C_P = \{a \in G \mid \mathrm{Ad}(a)\omega(\lambda) = \omega(\lambda)\}.$$

We have the identification

$$G/C_P = \{\mathrm{Ad}(a)\omega \in \Omega G \mid a \in G\},$$

and this is just the connected component of $\mathcal{F}$ containing ω. Now let $K = \{a \in G \mid \mathrm{Ad}(a)\omega(-1) = \omega(-1)\}$. Then we have a compact symmetric space G/K, totally geodesically embedded in G, which is a connected component of M. Let T be a maximal torus of G containing ω, and let $\mathbf{t}$ denote the maximal abelian subalgebra of $\mathbf{g}$ corresponding to T. Then $P \in \mathbf{t}$, and the linear endomorphism $\mathrm{ad}(P)$ on $\mathbf{g}^c$ has eigenvalues in $2\pi\sqrt{-1}\mathbf{Z}$. We denote by $\mathbf{g}_\ell$ the $(2\pi\sqrt{-1}\ell)$-eigenspace of $\mathrm{ad}(P)$. We have the eigenspace decomposition of $\mathbf{g}^c$ with respect to $\mathrm{ad}(P)$:

$$\mathbf{g}^c = \bigoplus_\ell \mathbf{g}_\ell.$$

Moreover,

$$\mathbf{c}_P^c = \mathbf{g}_0 \quad \text{and} \quad \mathbf{k}^c = \bigoplus_{\ell \text{ even}} \mathbf{g}_\ell,$$

where $\mathbf{c}_P$ and $\mathbf{k}$ denote the Lie algebras of C_P and K, respectively. The natural projection $\pi : G/C_P \longrightarrow G/K$ is a restriction of $\pi : \Omega(G, \sigma) \longrightarrow N^\sigma$, and may be considered as a "twistor subfibration". This point of view was established in [12]. Relative to the fibration $\pi : G/C_P \longrightarrow G/K$, the complexified vertical subspace corresponds to $\bigoplus_{\ell \text{ even}, \neq 0} \mathbf{g}_\ell$ and the complexified horizontal subspace corresponds to $\bigoplus_{\ell \text{ odd}} \mathbf{g}_\ell$. The superhorizontal subspace ([12]) corresponds to $\mathbf{g}_{-1} \oplus \mathbf{g}_1$. Let K^c be the complex Lie subgroup of G^c generated by $\mathbf{k}^c$. It is known that if $f : \Sigma \longrightarrow G/C_P$ is horizontal and holomorphic, then $\pi \circ f : \Sigma \longrightarrow G/K$ is harmonic (cf. [11, 12]).

Under the identifications $T_{eC_P}(G/C_P)^c \cong \bigoplus_{\ell \neq 0} \mathbf{g}_\ell$ and $T_\omega \Omega G^c \cong T_e \Omega G^c \cong \bigoplus_{\ell \neq 0} \mathbf{g}_i$, the derivative at the origin eC_P of the embedding $\iota : G/C_P \longrightarrow \Omega G$ identifies $\mathbf{g}_\ell$ with $(\lambda^{-\ell} - 1)\mathbf{g}_\ell$ (see [23], for example). Hence we see that if $f : \Sigma \longrightarrow G/C_P$ is superhorizontal and holomorphic, then $\iota \circ f$ is an extended solution. Let G_P be the parabolic subgroup of G^c generated by the parabolic subalgebra $\mathbf{g}_P = \bigoplus_{\ell \leq 0} \mathbf{g}_\ell$. It is well known that $G/C_P = G^c/G_P$, which is a complex manifold. The embedding $\iota : G/C_P = G^c/G_P \longrightarrow \Omega G = LG^c/L^+G^c$ is holomorphic, and G^c-equivariant with respect to the injective homomorphism $G^c \longrightarrow LG^c$ of the constant loops. The action of G^c on G/C_P preserves the complex structure and the superhorizontal subspaces, and the action of K^c preserves in addition the horizontal subspaces. Hence we obtain:

Proposition 9 *(i) If $a \in G^c$ and $f : \Sigma \longrightarrow G/C_P$ is superhorizontal and holomorphic, then $a \circ f : \Sigma \longrightarrow G/C_P$ is also superhorizontal and holomorphic. (ii) If $a \in K^c$ and $f : \Sigma \longrightarrow G/C_P$ is horizontal and holomorphic, then $a \circ f : \Sigma \longrightarrow G/C_P$ is also horizontal and holomorphic.*

In general, the action of G^c on G/C_P does not preserve the horizontal subspaces. We can characterize those twistor fibrations such that the action of G^c on G/C_P preserves the horizontal subspaces, as follows. Let $\Pi = \{\alpha_1, \ldots, \alpha_r\}$ be the fundamental root system of **g** with respect to **t** and let $\tilde{\alpha} = m_1\alpha_1 + \ldots + m_r\alpha_r$ be the highest root. Define

$$\begin{aligned}\Pi_0 &= \{\alpha \in \Pi \mid \alpha(P) = 0\}, \\ \Pi_1 &= \{\alpha \in \Pi \mid \alpha(P) \text{ is odd}\}, \\ \Pi_2 &= \{\alpha \in \Pi \mid 0 \neq \alpha(P) \text{ is even}\}.\end{aligned}$$

Assume that G is simple. If the action of G^c on G/C_P preserves the horizontal subspaces, then $\Pi_1 = \emptyset$ or $\Pi_2 = \emptyset$. If $\Pi_1 = \emptyset$, then G/K is a point, and if $\Pi_2 = \emptyset$, then $\Pi_1 = \{\alpha_j\}$, $m_j = 1$ or 2, or $\Pi_2 = \{\alpha_{j_1}, \alpha_{j_2}\}$, $m_{j_1} = 1, m_{j_2} = 1$. Such twistor fibrations were classified in [11].

7 Topology of spaces of harmonic maps

Up to now we have discussed group actions on harmonic maps. In this section we explain the method of Morse-Bott theoretic deformations of harmonic maps, which leads to results on the topology of spaces of harmonic maps.

Consider a finite dimensional generalized flag manifold of G, that is an orbit $\text{Ad}(G)P = G/C_P$ of a point P of **g** under the adjoint representation. Let Q be any element of **g**. Then one may define the height function $h^Q : \text{Ad}(G)P \longrightarrow \mathbf{R}$ by $h^Q(X) = \langle X, Q \rangle$. It is classical that this is a Morse-Bott function. Its nondegenerate critical manifolds and their stable or unstable manifolds can be described explicitly in Lie theoretic terms ([9]). Let ∇h^Q be the gradient of h^Q with respect to the natural Kähler metric on $\text{Ad}(G)P$. Then the flow line of $-\nabla h^Q$ is given by $t \longmapsto (\exp\sqrt{-1}tQ)^\natural X$.

The loop group ΩG is an infinite dimensional version of a generalized flag manifold. By using an identification $\Omega G \cong T\tilde{\times}LG/T \times G$, one may exhibit ΩG as the adjoint orbit of the group $T\tilde{\times}LG$ through $(\sqrt{-1}, 0)$ in $\sqrt{-1}\mathbf{R} \oplus L\mathbf{g}$. For any fixed $Q \in \mathbf{g}$, we consider the momentum functional

$$K^Q(\gamma) = \int_{S^1} \langle \gamma^{-1}\gamma', Q \rangle dt,$$

where $\langle \, , \, \rangle$ denotes an Ad-invariant inner product of **g**. It is a height function h^Q on the adjoint orbit ΩG in $\sqrt{-1}\mathbf{R} \oplus L\mathbf{g}$, with respect to the L^2-inner product. The set of critical points is ΩC_Q, and the flow of $-\nabla K^Q$ with respect to the standard Kähler metric is given by the "natural action" of $\{\exp\sqrt{-1}tQ\}$ on ΩG.

For applications to harmonic maps of finite uniton number, we are interested in the restriction of K^Q to the finite dimensional subvariety $F_{n,k}$, with $G = U(n)$. Let us consider the flow of $-\nabla K^Q$. It is given by the natural action of $\{\exp(\sqrt{-1}tQ)\} \subset L^+G^c$, so it preserves $F_{n,k}$. This flow can be used to establish deformations of harmonic maps:

Theorem 6 ([23]) *Let $\varphi : \Sigma \longrightarrow U(n)$ be a harmonic map of finite uniton number. Assume that φ admits an extended solution $\Phi = \sum_{\alpha=0}^{m} \lambda^{\alpha} T_{\alpha}$ such that rank $T_0(z) \geq 2$ for all $z \in \Sigma$. Then $\varphi : \Sigma \longrightarrow U(n)$ can be continuously deformed through harmonic maps of finite uniton number into a harmonic map $\psi : \Sigma \longrightarrow U(n-1)$ of finite uniton number.*

We shall explain the principle behind this theorem in the special case of harmonic maps into symmetric spaces of inner type. Consider the twistor fibration $\pi : G/C_P \longrightarrow G/K$ of the previous section. Let $\varphi : \Sigma \longrightarrow G/K$ be a harmonic map into a symmetric space G/K which can be lifted to an extended solution Φ into the twistor space $G/C_P \subset \Omega G$ over G/K. Suppose that the image of Φ is contained in the stable manifold $S(N)$ of a nondegenerate critical manifold $N = \text{Ad}(C_Q)X = C_Q/(C_Q)_X$ of h^Q. (Here we have $X = \text{Ad}(b)P$ for some $b \in G$.) Then $\{\Phi^t = (\exp\sqrt{-1}tQ)^\natural \Phi \mid 0 \leq t \leq \infty\}$ provides a continuous deformation of Φ to an extended solution $\Phi^\infty : \Sigma \longrightarrow N$. Note that $C_Q/(C_Q)_X \cap \text{Ad}(b)K$ is totally geodesically embedded in $G/\text{Ad}(b)K = G/K$, and the fibration $C_Q/(C_Q)_X \longrightarrow C_Q/(C_Q) \cap \text{Ad}(b)K$ is a twistor subfibration of $G/C_Q \longrightarrow G/K$. Thus we obtain a continuous deformation of the harmonic map φ to a harmonic map $\varphi^\infty = \pi \circ \Phi^\infty$, which maps into the smaller symmetric space $C_Q/(C_Q)_X \cap \text{Ad}(b)K$.

As applications of this we have the following results:

Theorem 7 ([23]) *If $n \geq 2$, then any isotropic harmonic map $\varphi : \Sigma \longrightarrow \mathbf{C}P^n$ can be deformed continuously through isotropic harmonic maps into an isotropic harmonic map $\psi : \Sigma \longrightarrow \mathbf{C}P^2$. In particular, if $n \geq 2$, then any harmonic map $\varphi : S^2 \longrightarrow \mathbf{C}P^n$ can be deformed continuously through harmonic maps into a harmonic map $\psi : S^2 \longrightarrow \mathbf{C}P^2$.*

Theorem 8 ([26],[23]) *If $n \geq 3$, then any isotropic harmonic map $\varphi : \Sigma \longrightarrow S^n$ can be deformed continuously through isotropic harmonic maps into a holomorphic map $\psi : \Sigma \longrightarrow S^2$. In particular, if $n \geq 3$, then any harmonic map $\varphi : S^2 \longrightarrow S^n$ can be deformed continuously through harmonic maps into a holomorphic map $\psi : S^2 \longrightarrow S^2$.*

Let $\text{Harm}_d(S^2, S^n)$ be the space of harmonic 2-spheres in the unit sphere of energy $4\pi d$, with $d > 0$. Then, since $\text{Hol}_d(S^2, S^2)$ is path-connected, we obtain:

Corollary 3 ([26]) *$\text{Harm}_d(S^2, S^n)$ is path-connected.*

In the case $n = 4$ this result was proved by Loo ([27]) and independently by Verdier ([44]).

It is known that $\text{Harm}_d(S^2, S^2)$ has two connected components, each of which is a copy of $\text{Hol}_d(S^2, S^2)$ and has fundamental group $\mathbf{Z}/2d\mathbf{Z}$ ([19]). In [20] the fundamental group of the space $\text{Harm}_d(S^2, S^n)$ was determined for $n > 2$, by using an extension of the above methods.

Theorem 9 ([20])

$$\pi_1 \text{Harm}_d(S^2, S^n) = \begin{cases} 0 & \text{if } n \geq 4, d \neq 2 \\ \mathbf{Z}/2\mathbf{Z} & \text{if } n \geq 4, d = 2 \\ \mathbf{Z}/2d\mathbf{Z} & \text{if } n = 3. \end{cases}$$

There is a similar result for $\text{Harm}_d(S^2, \mathbf{R}P^n)$ ([20]).

Finally we mention the space of harmonic 2-spheres in the quaternionic projective space $\mathbf{H}P^n$. From [4], we know that there are four classes of harmonic 2-spheres in $\mathbf{H}P^n (\subset Gr_2(\mathbf{C}^{2(n+1)}))$: (I) reducible, strongly isotropic, (II) irreducible, strongly isotropic, (III) reducible, finite isotropy order, (IV) irreducible, finite isotropy order. If the rank of the ∂'-Gauss bundle of φ is equal to the rank of φ, we call φ *irreducible*, and otherwise we call φ *reducible*. A quaternionic projective space $\mathbf{H}P^n$ has two natural twistor spaces, $\mathbf{C}P^{2n+1}$ and $\mathcal{T}_n$ ([11],[21]). Any harmonic 2-sphere in $\mathbf{H}P^n$ which is strongly isotropic (of class (I) or (II)) can be lifted to a horizontal holomorphic map into $\mathcal{T}_n$ ([21]), and any harmonic 2-sphere in $\mathbf{H}P^n$ which is of class (III) can be lifted to a horizontal holomorphic map into $\mathbf{C}P^{2n+1}$ ([4]). Though not every harmonic 2-sphere in $\mathbf{H}P^n$ of class (IV) can be lifted to a horizontal holomorphic map into $\mathbf{C}P^{2n+1}$ or $\mathcal{T}_n$, it can be transformed to a harmonic 2-sphere of class (III) by a finite number of "forward and backward replacements" ([4]).

Applying the argument of [23] to the twistor spaces $\mathbf{C}P^{2n+1}$ and $\mathcal{T}_n$ over $\mathbf{H}P^n$, M. Mukai [30] has recently obtained the following result:

Theorem 10 *Any harmonic 2-sphere in $\mathbf{H}P^n$ which is of class (I), (II) or (III) can be deformed continuously through harmonic maps to a harmonic 2-sphere in $S^4 = \mathbf{H}P^1$. Hence the space of all harmonic 2-spheres in $\mathbf{H}P^n$ of fixed energy which are of class (I), (II) or (III) is path-connected.*

Bibliography

[1] G. Arsenault, M. Jacques and Y. Saint-Aubin, *Collapse and exponentiation of infinite symmetry algebras of Euclidean projective and Grassmannian sigma models*, J. Math. Phys. **29** (1988), 1465–1471.

[2] G. Arsenault and Y. Saint-Aubin, *The hidden symmetry of $U(n)$ principal σ models revisited*: I, Nonlinearity **2** (1989), 571–591.

[3] G. Arsenault and Y. Saint-Aubin, *The hidden symmetry of $U(n)$ principal σ models revisited*: II, Nonlinearity **2** (1989), 593–607.

[4] A. Bahy-El-Dien and J. C. Wood, *The explicit construction of all harmonic two-spheres in quaternionic projective spaces*, Proc. London Math. Soc. **62** (1991), 202–224.

[5] J. L. M. Barbosa, *On minimal immersions of S^2 into S^{2m}*, Trans. Amer. Math. Soc. **210** (1975), 75–106.

[6] E. J. Beggs, *Solitons in the chiral equation*, Commun. Math. Phys. **128** (1990), 131–139.

[7] M. J. Bergvelt and M. A. Guest, *Actions of loop groups on harmonic maps*, Trans. Amer. Math. Soc. **326** (1991), 861–886.

[8] J. Bolton and L. M. Woodward, *Moduli spaces of harmonic 2-spheres*, Geometry and Topology of submanifolds, IV (Leuven, 1991), World Scientific, Singapore, 1992, 143–151.

[9] R. Bott, *An application of the Morse theory to the topology of Lie groups*, Bull. Soc. Math. France **84** (1956), 251–281.

[10] R. L. Bryant, *Conformal and minimal immersions of compact surfaces into the 4-sphere* J. Differential Geom. **17** (1982), 455–473.

[11] R. L. Bryant, *Lie groups and twistor spaces*, Duke Math. J. **52** (1985), 223–261.

[12] F. E. Burstall and J. H. Rawnsley, *Twistor Theory for Riemannian Symmetric Spaces*, Lect. Notes in Math. 1424, Springer-Verlag, 1990.

[13] E. Calabi, *Quelques applications de l'analyse complexe aux surfaces d'aire minima*, Topics in Complex Manifolds, Séminaire de Mathématiques Supérieures 30, University of Montreal, 1967, 58–81.

[14] B.-Y. Chen and T. Nagano, *Totally geodesic submanifolds of symmetric spaces II*, Duke Math. J. **44** (1978), 405–425.

[15] S.-S. Chern, *On the minimal immersions of the two-sphere in a space of constant curvature*, Problems in Analysis, Princeton University Press, 1970, 27–40.

[16] L. Dolan, *Kac-Moody algebra is hidden symmetry of chiral model*, Phys. Rev. Lett. **47** (1981), 1371–1374.

[17] J. Eells and L. Lemaire, *Another report on harmonic maps*, Bull. London Math. Soc. **20** (1988), 385–524.

[18] J. Eells and J. C. Wood, *Harmonic maps from surfaces to complex projective spaces*, Adv. Math. **49** (1983), 217–263.

[19] S. I. Épshtein, *Fundamental groups of spaces of coprime polynomials*, Funkts. Anal. i Prilozh. **7** (1973), 90–91; English transl.: Funct. Anal. Appl. **7** (1973), 82–83.

[20] M. Furuta, M. A. Guest, M. Kotani and Y. Ohnita, *On the fundamental group of the space of harmonic 2-spheres in the n-sphere*, Math. Z. (to appear).

[21] J. F. Glazebrook, *The construction of a class of harmonic maps to quaternionic projective spaces*, J. London Math. Soc. **30** (1984), 151–159.

[22] M. A. Guest, *Geometry of maps between generalized flag manifolds*, J. Differential Geom. **25** (1987), 223–247.

[23] M. A. Guest and Y. Ohnita, *Group actions and deformations for harmonic maps*, J. Math. Soc. Japan. (to appear).

[24] J. Harnad, Y. Saint-Aubin and S. Shnider, *Bäcklund transformations for nonlinear sigma models with values in Riemannian symmetric spaces*, Commun. Math. Phys. **92** (1984), 329–367.

[25] M. Jacques and Y. Saint-Aubin, *Infinite-dimensional Lie algebras acting on the solution space of various σ models*, J. Math. Phys. **28** (1987), 2463–2479.

[26] M. Kotani, *Connectedness of the space of minimal 2-spheres in $S^{2m}(1)$*, Proc. Amer. Math. Soc. (to appear).

[27] B. Loo, *The space of harmonic maps of S^2 into S^4*, Trans. Amer. Math. Soc. **313** (1989), 81–102.

[28] I. McIntosh, *Global solutions of the elliptic 2D periodic Toda lattice*, preprint, University of Newcastle, 1993.

[29] S. A. Mitchell, *A filtration of the loops on $SU(N)$ by Schubert varieties*, Math. Z. **193** (1986), 347–362.

[30] M. Mukai, *On connectedness of the space of harmonic 2-spheres in quaternionic projective spaces*, Tokyo J. Math. (to appear).

[31] T. Nagano, *The involutions of compact symmetric spaces*, Tokyo J. Math. **11** (1988), 57–79.

[32] T. Nagano, *The involutions of compact symmetric spaces II*, Tokyo J. Math. **15** (1992), 39–82.

[33] Y. Ohnita and G. Valli, *Pluriharmonic maps into compact Lie groups and factorisation into unitons*, Proc. London Math. Soc. **61** (1990), 546–570.

[34] K. Pohlmeyer, *Integrable Hamiltonian systems and interactions through quadratic constraints*, Commun. Math. Phys. **46** (1976), 207–221.

[35] A. N. Pressley and G. B. Segal, *Loop Groups*, Oxford University Press, 1986.

[36] J. H. Rawnsley, *Harmonic 2-spheres*, in: Quantum Theory and Geometry (Les Treilles 1987), Math. Phys. Studies **10**, Kluwer, Dordrecht, 1988, 175-189.

[37] M. Sato, *Soliton equations as dynamical systems on an infinite dimensional Grassmann manifold*, RIMS Kokyuroku **439** (1981), 30–46, Kyoto University.

[38] G. B. Segal, *Loop groups and harmonic maps*, Advances in Homotopy Theory, L.M.S. Lect. Notes 139, Cambridge University Press, 1989, 153–164.

[39] K. Takasaki, *A new approach to the self-dual Yang-Mills equations*, Commun. Math. Phys. **94** (1984), 35–59.

[40] K. Ueno, *Infinite dimensional Lie algebras acting on chiral fields and the Riemann-Hilbert problem*, Publ. RIMS, Kyoto University, **19** (1983), 59–82.

[41] K. Ueno and Y. Nakamura, *Infinite dimensional Lie algebras and transformation theories for nonlinear field equations*, Proc. of RIMS Symposium on Non-Linear Integrable Systems - Classical Theory and Quantum Theory, Kyoto, 1981 (M. Jimbo and T. Miwa, eds.), World Scientific (Singapore), 1983, 241–272.

[42] K. Uhlenbeck, *Harmonic maps into Lie groups (Classical solutions of the chiral model)*, J. Differential Geom. **30** (1989), 1–50.

[43] G. Valli, *Holomorphic maps from compact manifolds into loop groups as Blaschke products*, preprint, Università di Roma, 1991.

[44] J. L. Verdier, *Applications harmoniques de S^2 dans S^4:* II, in: Harmonic Mappings, Twistors and σ-models, Advanced Series in Math. Phys. 4 (P. Gauduchon, ed.), World Scientific (Singapore), 1988, 124–147.

[45] J.C. Wood, *Harmonic maps into symmetric spaces and integrable systems*, this volume.

[46] Y.S. Wu, *The group theoretic aspects of infinitesimal Riemann-Hilbert transform and hidden symmetry*, Commun. Math. Phys. **90** (1983), 461–472.

[47] I. Yokota, *On the homology of classical Lie groups*, Jour. Inst. Poly. Osaka City University **8** (1957), 93–120.

[48] V. E. Zakharov and A. V. Mikhailov, *Relativistically invariant two-dimensional models of field theory which are integrable by means of the inverse scattering problem method*, Zh. Eksp. Teor. Fiz. **74** (1978), 1953–1973; English transl.: Sov. Phys. JETP **47** (1987), 1017–1027.

[49] V. E. Zakharov and A. B. Shabat, *Integration of non-linear equations of mathematical physics by the inverse scattering method* II, Funkt. Anal. i Prilozh. **13** (1979), 13–22; English transl.: Funct. Anal. Appl. **13** (1979), 166–174.

THE TWISTOR APPROACH

Twistors, nilpotent orbits and harmonic maps

P.Z. Kobak

1 Introduction

Twistor theory provides a useful tool which has applications in the theory of harmonic maps. A good example is the Calabi-Penrose twistor fibration $\mathbb{CP}^3 \to S^4$. All harmonic spheres in S^4 can be obtained from projections of holomorphic horizontal curves in $\mathbb{CP}^3$ (a holomorphic curve is horizontal if it is tangent to the complex contact distribution $\mathcal{H} \subset T\mathbb{CP}^3$ which is perpendicular to the fibres of the twistor fibration with respect to the Fubini-Study metric). In order to construct holomorphic curves tangent to the distribution $\mathcal{H}$ one can use the Bryant correspondence which maps $\mathbb{CP}^3$ birationally to $\mathbb{P}T^*\mathbb{CP}^2$ and maps $\mathcal{H}$ to the canonical complex contact distribution on $\mathbb{P}T^*\mathbb{CP}^2$ (see [6] and [27]). The flag manifold $F_{12}(\mathbb{C}^3) \simeq \mathbb{P}T^*\mathbb{CP}^2$ is the twistor space of $\mathbb{CP}^2$ and Burstall shows in [11] that in fact all twistor spaces of compact quaternion-Kähler symmetric spaces of the same dimension are birationally equivalent as complex contact manifolds.

An alternative approach was used by Loo in [29]. He describes the moduli space of harmonic 2-spheres in S^4 by constructing a 2:1 covering map from $\mathbb{CP}^3 \setminus \{Z_0 \cup Z_\infty\}$ to $\mathbb{P}T^*(\mathbb{CP}^1 \times \mathbb{CP}^1)$ which maps the horizontal distribution $\mathcal{H}$ on $\mathbb{CP}^3$ to the canonical contact distribution on $\mathbb{P}T^*(\mathbb{CP}^1 \times \mathbb{CP}^1)$ (here Z_0 and Z_∞ denote twistor fibres over two antipodal points $0, \infty \in S^4$). This map is then used to characterise harmonic spheres in S^4 in terms of pairs of meromorphic functions on S^2 with the same ramification divisor.

Let $\mathbb{O}$ denote the nonassociative algebra of Cayley numbers. As a vector space $\mathbb{O}$ is isomorphic to $\mathbb{H}^2$. The 4-sphere S^4 can be identified with the set of quaternionic lines in $\mathbb{H}^2$ i.e. $S^4 = \mathbb{HP}^1$. If, instead of lines, we consider subalgebras of $\mathbb{O}$, isomorphic to $\mathbb{H}$ we get the 8-dimensional symmetric space $G_2/SO(4)$. It turns out that there is a class of harmonic spheres in $G_2/SO(4)$ which have a strikingly similar description to the one given by Loo in the case of harmonic spheres in S^4 [24, 25]. This is not a coincidence and this work originated as an attempt to understand the algebraic reasons which account for the similarities. In order to put S^4 and $G_2/SO(4)$ on equal footing, note that

1. Both S^4 and $G_2/SO(4)$ are Wolf (i.e. quaternion-Kähler compact symmetric) spaces and their twistor spaces are projectivised minimal nilpotent orbits in $\mathfrak{sp}(2,\mathbb{C})$ and $\mathfrak{g}_2^\mathbb{C}$ respectively.

2. The minimal nilpotent orbit in $\mathfrak{sp}(2,\mathbb{C})$ (respectively $\mathfrak{g}_2^\mathbb{C}$) is a finite branched covering of the nilpotent variety in $\mathfrak{sp}(1,\mathbb{C}) \oplus \mathfrak{sp}(1,\mathbb{C})$ (respectively $\mathfrak{sl}(3,\mathbb{C})$).

There is precisely one Wolf space corresponding to each complex simple Lie algebra $\mathfrak{g}^\mathbb{C}$. The associated twistor space is the projectivised minimal nilpotent orbit in $\mathfrak{g}^\mathbb{C}$. The cases when a nilpotent orbit is a finite cover of another nilpotent orbit have been classified by Brylinski and Kostant in [8]. One may therefore expect to find other examples of Wolf spaces with families of harmonic spheres which follow the pattern for S^4 and $G_2/SO(4)$.

The article begins with a brief survey of twistor methods for harmonic maps. The next section contains an outline of a construction of harmonic spheres in S^4 developed in [29]. We shall see later that this approach, when suitably rephrased, can be applied to other Wolf spaces. In the next section we describe Wolf spaces and their associated geometry which can be used to construct harmonic maps. Sections 5–7 are concerned with the contact geometry of nilpotent orbits. The relevant theory is well known to the algebraists, we present it in a more geometric way and we give many examples in an attempt to make it more accessible. We introduce contact structures on nilpotent orbits in Section 5. In the next section we describe certain holomorphic fibrations of nilpotent orbits over flag manifolds. We show in Section 7 that these fibrations have contact fibres and, consequently, they can be used to define contact maps from nilpotent orbits to projectivised cotangent bundles of flag manifolds. Such maps, together with finite coverings of nilpotent orbits described in Section 8 can be applied to construct holomorphic horizontal curves in quaternionic twistor spaces and, consequently, a class of minimal surfaces in Wolf spaces. The last section is devoted to examples of applications of the described theory to harmonic maps. The section begins with a Lie-algebraic interpretation of the construction of harmonic spheres in S^4. We then describe the construction for real Grassmannians $Gr_4(\mathbb{R}^n)$ and for $G_2/SO(4)$.

2 Twistor fibrations

One of the methods of constructing harmonic maps into Riemannian manifolds is by use of twistor fibrations. In this section we shall give a brief description of the relevant results. We refer the reader to the surveys [41, 10, 30] and to [18] for more details and for further references.

Definition 1 [13] Let M be a Riemannian manifold. A fibration $\pi : Z \to M$ is called a *twistor fibration* (with *twistor space* Z) if Z is an almost complex manifold and for any holomorphic map $\psi : S \to Z$ of an almost Hermitian manifold S with co-closed Kähler form the composition $\pi \circ \psi : S \to M$ is harmonic.

A standard example of a twistor space is the bundle $J(M)$ where M is an oriented $2n$-dimensional Riemannian manifold and $J_x(M)$ consists of Hermitian structures in T_xM compatible with the orientation (cf. [33, 19]). There are two natural almost complex structures J_1 and J_2 on $J(M)$, defined as follows. The Levi-Civita connection on M gives a splitting of the tangent space of $J(M)$ into vertical and horizontal parts, $T_zJ(M) = V_z \oplus H_z$. A complex structure z on T_xM induces a complex structure on H_z, and V_z is naturally complex since $J_x(M) \cong SO(2n)/U(n)$ is a Hermitian symmetric space. Then, with a careful choice of orientation, one can define $J_1 = J_H \oplus J_V$ and $J_2 = J_H \oplus (-J_V)$. If $\dim M > 4$ then J_1 is integrable if and only if M is conformally flat (see [33]). If M is a 4-manifold then one can use the Hodge $*$-operator to split $\Lambda^2 T^*M$ into ± 1 eigenspaces $\Lambda^2_\pm T^*M$. A Hermitian structure is determined by its Kähler form and this gives an identification of the twistor space $J(M)$ with the sphere bundle $S(\Lambda^2_+ T^*M)$. The Weyl tensor can be also split into two components W_+ and W_- and the J_1 structure on $S(\Lambda^2_+ T^*M)$ is integrable if and only if M is anti-self-dual i.e. $W_+ = 0$ (see [2] or [33, 34]).

The almost complex structure J_2 is never integrable but its importance comes from the fact that $(J(M), J_2)$ is a twistor fibration [33, 19]. The twistor space $(J(M), J_2)$ can be used to construct and, in some cases, to parametrise harmonic maps in M. For example we have the following

Theorem 1 [19] *If M is an orientable 4-manifold then there is a bijective correspondence between nonconstant weakly conformal harmonic maps from a Riemann surface Σ to M and nonvertical J_2-holomorphic curves $\Sigma \to S(\Lambda^2_+ T^*M)$.*

Recall that a map $\phi : \Sigma \to M$ is *weakly conformal* if $h(\phi_*\frac{\partial}{\partial z}, \phi_*\frac{\partial}{\partial z})^{2,0} = 0$, i.e. ϕ is conformal away from the points where the differential $d\phi$ vanishes (h denotes the complexification of the Riemannian metric on M), see [43].

One can also consider all Hermitian structures on M (not necessarily compatible with the orientation), this gives the twistor space $\tilde{J}(M)$ with fibre $O(2n)/U(n)$. Note that if M is an oriented 4-manifold then $\tilde{J}(M)$ splits into two components: $J(M) = S(\Lambda^2_+ T^*M)$ and $S(\Lambda^2_- T^*M)$. We will denote these components simply by $S(\Lambda^2_+)$ and $S(\Lambda^2_-)$.

In general harmonic maps $\phi : \Sigma \to M$ which come from J_2-holomorphic curves in the twistor space $\tilde{J}(M)$ are characterised by the following theorem (w_1 denotes the first Stiefel-Whitney class).

Theorem 2 [13] *A map $\phi : \Sigma \to M$ of a Riemann surface is a projection of a J_2-holomorphic curve $\Sigma \to \tilde{J}(M)$ if and only if it is weakly conformal, harmonic and $\phi_*w_1(M) = 0$.*

It is of course easier to construct holomorphic curves in complex rather than almost complex manifolds. Since the almost complex structures J_1 and J_2 coincide on the horizontal distribution, a J_1-holomorphic horizontal map $\phi : S \to \tilde{J}(M)$ is holomorphic also with respect to J_2. Therefore if J_1 is integrable one can construct harmonic surfaces in M from holomorphic horizontal curves in the twistor space.

There is a rich supply of complex twistor spaces (with integrable J_1) which fibre over inner symmetric spaces. They come from orbits of the isometry group action on the zero locus of the Nijenhuis tensor of J_1 and it turns out that they are flag manifolds. Their importance comes from the following

Theorem 3 [13] *Let $\psi: S^2 \to M$ be a harmonic map into a compact inner symmetric space. Then there exists a flag manifold $\mathcal{F}$ which is a subbundle of $\tilde{J}(M)$ (with compatible J_1 and J_2) and a J_2-holomorphic curve $S^2 \to \mathcal{F}$ which projects to ψ.*

Let $\mathcal{F} \to M$ be a twistor fibration of a flag manifold over an inner symmetric space. In general the horizontal distribution on $\mathcal{F}$ may not be holomorphic (twistor spaces with holomorphic horizontal distributions were classified in [7], see also [33]). But it contains a natural holomorphic subdistribution, called the *superhorizontal distribution* (see Chapter 4, C in [13]). It follows that superhorizontal holomorphic curves project to harmonic maps. Such harmonic maps have to be weakly conformal (in fact real-isotropic, see [18, §7.7 and §7.8]) and are called *superminimal*. One can obtain all harmonic 2-spheres from holomorphic superhorizontal curves precisely when $M = \mathbb{CP}^n$ or S^{2n} (in this case the superhorizontal and the horizontal distributions coincide). Harmonic spheres in $\mathbb{CP}^n$ and S^{2n} have to be superminimal since one can construct certain holomorphic differentials which vanish precisely when a harmonic map is horizontal, and S^2 has no nonzero holomorphic differentials, see [43].

Let us have a look at two simple examples of twistor fibrations of flag manifolds over Riemannian symmetric spaces. We will consider the symmetric spaces $S^4 = Sp(2)/(Sp(1) \times Sp(1))$ and $\mathbb{CP}^2 = SU(3)/S(U(2) \times U(1))$. Note that $S^4 = \mathbb{HP}^1$ and $\mathbb{CP}^2 = Gr_2(\mathbb{C}^3)$ are the simplest examples of quaternion-Kähler symmetric spaces described in Section 4 and $S(\Lambda_+^2 T^* S^4)$ (respectively $S(\Lambda_+^2 T^* \mathbb{CP}^2)$) are by definition their *quaternionic twistor spaces*. The corresponding superminimal maps will be called *quaternionic superminimal* (see Definition 3).

Example 1 Since the 4-sphere S^4 is conformally flat both twistor spaces $S(\Lambda_+^2)$ and $S(\Lambda_-^2)$ are complex. Consequently, there are two types of harmonic spheres in S^4. The quaternionic twistor space $S(\Lambda_+^2)$ can be described as the Calabi-Penrose twistor fibration $\pi : \mathbb{CP}^3 \to S^4$. In projective coordinates ($S^4 = \mathbb{HP}^1$) it can be written as the Hopf map

$$\mathbb{CP}^3 \ni [x_0, x_1, x_2, x_3] \to [x_0 + jx_1, x_2 + jx_3] \in S^4$$

and the horizontal distribution is perpendicular to the fibres of the twistor projection with respect to the Fubini-Study metric on $\mathbb{CP}^3$. The bundle $S(\Lambda_-^2)$ on the other hand is isomorphic to the fibration $A \circ \pi : S(\Lambda_+^2) \to S^4$ where A is the antipodal map of S^4. As a result if $\gamma : S^2 \to S^4$ is harmonic then either γ or $A \circ \gamma$ is a projection of a horizontal holomorphic curve in $\mathbb{CP}^3$ i.e. is a quaternionic superminimal sphere (Definition 3) or, in the terminology of Bryant, a positive spin superminimal sphere, see [6, Theorems C, D].

The next step is to find a way of characterising horizontal holomorphic curves in $\mathbb{CP}^3$. This was done by Bryant who found explicit formulae for horizontal holomorphic curves $\Sigma \to \mathbb{CP}^3$ in terms of pairs of meromorphic functions on Σ. Lawson [27] gave a geometric interpretation of this construction by defining a contact birational correspondence (known as the Bryant correspondence) between $\mathbb{CP}^3$ and the flag manifold $F_{12}(\mathbb{C}^3)$ consisting of lines in 2-planes in $\mathbb{C}^3$. An alternative construction (for $\Sigma = S^2$) which makes use of a contact birational correspondence between $\mathbb{CP}^3/\mathbb{Z}_2$ and $\mathbb{P}T^*(\mathbb{CP}^1 \times \mathbb{CP}^1)$ is described in the next section.

Example 2 Let us consider the projective 2-plane $\mathbb{CP}^2$. There are three $SU(3)$ flag manifolds: $\mathbb{CP}^2$, the dual projective plane $\mathbb{CP}^{2*} = Gr_2(\mathbb{C}^3)$ and $F_{12}(\mathbb{C}^3)$. Since $\mathbb{CP}^2$ is anti-self-dual, $S(\Lambda_+^2)$ is integrable (we assume that the orientation of $\mathbb{CP}^2$ is opposite to that given by its Kähler structure). As a complex manifold $S(\Lambda_+^2)$ can be identified with the flag manifold $F_{12}(\mathbb{C}^3)$ and the map

$$F_{12}(\mathbb{C}^3) \ni (l \subset V) \to l^\perp \cap V \in \mathbb{CP}^2$$

is the twistor projection. The flag manifolds $\mathbb{CP}^2$ and $\mathbb{CP}^{2*}$ correspond to the sections ω and $-\omega$ of $S(\Lambda_-^2)$ where ω is the Kähler form of $\mathbb{CP}^2$. They fibre trivially over $\mathbb{CP}^2$ but the first fibration is holomorphic whereas the other one is antiholomorphic. As a result there are three types of harmonic spheres in $\mathbb{CP}^2$: holomorphic and antiholomorphic maps from S^2 into $\mathbb{CP}^2$ and quaternionic superminimal spheres (mixed pairs in the terminology of [15]).

Harmonic spheres in complex projective spaces are well understood, we refer the reader to [16, 9, 20, 43]. As a Riemannian symmetric space $\mathbb{CP}^2$ is identical with $Gr_2(\mathbb{C}^3)$ which is the first element of the infinite sequence $Gr_2(\mathbb{C}^n)$ of Wolf spaces corresponding to the unitary groups (cf. Table 4.1). There is abundant literature on harmonic maps in complex Grassmannians, see for example [15, 14, 42, 40] and also [3] for the results on harmonic 2-spheres in quaternionic projective spaces.

3 Example: harmonic 2-spheres in S^4

The following construction was used by Loo in [29] to study the moduli space of harmonic 2-spheres in S^4. Let us consider the twistor fibration $\mathbb{CP}^3 \to S^4$ and define the map

$$\psi : \mathbb{CP}^3 \setminus \{Z_0 \cup Z_\infty\} \ni [x_0, x_1, x_2, x_3] \to ([x_0, x_1], [x_2, x_3]) \in \mathbb{CP}^1 \times \mathbb{CP}^1,$$

where Z_0 and Z_∞ are the twistor fibres over two antipodal points $0 = [0, 1]$ and $\infty = [1, 0]$ in S^4. The map ψ is a submersion with contact fibres and it follows that the map

$$\tilde{\psi} : \mathbb{CP}^3 \setminus \{Z_0 \cup Z_\infty\} \ni z \to \psi_*(\mathcal{H}_z) \in \mathbb{P}T^*(\mathbb{CP}^1 \times \mathbb{CP}^1)$$

is contact (Lemma 1). One can show that $\tilde{\psi}$ is a 2:1 covering and the image of $\tilde{\psi}$ consists of lines in $T(\mathbb{CP}^1 \times \mathbb{CP}^1)$ which are transversal to the fibres of the projections $p_1, p_2 : \mathbb{CP}^1 \times \mathbb{CP}^1 \to \mathbb{CP}^1$. This gives Diagram 3.1 (all maps in the diagram, except for the twistor fibration π, are holomorphic).

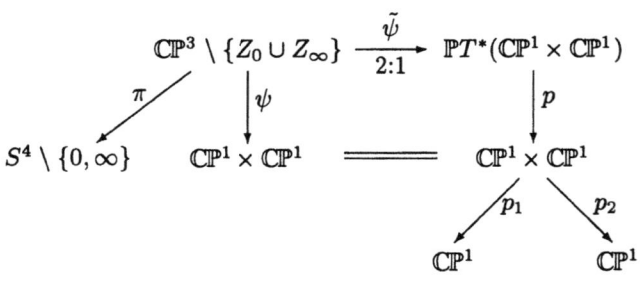

(3.1)

Now if γ is a quaternionic superminimal sphere in S^4 which avoids the points 0 and ∞, then the corresponding holomorphic horizontal curve $\tilde{\gamma} : S^2 \to \mathbb{CP}^3$ stays away from the twistor fibres Z_0 and Z_∞. Therefore the curve $p \circ \tilde{\psi} \circ \tilde{\gamma}$ is transversal to the fibres of the projections p_1, p_2, i.e. the functions $p_i \circ p \circ \tilde{\psi} \circ \tilde{\gamma}$, $i = 1, 2$ have the same ramification divisor. Conversely, let w be a curve in $\mathbb{CP}^1 \times \mathbb{CP}^1$ defined by a pair of meromorphic functions on S^2 with the same ramification divisor. The lines tangent to w belong to $\mathbb{P}T^*(\mathbb{CP}^1 \times \mathbb{CP}^1)$ and we get a curve $w' : S^2 \to \mathbb{P}T^*(\mathbb{CP}^1 \times \mathbb{CP}^1)$ (the Gauss lift of w). The curve w' is tangent to the canonical contact structure on $\mathbb{P}T^*(\mathbb{CP}^1 \times \mathbb{CP}^1)$ and since w cannot be tangent to fibres of p_1 and p_2 (because $p_1 \circ w$ and $p_2 \circ w$ have the same ramification divisor), it lies in the image of the map $\tilde{\psi}$. Since $\tilde{\psi}$ is 2:1, we get a pair of holomorphic horizontal curves in $\mathbb{CP}^3$ and then a pair of harmonic 2-spheres in S^4 (such spheres are called *conjugate* in [29], they differ by the action of the involution $[z_0, z_1] \to [z_0, -z_1]$). As a result one gets the following characterisation of minimal spheres in S^4.

Theorem 4 [29] *There is a 2:1 correspondence between quaternionic superminimal maps $\gamma : S^2 \to S^4$ avoiding two antipodal points $0, \infty \in S^4$ and pairs of meromorphic functions on S^2 with the same ramification divisor.*

4 Quaternion-Kähler manifolds

There is a particularly important class of Riemannian manifolds with complex twistor spaces and holomorphic horizontal distributions [32].

Definition 2 An oriented Riemannian manifold M is *quaternion-Kähler* if $J(M)$ contains a proper 2-sphere subbundle Z preserved by parallel translations with respect to the Levi-Civita connection on M and locally spanned by three orthogonal almost complex structures I, J, K which satisfy the relations $IJ = -JI = K$.

It follows from the definition that each tangent space of a quaternion-Kähler manifold can be thought of as an $\mathbb{H}$-module, so $\dim M = 4n$. If M_4 is a 4-manifold then $Z(M_4) = S(\Lambda_+^2 T^* M_4)$ is already a 2-sphere bundle so $n \geq 2$. We shall call the bundle Z the *quaternionic twistor space* of M (or simply twistor space, if there is no confusion). Quaternion-Kähler manifolds are Einstein and their definition can be naturally extended to the case when $n = 1$. A 4-dimensional manifold is called quaternion-Kähler if it is Einstein and anti-self-dual. We recall that the anti-self-duality condition ensures that J_1 is integrable. One usually assumes that quaternion-Kähler manifolds have nonzero scalar curvature, otherwise they would be locally hyperKähler. We refer the reader to [4, Chapter 14] for more information on quaternion-Kähler and hyperKähler geometry.

The horizontal distribution on a quaternionic twistor space has complex codimension 1 and is holomorphic. As a result it endows the twistor space with a complex contact structure (see [32]) and provides a class of minimal surfaces in the corresponding quaternion-Kähler manifold.

Definition 3 A harmonic map of a Riemann surface to a quaternion-Kähler manifold will be called *quaternionic superminimal* if and only if it is a projection of a horizontal holomorphic curve in the quaternionic twistor space.

Quaternionic superminimal maps are very special. They are of course superminimal (and hence weakly conformal), they also have to be *inclusive* [33], see also [15, §5B, C]. A map $\gamma : \Sigma \to M$ of a Riemann surface to a quaternion-Kähler manifold is called inclusive if for each $x \in \Sigma$ there exists an almost complex structure J_x in the twistor fibre Z_x such that the differential $d_x\gamma : T_x\Sigma \to (T_{\gamma(x)}M, J_x)$ is complex. Inclusive maps can be viewed as quaternionic analogues of holomorphic curves. Note that the map $\Sigma \ni x \to J_x \in Z$ is a lift of the inclusive curve γ to the quaternionic twistor space Z.

In this paper we shall be interested in compact homogeneous quaternion-Kähler manifolds. They were classified by Wolf and Alekseevskiĭ ([39, 1]). They are symmetric and are known as *Wolf spaces*. There is one such manifold for each compact simple Lie algebra $\mathfrak{g}$ and it can be defined in the following way. Let $\mathfrak{t}$ be a maximal torus in $\mathfrak{g}$. Take the $\mathfrak{su}(2)$ in $\mathfrak{g}$ with the root system $\{\rho, -\rho\}$ where ρ is the highest root. The conjugacy class $M = G/N_G(\mathfrak{su}(2))$ of such $\mathfrak{su}(2)$ is the Wolf space corresponding to the Lie algebra $\mathfrak{g}$ (G denotes a compact simple Lie group with Lie algebra $\mathfrak{g}$ and $N_G(\mathfrak{su}(2))$ is the normaliser of $\mathfrak{su}(2)$ in G). One can write M as $G/(HSp(1))$ whereas Z is equal to $G/(HU(1))$, for some subgroup $H \subset G$.

The three families of classical Wolf spaces together with associated twistor bundles are listed below.

Z	M	$=$	$G/N_G(su(2))$
$\mathbb{CP}^{2n+1}$	$\mathbb{HP}^n$	$=$	$Sp(n+1)/(Sp(n) \times Sp(1))$
$F_{1,n-1}(\mathbb{C}^n)$	$Gr_2(\mathbb{C}^n)$	$=$	$SU(n)/S(U(n-2) \times U(2))$
Z_n	$Gr_4(\mathbb{R}^n)$	$=$	$SO(n)/(SO(n-4) \times SO(4))$

(4.1)

In the list above $F_{1,n-1}(\mathbb{C}^n)$ is the flag manifold of lines in hyperplanes in $\mathbb{C}^n$ and $Z_n \subset Gr_2(\mathbb{C}^n)$ consists of totally isotropic 2-planes. $Gr_4(\mathbb{R}^n)$ denotes the Grassmannian of oriented 4-planes in $\mathbb{R}^n$. There are five more Wolf spaces corresponding to exceptional Lie algebras:

$$E_6/SU(6)Sp(1), \quad E_7/Spin(12)Sp(1), \quad E_8/E_7Sp(1)$$
$$F_4/Sp(3)Sp(1) \quad \text{and} \quad G_2/SO(4).$$

There is another description of quaternionic twistor spaces which will be essential in the sequel. Let M be a quaternion-Kähler manifold with nonzero scalar curvature and let Z denote the corresponding twistor space. We denote by G the connected component of the isometry group of M and by $\mathcal{L}$ the line bundle over Z given by the complex contact structure. The action of G on M induces a complex symplectic action of $G^{\mathbb{C}}$ on $\mathcal{L}^*$ and one can use the associated moment map to construct a map $Z \to \mathbb{P}\mathfrak{g}^{\mathbb{C}}$. Under suitable conditions this gives an identification of Z with a projectivised nilpotent orbit in $\mathfrak{g}^{\mathbb{C}}$ (see [36, 37] for further details). For example one has the following theorem.

Theorem 5 [37] *If Z is a twistor space of a quaternion-Kähler manifold of positive scalar curvature such that Z is $G^{\mathbb{C}}$-homogeneous as a complex contact manifold and the symmetry group $G^{\mathbb{C}}$ is reductive, then, up to finite covers, Z is the projectivisation of a nilpotent orbit of the semisimple part of $\mathfrak{g}^{\mathbb{C}}$.*

It is therefore natural to consider the *nilpotent variety* $\mathcal{N} = \{A \in \mathfrak{g}^{\mathbb{C}} : (\mathrm{ad}_A)^k = 0 \text{ for some } k\}$. It breaks down under the action of $G^{\mathbb{C}}$ into a finite disjoint union of conjugacy classes (nilpotent orbits). It turns out that all projectivised nilpotent orbits are quaternionic twistor spaces [36, 38]. In particular, if $z \in \mathfrak{g}^{\mathbb{C}}_\rho$ lies in the root space of $\mathfrak{g}^{\mathbb{C}}$ corresponding to the highest root ρ then the projectivisation of the nilpotent orbit $\mathcal{N}_m = G^{\mathbb{C}} \cdot z$ is the twistor space of the Wolf space $G/N_G(\mathfrak{sp}(1))$. We shall call the orbit $\mathcal{N}_m$ the *highest root orbit* or the *minimal orbit* — it is the nilpotent orbit of the smallest dimension in $\mathcal{N} \setminus \{0\}$. On the other hand generic nilpotent elements in $\mathfrak{g}^{\mathbb{C}}$ form the *regular nilpotent orbit* $\mathcal{N}_r$ and $\mathcal{N} = \bar{\mathcal{N}}_r$. The regular nilpotent orbit is the orbit of the highest dimension in $\mathcal{N}$ and its complex dimension is equal to $\dim \mathfrak{g}^{\mathbb{C}} - \mathrm{rank}\, \mathfrak{g}^{\mathbb{C}}$.

5 Contact distributions

Let $\mathcal{N} \subset \mathfrak{g}^{\mathbb{C}}$ be a (nonzero) nilpotent orbit. Its projectivisation carries a natural $G^{\mathbb{C}}$-invariant contact structure which is precisely the contact structure on $\mathbb{P}\mathcal{N}$ viewed as a quaternionic twistor space [36]. It can be defined explicitly in the following way. Let e be an element of the orbit $\mathcal{N}$. It follows from the Jacobson-Morosov Theorem [26] that there exist $x, e_- \in \mathfrak{g}^{\mathbb{C}}$ which together with $e_+ = e$ form an $\mathfrak{sl}(2,\mathbb{C})$-*triple,* that is, they span a three-dimensional subalgebra of $\mathfrak{g}^{\mathbb{C}}$, isomorphic to $\mathfrak{sl}(2,\mathbb{C})$, and satisfy the relations

$$[x, e_+] = e_+, \quad [x, e_-] = -e_-, \quad [e_+, e_-] = x. \tag{5.1}$$

We can use the surjective map $\tau : \mathfrak{g}^{\mathbb{C}} \ni X \to [e, X] \in T_e\mathcal{N}$ to identify $T_e\mathcal{N}$ with $\mathfrak{g}^{\mathbb{C}}/Z(e)$ where $Z(e) = \ker \tau$ is the centraliser of e in $\mathfrak{g}^{\mathbb{C}}$. One can use the Killing form on $\mathfrak{g}^{\mathbb{C}}$ to define an invariant 1-form θ on $\mathcal{N}$,

$$\theta([e, X]) = \langle e, X \rangle. \tag{5.2}$$

If $Y \in Z(e)$ then $[e, Y] = 0$ and $\langle x, [e, Y] \rangle = 0$ so $\langle e, Y \rangle = \langle Y, [x, e] \rangle = 0$ and θ is well defined. The distribution $D = p_* \ker \theta$ is a complex contact structure on $\mathbb{P}\mathcal{N}$ (where $p : \mathcal{N} \to \mathbb{P}\mathcal{N}$ is the obvious projection). It follows from Formula 5.2 that $\ker \theta_e = e^\perp$ so for $A \in \mathbb{P}\mathcal{N}$ the contact hyperplane $D_A \subset T_A\mathbb{P}\mathcal{N}$ can be written as

$$D_A = p_*(A^\perp/Z(A)). \tag{5.3}$$

The 2-form $\omega = d\theta$ endows $\mathcal{N}$ with a complex symplectic structure. This is the Kostant-Kirillov-Soriau symplectic structure. We have

$$\begin{aligned}\omega([e, X], [e, Y]) &= [e, X]\langle e, Y \rangle - [e, Y]\langle e, X \rangle - \theta(e, [X, Y]) \\ &= -\langle e, [X, Y] \rangle.\end{aligned}$$

We have seen that nilpotent orbits and their projectivisations give very natural examples of complex symplectic and complex contact manifolds. Manifolds T^*M and $\mathbb{P}T^*M$, where M is a complex manifold provide an even more basic class of examples. We shall see later that in the case when M is a flag manifold these two classes are closely related.

Let us recall that the contact structure on $\mathbb{P}T^*M$ is defined in the following way. If π denotes the projection $T^*M \to M$ then the formula $\theta_A : T_AT^*M \ni X \to A(\pi_*X)$ defines a canonical 1-form θ on T^*M. Then $d\theta$ is a complex symplectic form on T^*M and $p_* \ker \theta$ is a contact distribution on $\mathbb{P}T^*M$ (again $p : T^*M \to \mathbb{P}T^*M$ is the obvious projection). In other words points of $\mathbb{P}T_x^*M$ parametrise hyperplanes in T_xM and if $A \subset T_xM$ is a hyperplane then $D_A = (d\pi')^{-1}(A)$ is the contact hyperplane in $T_A\mathbb{P}T^*M$, where π' is the projection $\mathbb{P}T^*M \to M$. It is therefore clear that the fibres of π' are tangent to the contact distribution on $\mathbb{P}T^*M$. Conversely, if M is a contact manifold and $\psi : M \to N$ is a submersion with contact fibres, then the projection $\psi_*(D_A)$ of a contact hyperplane in TM is a hyperplane in TN. This gives a map

$$\tilde{\psi} : M \ni A \to \psi_*(D_A) \in \mathbb{P}T^*N. \tag{5.4}$$

We have the following lemma which is implicit in [29].

Lemma 1 *Let M be a contact manifold and let $\psi : M \to N$ be a submersion with contact fibres. Then the map $\tilde{\psi} : M \to \mathbb{P}T^*N$ defined as in Formula 5.4 is contact, i.e. it maps the contact distribution on M to the canonical contact distribution on $\mathbb{P}T^*N$.*

Proof. The lemma follows immediately from the commutative diagram

$$\begin{array}{ccc} M & \xrightarrow{\tilde{\psi}} & \mathbb{P}T^*N \\ \downarrow \psi & & \downarrow \pi' \\ N & = & N \end{array}$$

If D denotes the complex contact distribution on M, $A \in M$ and $B = \tilde{\psi}(A)$ then $\pi'_*(\tilde{\psi}_*(D_A)) = \psi_*(D_A) = B$. This means that $\tilde{\psi}_*(D_A)$ is the contact hyperplane in $T_B \mathbb{P}T^*\mathcal{N}$. □

We shall call $\tilde{\psi}$ the contact map *associated* to the fibration ψ.

Remark 1 The lemma, although elementary, can be very useful. For example the projection

$$\phi : \mathbb{CP}^3 \setminus B \ni [x_0, x_1, x_2, x_3] \mapsto [x_0^2, x_0 x_1 + x_2 x_3, x_0 x_2] \in \mathbb{CP}^2,$$

where B is the base locus given by the equations $x_0 = 0$ and $x_2 x_3 = 0$, has contact fibres, and the associated map $\tilde{\phi} : \mathbb{CP}^3 \to \mathbb{P}T^*\mathbb{CP}^2 \simeq F_{12}(\mathbb{C}^3)$ is the Bryant correspondence.

6 Canonical fibrations over flag manifolds

Flag manifolds are defined as conjugacy classes of *parabolic subalgebras* in complex semisimple Lie algebras. To define a parabolic subalgebra $\mathfrak{p}$ of a complex semisimple Lie algebra $\mathfrak{g}^{\mathbb{C}}$ one can choose a vector $x \in \text{span}_{\mathbb{R}} \Delta$ where Δ denotes the root system of $\mathfrak{g}^{\mathbb{C}}$ with respect to a Cartan subalgebra $\mathfrak{j}$ and put

$$\mathfrak{p} = \mathfrak{j} \oplus \bigoplus_{\substack{\alpha \in \Delta \\ (\alpha, x) \geq 0}} \mathfrak{g}^{\mathbb{C}}_{\alpha} \tag{6.1}$$

where $\mathfrak{g}^{\mathbb{C}}_{\alpha}$ denotes the root space of $\alpha \in \Delta$.

If there are no roots perpendicular to x then $\mathfrak{p}$ is called a *Borel subalgebra*. Borel subalgebras can be also defined invariantly as maximal solvable subalgebras in $\mathfrak{g}^{\mathbb{C}}$ whereas a subalgebra $\mathfrak{p} \subset \mathfrak{g}^{\mathbb{C}}$ is parabolic if and only if it contains a Borel subalgebra.

Parabolic algebras are self-normalising: if P is the normaliser of $\mathfrak{p}$ in $G^\mathbb{C}$ then $\mathfrak{p} = \mathrm{Lie}(P)$ (P is called a parabolic group). Therefore $\mathcal{F} = G^\mathbb{C}/P \simeq G^\mathbb{C} \cdot \mathfrak{p}$ is the flag manifold determined by the parabolic algebra $\mathfrak{p}$. Flag manifolds which correspond to Borel subalgebras will be called *complete flag manifolds*.

Remark 2 Let $\rho \in \Delta$ be the highest root and let $z \in \mathfrak{g}_\rho^\mathbb{C}$, $z \neq 0$. The stabiliser of the line $\mathbb{C}z \in \mathbb{P}\mathfrak{g}^\mathbb{C}$ can be written as $\mathfrak{p} = \bigoplus_{\alpha \in A} \mathfrak{g}_\alpha^\mathbb{C}$, where $A = \{\alpha \in \Delta : \langle \rho, \alpha \rangle \geq 0\}$. This shows that $\mathfrak{p}$ is a parabolic subalgebra of $\mathfrak{g}^\mathbb{C}$ and the projectivised minimal nilpotent orbit $\mathbb{P}\mathcal{N}_m$ is a flag manifold.

It will be useful to have an algebraic description of the cotangent space of a flag manifold. For a parabolic algebra $\mathfrak{p}$ one can write $\mathfrak{p} = \mathfrak{l} \oplus \mathfrak{n}$ where $\mathfrak{l}$ is the reductive factor of $\mathfrak{p}$ and $\mathfrak{n} = \mathfrak{p}^\perp$ is its nilradical:

$$\mathfrak{l} = \mathfrak{j} \oplus \bigoplus_{\substack{\alpha \in \Delta \\ \langle \alpha, x \rangle = 0}} \mathfrak{g}_\alpha^\mathbb{C} \quad \text{and} \quad \mathfrak{n} = \bigoplus_{\substack{\alpha \in \Delta \\ \langle \alpha, x \rangle > 0}} \mathfrak{g}_\alpha^\mathbb{C}. \tag{6.2}$$

The holomorphic tangent bundle of $\mathcal{F} = G^\mathbb{C}/P$ can be written as the associated bundle $T\mathcal{F} = G^\mathbb{C} \times_P \mathfrak{g}^\mathbb{C}/\mathfrak{p}$. The Killing form on $\mathfrak{g}^\mathbb{C}$ induces a nondegenerate bilinear map $\mathfrak{g}^\mathbb{C}/\mathfrak{p} \times \mathfrak{p}^\perp \to \mathbb{C}$ which gives a pairing between $T\mathcal{F}$ and $G^\mathbb{C} \times_P \mathfrak{p}^\perp$. We can therefore write $T^*\mathcal{F} = G^\mathbb{C} \times_P \mathfrak{p}^\perp$ and points of $T^*\mathcal{F}$ can be regarded as pairs $(\mathfrak{p}, x)$ where $\mathfrak{p} \in \mathcal{F}$ and $x \in \mathfrak{p}^\perp$.

We shall use Lemma 1 to relate contact structures on projectivised nilpotent orbits to canonical contact structures on projectivised cotangent spaces of flag manifolds. First we need a suitable fibration with contact fibres. There is a natural way of assigning parabolic algebras to nilpotent elements of complex semisimple Lie algebras (see for example [35, Remark III.4.18]). In other words projectivised nilpotent orbits fibre homogeneously in a canonical way over flag manifolds. We shall describe here these fibrations and their basic properties following the exposition of Burstall [12].

Let $e \in \mathfrak{g}^\mathbb{C}$ be a nilpotent element which together with $x, e_- \in \mathfrak{g}^\mathbb{C}$ forms a $\mathfrak{sl}(2, \mathbb{C})$ triple (Formula 5.1). The element x is semisimple in $\mathfrak{g}^\mathbb{C}$ and one can decompose $\mathfrak{g}^\mathbb{C}$ into ad_x eigenspaces,

$$\mathfrak{g}^\mathbb{C} = \bigoplus_{i \in \frac{1}{2}\mathbb{Z}} \mathfrak{s}_i.$$

Now $[\mathfrak{s}_i, \mathfrak{s}_j] \subset \mathfrak{s}_{i+j}$ and therefore

$$\mathfrak{p}_e = \bigoplus_{i \geq 0} \mathfrak{s}_i \tag{6.3}$$

is a subalgebra of $\mathfrak{g}^\mathbb{C}$. Its polar

$$\mathfrak{n} = \mathfrak{p}_e^\perp = \bigoplus_{i \geq \frac{1}{2}} \mathfrak{s}_i$$

is nilpotent so it follows from Grothendieck's Lemma (see [13], Proposition 4.2) that $\mathfrak{p}_e$ is a parabolic algebra. If (x', e'_-, e) is another $\mathfrak{sl}(2, \mathbb{C})$-triple, then $x - x' \in \mathrm{im}\,\mathrm{ad}_e \cap \ker\mathrm{ad}_e$ ([26], 3.6) and one can show that $\mathfrak{p}_e$ does not depend on the choice of x. Consequently, we have a homogeneous (i.e. $G^{\mathbb{C}}$-equivariant) map from the nilpotent orbit $\mathcal{N} = G^{\mathbb{C}} \cdot e$ to the flag manifold $\mathcal{F}$. Rescaling e does not affect $\mathfrak{p}_e$ so we also have a fibration $\mathbb{P}\mathcal{N} \to \mathcal{F}$. Summarizing, we have the following theorem.

Theorem 6 *Let $\mathcal{N}$ be a nilpotent orbit in a complex semisimple Lie algebra. Then there is a flag manifold $\mathcal{F}$ such that $\mathcal{N}$ fibres homogeneously and holomorphically in a canonical way over $\mathcal{F}$. This fibration induces a fibration of the projectivised nilpotent orbit $\mathbb{P}\mathcal{N} \to \mathcal{F}$. Moreover, if $\mathfrak{p}_e$ is the image of $e \in \mathcal{N}$ then $e \in \mathfrak{p}_e^\perp$.*

For conciseness we shall call the fibration $\mathcal{N} \to \mathcal{F}$ defined above the *canonical fibration*.

Remark 3 Let $\mathfrak{p} \subset \mathfrak{g}^{\mathbb{C}}$ be a parabolic algebra, $x \in \mathfrak{p}^\perp$ a nilpotent element. This gives a nilpotent orbit $\mathcal{N} = G^{\mathbb{C}} \cdot x$ and a flag manifold $\mathcal{F} = G^{\mathbb{C}}/P$. If the condition $\dim \mathcal{N} = 2\dim F$ is satisfied then x is called a *Richardson element* and the algebra $\mathfrak{p}$ is called a *polarisation* of x. In this case the orbit $P \cdot x$ is open and dense in $\mathfrak{p}^\perp$. It is proved in [31] that the nilradical of every parabolic algebra contains Richardson elements. On the other hand not every nilpotent element has polarisations. For a nilpotent element x the set $\mathrm{Pol}(x)$ of its polarisations is finite (see [21, 23] for classification results).

If $x \in \mathfrak{p}^\perp$ is a Richardson element then the identity component of its centraliser $Z(x)^0$ is contained in the parabolic group P. A polarisation is called *stable* if $Z(x) \subset P$. This means that one can define a $G^{\mathbb{C}}$-homogeneous map $\mathcal{N} = G^{\mathbb{C}} \cdot x \to G^{\mathbb{C}}/P$. In particular, the canonical projection $s : \mathcal{N} \to \mathcal{F}$, in the case when $\dim \mathcal{N} = 2\dim \mathcal{F}$, can be viewed as a map which assigns a distinguished stable polarisation to each element of $\mathcal{N}$.

Example 3 Nilpotent orbits in $\mathfrak{sl}(n, \mathbb{C})$ are in 1:1 correspondence with *partitions* $\lambda = (\lambda_1 \geq \lambda_2 \geq \ldots \geq \lambda_r \geq 0)$, $\lambda_1 + \lambda_2 + \ldots + \lambda_r = n)$. A partition λ determines a nilpotent element e which consists of Jordan blocks of size $\lambda_k \times \lambda_k$, $k = 1 \ldots r$:

$$e^i = \begin{pmatrix} 0 & & & \\ 1 & 0 & & \\ & \ddots & \ddots & \\ & & 1 & 0 \end{pmatrix} \quad (i \times i \text{ matrix}), \quad e = \begin{pmatrix} e^{\lambda_1} & & \\ & \ddots & \\ & & e^{\lambda_r} \end{pmatrix}.$$

Let us define matrices

$$e^i_- = \begin{pmatrix} 0 & 1(i-1) & & & \\ & 0 & 2(i-2) & & \\ & & \ddots & \ddots & \\ & & & 0 & (i-1)1 \\ & & & & 0 \end{pmatrix}, \quad e_- = \begin{pmatrix} e^{\lambda_1}_- & & \\ & \ddots & \\ & & e^{\lambda_r}_- \end{pmatrix}$$

and

$$x^i = \begin{pmatrix} 1-i & & & \\ & 3-i & & \\ & & \ddots & \\ & & & i-1 \end{pmatrix}, \quad x = \begin{pmatrix} x^{\lambda_1} & & \\ & \ddots & \\ & & x^{\lambda_r} \end{pmatrix}.$$

Then the matrices $e/\sqrt{2}, e_-/\sqrt{2}, x/2$ form an $\mathfrak{sl}(2,\mathbb{C})$-triple. By choosing a skew or symmetric nondegenerate bilinear form on $\mathbb{C}^n$ and a suitable basis (see [35, 23]) one can use the same matrices to construct $\mathfrak{sl}(2,\mathbb{C})$-triples in other classical Lie algebras.

The parabolic algebra $\mathfrak{p}_e$ can be found from Formula 6.3. For example if e is a regular nilpotent element then it consists of just one Jordan block e^n. Diagonal matrices in $\mathfrak{sl}(n,\mathbb{C})$ form a Cartan algebra with root system $\lambda_i - \lambda_j, i \neq j \in \{1,\ldots n\}$ and root spaces $\mathfrak{g}_{ij}$ consist of matrices in which only the (i,j) entry can be nonzero. Since $(\lambda_i - \lambda_j)x = 2(i-j)$, the parabolic algebra consists of lower-triangular matrices so the corresponding flag manifold $\mathcal{F}_r$ consists of complete flags in $\mathbb{C}^n$. The canonical projection can be written as the map

$$\mathcal{N}_r \ni A \;\to\; (\operatorname{im} A^n \subset \operatorname{im} A^{n-1} \subset \ldots \subset \operatorname{im} A \subset \operatorname{im} A^0) \;\in \mathcal{F}_r$$
$$\phantom{\mathcal{N}_r \ni A \;\to\; (}\|\phantom{\operatorname{im} A^n \subset \operatorname{im} A^{n-1} \subset \ldots \subset \operatorname{im} A \subset }\|$$
$$\phantom{\mathcal{N}_r \ni A \;\to\; (}0\phantom{\operatorname{im} A^n \subset \operatorname{im} A^{n-1} \subset \ldots \subset \operatorname{im} A \subset }\mathbb{C}^n$$

Flags corresponding to nonregular nilpotent elements $A \in \mathfrak{sl}(n,\mathbb{C})$ can be built from blocks of the form $\ker A^p \cap \operatorname{im} A^q$, $p, q \in \mathbb{N}$.

Let us have a look at the simplest case (which, as we shall see later, has nontrivial applications). The nilpotent variety $\mathcal{N} = \{A \neq 0 \in \mathfrak{sl}(3,\mathbb{C}) : A^3 = 0\}$ in $\mathfrak{sl}(3,\mathbb{C})$ consists of three nilpotent orbits:

$$\mathcal{N}_m = \{A \in \mathfrak{sl}(3,\mathbb{C}) : A^2 = 0, A \neq 0\} \quad \text{(minimal nilpotent orbit)}$$
$$\mathcal{N}_r = \{A \in \mathfrak{sl}(3,\mathbb{C}) : A^3 = 0, A^2 \neq 0\} \quad \text{(regular nilpotent orbit)}$$

and the zero orbit $\mathcal{N}_0 = \{0\}$. The canonical fibrations in this case are simply the maps

$$\mathcal{N}_m \ni A \;\to\; (\operatorname{im} A \subset \ker A \subset \mathbb{C}^3) \in F_{12}(\mathbb{C}^3)$$
$$\mathcal{N}_r \ni A \;\to\; (\ker A \subset \operatorname{im} A \subset \mathbb{C}^3) \in F_{12}(\mathbb{C}^3)$$

so the projectivised orbit $\mathbb{P}\mathcal{N}_m$ can be identified with the flag manifold $F_{12}(\mathbb{C}^3)$. Under this identification the canonical map for the projectivised regular orbit is simply the map $\mathbb{P}\mathcal{N}_r \ni [A] \to [A^2] \in \mathbb{P}\mathcal{N}_m$.

In general flag manifolds in a complex semisimple Lie algebra $\mathfrak{g}^{\mathbb{C}}$ are parametrised by subsets $S \subset R$ of the set of simple roots, whereas nilpotent orbits are represented by assigning numbers $0, \frac{1}{2}$ or 1 to simple roots of $\mathfrak{g}^{\mathbb{C}}$. More precisely, let $\mathfrak{j} \subset \mathfrak{g}^{\mathbb{C}}$ be a Cartan subalgebra with the root system Δ and positive roots Δ_+. We can use the Killing form on $\mathfrak{g}^{\mathbb{C}}$ to identify $\operatorname{span}_{\mathbb{R}} \Delta$ with a real subspace $\mathfrak{j}^{\sharp} \subset \mathfrak{j}$. Let $\mathcal{C} \subset \mathfrak{j}^{\sharp}$ be the fundamental Weyl chamber, $\mathcal{C} = \{x \in \mathfrak{j}^{\sharp} \mid \langle \phi, x \rangle \geq 0 \;\forall \phi \in \Delta_+\}$. If a parabolic algebra $\mathfrak{p}$ is determined by $x \in \mathcal{C}$ then $\mathfrak{p}$ contains the Borel algebra

$$\mathfrak{b} = \mathfrak{j} \oplus \bigoplus_{\alpha \in \Delta_+} \mathfrak{g}_\alpha^{\mathbb{C}} \qquad (6.4)$$

and it corresponds to the subset $S = x^\perp \cap \Delta_+ \subset R$:

$$\mathfrak{p} = \mathfrak{j} \oplus \underbrace{\bigoplus_{\alpha \in (\text{span } S) \cap \Delta} \mathfrak{g}_\alpha^{\mathbb{C}}}_{\mathfrak{l}} \oplus \underbrace{\bigoplus_{\alpha \in \Delta_+ \setminus (\text{span } S)} \mathfrak{g}_\alpha^{\mathbb{C}}}_{\mathfrak{n}} \qquad (6.5)$$

Flag manifolds can therefore be represented by Dynkin diagrams of $\mathfrak{g}^{\mathbb{C}}$ with crosses through the nodes corresponding to roots in $R \setminus S$.

Let $\mathcal{N} \subset \mathfrak{g}^{\mathbb{C}}$ be a nilpotent orbit. There exists an $\mathfrak{sl}(2, \mathbb{C})$-triple (e, e_-, x) such that $e \in \mathcal{N}$ and $x \in \mathcal{C}$. If α is a simple root one can assign the number $a_\alpha = \langle \alpha, x \rangle$ to the corresponding node in the Dynkin diagram. It turns out that the numbers $\{a_\alpha\}_{\alpha \in R}$ uniquely determine $\mathcal{N}$ (they are called *characteristic numbers* of the nilpotent orbit $\mathcal{N}$, see [17, 26]). The possible values for a_α are 0, $\frac{1}{2}$ or 1, but not all combinations can occur and the admissible ones can be found in [17]. It follows from Formula 6.3 that the flag manifold $\mathcal{F}$ over which $\mathcal{N}$ fibres corresponds to the Dynkin diagram with crosses through nonzero nodes.

Example 4 There are 3 flag manifolds and 4 (nonzero) nilpotent orbits in the exceptional Lie algebra $\mathfrak{g}_2^{\mathbb{C}}$. The canonical fibrations $\mathcal{N} \to \mathcal{F}$ are listed in the following table.

$\mathcal{N}$	a_1	a_2	$\dim \mathcal{N}$	$\mathcal{F}$	$\dim \mathcal{F}$
highest root orbit	$\frac{1}{2}$	0	6	$\mathbb{P}\mathcal{N}_m$	5
short root orbit	0	$\frac{1}{2}$	8	Q^5	5
subregular orbit	1	0	10	$\mathbb{P}\mathcal{N}_m$	5
regular orbit	1	1	12	$\mathcal{F}_r$	6

In the table $\mathcal{F}_r$ denotes the complete flag manifold ($\mathcal{F}_r = G_2^{\mathbb{C}}/B$ where B is a Borel subgroup), $\mathbb{P}\mathcal{N}_m$ is the quaternionic twistor space of $G_2/SO(4)$ ($\mathcal{N}_m$ denotes the highest root orbit in $\mathfrak{g}_2^{\mathbb{C}}$) and $Q^5 \cong \mathbb{P}(T^{1,0}S^6)$ can be identified with the 5-dimensional nondegenerate quadric in $\mathbb{CP}^6$. Note that only the regular and the subregular orbits have polarisations.

7 Relations between contact structures

We shall now use the canonical fibration $\mathbb{P}\mathcal{N} \to \mathcal{F}$ to construct a contact map $\psi : \mathbb{P}\mathcal{N} \to \mathbb{P}T^*\mathcal{F}$. Note that the last statement in Theorem 6 has a simple geometric interpretation.

Lemma 2 *Let $f : \mathbb{P}\mathcal{N} \to \mathcal{F}$ be a holomorphic homogeneous fibration such that if $l \in \mathbb{P}\mathcal{N}$ then $l \subset f(l)^{\perp}$. Then the fibres of f are contact.*

Proof. Let p denote the projection $\mathcal{N} \to \mathbb{P}\mathcal{N}$. Let $l = \mathbb{C} \cdot z \in \mathbb{P}\mathcal{N}$, $z \in \mathcal{N}$ and take $\mathfrak{p} = f(l)$. Since $f \circ p$ is $G^{\mathbb{C}}$-equivariant, the stabiliser of z must be contained in the stabiliser of $f(p(z))$. Therefore the centraliser $Z(z)$ of z in $\mathfrak{g}^{\mathbb{C}}$ is contained in $\mathfrak{p}$. The tangent space $T_z(f \circ p)^{-1}(\mathfrak{p})$ can be identified with $\mathfrak{p}/Z(z)$. According to Formula 5.3 the contact hyperplane $\mathcal{H}_l \subset T_l \mathbb{P}\mathcal{N}$ is a projection of $l^{\perp}/Z(z) \subset T_z \mathcal{N}$. We want to show that $T_z(f \circ p)^{-1}(\mathfrak{p})$ is contained in $\mathcal{H}_l$, i.e. $\mathfrak{p}/Z(z) \subset l^{\perp}/Z(z)$. But $l \subset \mathfrak{p}^{\perp}$ implies $\mathfrak{p} \subset l^{\perp}$ and this ends the proof. □

Let $\psi : \mathbb{P}\mathcal{N} \to \mathcal{F}$ be a canonical fibration. Lemmas 1, 2 and Theorem 6 imply that the associated map $\tilde{\psi} : \mathbb{P}\mathcal{N} \to \mathbb{P}T^*\mathcal{F}$ defined by Formula 5.4 is contact. We will show that it is also injective.

Lemma 3 *If $l \in \mathbb{P}\mathcal{N}$ then it follows from Theorem 6 that $l \subset \psi(l)^{\perp}$. Therefore $\{(\psi(l), z) : z \in l\}$ is a line in $T^*_{\psi(l)}\mathcal{F}$ and we can define a holomorphic $G^{\mathbb{C}}$-equivariant fibre map*

$$\mathbb{P}\mathcal{N} \ni l \to (\psi(l), l) \in \mathbb{P}T^*\mathcal{F} \tag{7.1}$$

This is precisely the associated map $\tilde{\psi}$.

Proof. Let $\mathcal{H}$ denote the contact structure on $\mathbb{P}\mathcal{N}$. Then

$$\tilde{\psi}(l) = \psi_*(\mathcal{H}_l) = \psi_*(p_*(l^{\perp}/Z(l)) = l^{\perp}/\psi(l)$$

where p denotes the projection $\mathcal{N} \to \mathbb{P}\mathcal{N}$. The hyperplane $l^{\perp}/\psi(l) \subset T_{\psi(l)}\mathcal{F}$ is annihilated by $(\psi(l), l)$ and Formula 7.1 follows. □

The lemma implies that the map s defined by the formula

$$s : \mathbb{P}T^*\mathcal{F} \ni (\mathfrak{p}, n) \to n \in \mathbb{P}\bar{\mathcal{N}} \tag{7.2}$$

is the left inverse of $\tilde{\psi}$, $s \circ \tilde{\psi} = \mathrm{id}_{\mathcal{N}}$. Here $\bar{\mathcal{N}}$ denotes the closure of $\mathcal{N}$ in $\mathfrak{g}^{\mathbb{C}}$. Note that if $\dim \mathcal{N} = 2 \dim \mathcal{F}$ then, according to Remark 3, the image of $\tilde{\psi}$ is equal to the open dense orbit $G^{\mathbb{C}} \cdot (\mathfrak{p}, r) \subset \mathbb{P}T^*\mathcal{F}$ where $r \in \mathfrak{p}^{\perp}$ is a Richardson element. To summarize, we have

Theorem 7 *Let $\psi : \mathbb{P}\mathcal{N} \to \mathcal{F}$ be a canonical fibration of a nilpotent orbit over a flag manifold. Then the associated map $\tilde{\psi} : \mathbb{P}\mathcal{N} \to \mathbb{P}T^*\mathcal{F}$ defined by Formula 5.4 is injective and contact. Moreover, if $\dim \mathcal{N} = 2 \dim \mathcal{F}$ then the image of $\tilde{\psi}$ is an open dense $G^{\mathbb{C}}$-orbit in $\mathbb{P}T^*\mathcal{F}$.*

Remark 4 The map $s : \mathbb{P}T^*\mathcal{F} \to \mathbb{P}\bar{\mathcal{N}}$ defined in Formula 7.2 is a projectivisation of a map $T^*\mathcal{F} \to \bar{\mathcal{N}}$ which is known in the literature as the *generalised Springer resolution* of the nilpotent variety $\bar{\mathcal{N}}$ (see [5]). The generalised Springer resolution can be defined for any flag manifold $\mathcal{F} = G^{\mathbb{C}}/P$ as the moment map for the action of $G^{\mathbb{C}}$ on $\mathcal{F}$ and it is given by the formula

$$S : T^*\mathcal{F} = G^{\mathbb{C}} \times_P \mathfrak{p}^\perp \ni (p,x) \to x \in G^{\mathbb{C}} \cdot \mathfrak{p}^\perp.$$

The orbit $G^{\mathbb{C}} \cdot \mathfrak{p}^\perp$ is equal to $\bar{\mathcal{N}}$ where $\mathcal{N} = G^{\mathbb{C}} \cdot r$ is the orbit of a Richardson element $r \in \mathfrak{p}^\perp$ whereas $G^{\mathbb{C}} \cdot (\mathfrak{p}, r)$ is an open dense orbit in $\mathbb{P}T^*\mathcal{F}$ on which the Springer resolution is a finite covering.

If $\mathcal{N} \to \mathcal{F}$ is a canonical fibration with associated map $\tilde{\psi} : \mathbb{P}\mathcal{N} \to \mathbb{P}T^*\mathcal{F}$ and $\dim \mathcal{N} = 2 \dim \mathcal{F}$ then $\tilde{\psi}$ is a bijection between the open dense $G^{\mathbb{C}}$-orbits $\mathcal{N} \subset \bar{\mathcal{N}}$ and $G^{\mathbb{C}} \cdot (\mathfrak{p}, r) \subset \mathbb{P}T^*\mathcal{F}$. It is an inverse of S restricted to $G^{\mathbb{C}} \cdot (\mathfrak{p}, r)$ so the Springer resolution is 1:1 and hence birational. But in general, if $S : \mathcal{F} \to \mathcal{N}$ is a Springer resolution, it is possible that the canonical fibration maps the nilpotent orbit $\mathcal{N}$ to a flag manifold $\mathcal{F}' \neq \mathcal{F}$ (see the next example).

Example 5 If $\mathfrak{g}^{\mathbb{C}} = \mathfrak{sl}(n, \mathbb{C})$ then every Springer resolution is birational and every nilpotent element has polarisations. For example if $\mathcal{N}_m \subset \mathfrak{sl}(3, \mathbb{C})$ is the minimal nilpotent orbit then the canonical projection is trivial (Examples 3 and 6) but one can check directly that $\mathbb{P}\mathcal{N}_m \simeq F_{12}(\mathbb{C}^3) \simeq \mathbb{P}T^*\mathbb{CP}^2 \simeq \mathbb{P}T^*\mathbb{CP}^{2*}$. These identifications can be obtained for example as the associated maps for the homogeneous fibrations $\mathcal{F}_{12}(\mathbb{C}^3) \to \mathbb{CP}^2$ and $\mathcal{F}_{12}(\mathbb{C}^3) \to \mathbb{CP}^{2*}$ since these fibrations have contact fibres. Consequently, $\mathcal{N}_m$ is an adjoint orbit of Richardson elements for the $Sl(3, \mathbb{C})$ flag manifolds $\mathbb{CP}^2$ and $\mathbb{CP}^{2*}$.

Example 6 The case when $\mathcal{N}_m \subset \mathfrak{g}^{\mathbb{C}}$ is the minimal (i.e. the highest root) orbit is very special. We have seen in Remark 2 that $\mathbb{P}\mathcal{N}_m$ is a flag manifold. It follows from the remark and from Formula 6.3 that the canonical projection in this case is the trivial map $\mathcal{N}_m \to \mathcal{F} = \mathbb{P}\mathcal{N}_m$, so the associated map $\mathcal{F} \to \mathbb{P}T^*\mathcal{F}$ is a section of the projection $\mathbb{P}T^*\mathcal{F} \to \mathcal{F}$ (it is simply the contact structure on $\mathcal{F}$) and of course is not birational.

If $\mathcal{N}_r \subset \mathfrak{g}^{\mathbb{C}}$ is the regular nilpotent orbit, then all characteristic numbers (a_α) in the Dynkin diagram of $\mathfrak{g}^{\mathbb{C}}$ are equal to one, so $\mathcal{N}_r$ fibres over the complete flag manifold $\mathcal{F}_r = G^{\mathbb{C}}/B$ where $B \subset G^{\mathbb{C}}$ is a Borel subgroup. Then $\dim \mathcal{N}_r = \dim \mathfrak{g}^{\mathbb{C}} - \text{rank}\, \mathfrak{g}^{\mathbb{C}} = 2 \dim \mathcal{F}_r$ so the associated map $\mathbb{P}\mathcal{N}_r \to \mathcal{F}_r$ is always birational. The image of the map $\tilde{\psi} : \mathcal{N}_r \to \mathbb{P}T^*\mathcal{F}_r$ has the following geometric description in terms of homogeneous fibrations $\mathcal{F}_r \to \mathcal{F}$ with fibre $\mathbb{CP}^1$.

Proposition 1 *Let $\psi : \mathcal{N}_r \to \mathcal{F}_r$ be the canonical fibration of the regular nilpotent orbit in $\mathfrak{g}^{\mathbb{C}}$ over the complete flag manifold $\mathcal{F}_r$. Then the image of the associated fibration $\tilde{\psi} : \mathcal{N}_r \to \mathbb{P}T^*\mathcal{F}_r$ consists of hyperplanes transversal to the fibres of all homogeneous fibrations $\mathcal{F}_r \to \mathcal{F}$ with 1-dimensional fibre.*

Remark 5 The complete flag manifold $\mathcal{F}_r$ corresponds to the Dynkin diagram with crosses through all nodes. Therefore the Dynkin diagram of the flag manifold $\mathcal{F}$ has one uncrossed node and the number of fibrations $\mathcal{F}_r \to \mathcal{F}$ with fibre $\mathbb{CP}^1$ is equal to rank $\mathfrak{g}^{\mathbb{C}}$.

Proof. Let $f : \mathcal{F}_r \to \mathcal{F}$ be a homogeneous fibration with 1-dimensional fibre and let $\mathfrak{p} = f(\mathfrak{b})$, where $\mathfrak{b}$ is a Borel algebra. Then $f(\mathfrak{b}) \subset \mathfrak{p}$ since nontrivial homogeneous fibrations increase stabilisers. Let us choose a root system Δ with positive roots Δ_+ and simple roots R so that $\mathfrak{p}$ and $\mathfrak{b}$ are as in Formulae 6.4 and 6.5, for some subset $S \subset R$. The vertical space $T_\mathfrak{b}(f^{-1}(\mathfrak{p}))$ can be identified with $\mathfrak{b}/\mathfrak{p} \simeq \mathfrak{g}^\mathbb{C}_{-\alpha}$ where α is the uncrossed root in the Dynkin diagram of $\mathcal{F}$ (so $S = \{\alpha\}$). The vertical space is contained in the hyperplane $x \in \mathfrak{b}^\perp \simeq T^*_\mathfrak{b}\mathcal{F}_r$ precisely when $x \perp \mathfrak{g}^\mathbb{C}_{-\alpha}$ i.e. when the $\mathfrak{g}^\mathbb{C}_\alpha$ component of x vanishes. The lemma now follows since an element $x \in \mathfrak{b}^\perp$ is regular if and only if all its components in $\mathfrak{g}^\mathbb{C}_\alpha$, $\alpha \in R$, are nonzero ([35], III 3.5). □

8 Finite covers of nilpotent orbits

We have seen in the previous section that one can use the canonical projection to relate the contact structure on projectivised nilpotent orbits and thus the original geometry of corresponding quaternion-Kähler manifolds to the canonical contact structure on projectivised cotangent bundles of flag manifolds. This approach fails, however, when we need it most — Example 6 shows that the canonical fibration is trivial for minimal nilpotent orbits. Fortunately, in some cases one can overcome this difficulty by considering nilpotent orbits which are finitely covered by minimal nilpotent orbits. Brylinski and Kostant determined pairs of complex semisimple Lie algebras $\mathfrak{g}^\mathbb{C} \subset \mathfrak{g}'^\mathbb{C}$ and nilpotent orbits $\mathcal{N} \subset \mathfrak{g}^\mathbb{C}$, $\mathcal{N}' \subset \mathfrak{g}'^\mathbb{C}$ such that $\mathcal{N}'$ is an equivariant finite cover of $\mathcal{N}$. Many such examples were known earlier in the literature (see for example [28, 36]). We will be interested in the cases where $\mathcal{N}'$ is the highest root orbit in $\mathfrak{g}'^\mathbb{C}$. Pairs $\mathfrak{g}^\mathbb{C}$, $\mathfrak{g}'^\mathbb{C}$ in the table below come from [8]; $\mathcal{N}'$ is a k:1 cover of a nilpotent orbit $\mathcal{N}$. We denote by $\mathcal{F}$ the flag manifold which is the target of the canonical fibration $\mathcal{N} \to \mathcal{F}$ and $W_{\mathfrak{g}'}$ is the Wolf space which corresponds to $\mathfrak{g}'$.

$W_{\mathfrak{g}'}$	$\mathfrak{g}'$	$\mathfrak{g}$	k	dim $\mathcal{N}$	dim $\mathcal{F}$
$Gr_4(\mathbb{R}^7)$	$\mathfrak{so}(7)$	$\mathfrak{g}_2$	1	8	5
$Gr_4(\mathbb{R}^{2n+2})$	$\mathfrak{so}(2n+2)$	$\mathfrak{so}(2n+1)$	2	$4n-2$	$2n-1$
$Gr_2(\mathbb{C}^{2n})$	$\mathfrak{sl}(2n)$	$\mathfrak{sp}(n)$	2	$4n-2$	$4n-5$
$E_6/SU(6)Sp(1)$	$\mathfrak{e}_6$	$\mathfrak{f}_4$	2	22	15
$G_2/SO(4)$	$\mathfrak{g}_2$	$\mathfrak{sl}(3)$	3	6	3
$Gr_4(\mathbb{R}^{2n+1})$	$\mathfrak{so}(2n+1)$	$\mathfrak{so}(2n)$	2	$4n-4$	$2n-2$
$F_4/Sp(3)Sp(1)$	$\mathfrak{f}_4$	$\mathfrak{so}(9)$	2	16	10
$F_4/Sp(3)Sp(1)$	$\mathfrak{f}_4$	$\mathfrak{so}(8)$	4	16	11
$Gr_4(\mathbb{R}^8)$	$\mathfrak{so}(8)$	$\mathfrak{g}_2$	6	10	5
$\mathbb{HP}^{n-1}$	$\mathfrak{sp}(n)$	$\oplus_1^k \mathfrak{sp}(n_i)$	2^{k-1}	$2n$	$2n-k$

(8.1)

In the last row $\sum_{i=1}^k n_i = n$ and $k \geq 2$. The orbit $\mathcal{N}$ is a product of the highest root orbits in $\mathfrak{sp}(n_i, \mathbb{C})$.

Example 7 Let $\mathcal{N}'$ denote the highest root orbit in $\mathfrak{so}(n, \mathbb{C})$. The orthogonal projection $\mathfrak{so}(n, \mathbb{C}) \to \mathfrak{so}(n-1, \mathbb{C})$ gives a 2:1 covering $\mathcal{N}' \to \bar{\mathcal{N}}$. Let $\mathbb{Z}_2 = \{1, \sigma\}$ where σ is a diagonal $n \times n$ matrix with $\sigma_{11} = -1$ and $\sigma_{22} = \ldots = \sigma_{nn} = 1$. Then $\mathfrak{so}(n-1, \mathbb{C})$ consists of $\mathbb{Z}_2$-invariant elements in $\mathfrak{so}(n, \mathbb{C})$ and the covering is simply the averaging over $\mathbb{Z}_2$, $x \to \frac{1}{2}(x + \text{Ad}_\sigma x)$. Let $\phi : \mathbb{P}\mathcal{N}' \to \mathbb{P}\bar{\mathcal{N}}$ be the projectivised covering. If $\mathcal{N}_m$ denotes the minimal nilpotent orbit in $\mathfrak{so}(n-1, \mathbb{C})$ then $\mathbb{P}\bar{\mathcal{N}} = \mathbb{P}\mathcal{N} \cup \mathbb{P}\mathcal{N}_m$. Since $\mathcal{N}_m = \mathcal{N}' \cap \mathfrak{so}(n-1, \mathbb{C})$, ϕ is 1:1 on $\mathbb{P}\mathcal{N}_m$ and 2:1 on $\mathbb{P}\mathcal{N}' \setminus \mathbb{P}\mathcal{N}_m$. Moreover, ϕ maps $\mathbb{P}\mathcal{N}' \setminus \mathbb{P}\mathcal{N}_m$ to $\mathbb{P}\mathcal{N}$ which is the twistor space of $(Gr_4(\mathbb{R}^n) \setminus Gr_4(\mathbb{R}^{n-1}))/\mathbb{Z}_2$.

Example 8 If $\mathcal{N}' \subset \mathfrak{g}_2^\mathbb{C}$ is the highest root orbit and $\mathcal{N}_r \subset \mathfrak{sl}(3, \mathbb{C})$ is the regular nilpotent orbit then the orthogonal projection $\mathfrak{g}_2^\mathbb{C} \to \mathfrak{sl}(3, \mathbb{C})$ gives a 3:1 covering $\mathcal{N}' \to \bar{\mathcal{N}}_r$. This map can be also viewed as the averaging over the centre $\mathbb{Z}_3$ of $SU(3) \subset G_2$, $x \to \frac{1}{3}(x + \text{Ad}_\sigma x + \text{Ad}_{\sigma^2} x)$. The projectivised map $\phi : \mathbb{P}\mathcal{N}' \to \mathbb{P}\bar{\mathcal{N}}_r$ is a 3:1 branched covering. Let us denote by $\mathcal{N}_m$ the minimal nilpotent orbit in $\mathfrak{sl}(3, \mathbb{C})$. Since $\mathfrak{sl}(3, \mathbb{C}) \cap \mathcal{N}' = \mathcal{N}_m$ is the fixed point set of the $\mathbb{Z}_3$ action, ϕ is 1:1 on $\mathbb{P}\mathcal{N}_m$ and 3:1 on $\mathcal{N}' \setminus \mathcal{N}_m$. The $\mathbb{Z}_3$ action on the twistor space induces an action on $G_2/SO(4)$ with fixed point set $\mathbb{CP}^2$. Note that $\mathbb{P}\mathcal{N}_m = F_{12}(\mathbb{C}^3)$ is the quaternionic twistor space of $\mathbb{CP}^2$. The covering ϕ maps $\mathbb{P}\mathcal{N}' \setminus \mathbb{P}\mathcal{N}_m$ to $\mathcal{N}_r$ which is the twistor space of $(G_2/SO(4) \setminus \mathbb{CP}^2)/\mathbb{Z}_3$ (see [25] for a more detailed analysis of the relation between the quaternionic geometry of $G_2/SO(4)$ and the regular nilpotent orbit $\mathcal{N}_r$).

9 Applications to harmonic maps

In this section we shall show examples of applications of the developed theory to harmonic maps. We begin with the Lie-algebraic interpretation of the construction in Section 3.

Quaternionic projective spaces. According to the last row in the table in Section 8, in the simplest case ($n = 2$, $k = 1$, $n_1 = n_2 = 1$), there is a 2:1 branched covering map from the twistor space $Z = \mathbb{CP}^3$ of $S^4 = \mathbb{HP}^1$ to the projectivised nilpotent variety in $\mathfrak{sp}(1,\mathbb{C}) \oplus \mathfrak{sp}(1,\mathbb{C})$. Nilpotent elements in $\mathfrak{sp}(1,\mathbb{C}) \oplus \mathfrak{sp}(1,\mathbb{C})$ are pairs (A, B) where A, B are nilpotent in $\mathfrak{sp}(1,\mathbb{C})$. Pairs with $A \neq 0$ and $B \neq 0$ belong to the regular orbit $\mathcal{N}_r$. There is only one projective nilpotent orbit in $\mathfrak{sp}(1,\mathbb{C})$. It is equal to $\mathbb{CP}^1$ and the canonical projection $\mathcal{N}_r \to \mathbb{CP}^1 \times \mathbb{CP}^1$ is simply the map

$$\psi : \mathbb{P}\mathcal{N}_r \ni [A, B] \to ([A], [B]) \in \mathbb{CP}^1 \times \mathbb{CP}^1.$$

We can use the induced map $\tilde{\psi}$ to identify $\mathbb{P}\mathcal{N}_r$ with an open dense orbit in $\mathbb{P}T^*(\mathbb{CP}^1 \times \mathbb{CP}^1)$. According to Proposition 1 the image of $\tilde{\psi}$ consists of the lines in $T(\mathbb{CP}^1 \times \mathbb{CP}^1)$ which are transversal to fibres of the two projections of the product $\mathbb{CP}^1 \times \mathbb{CP}^1$ to its factors. It remains to find the branching set of the covering map $Z \to \mathbb{P}\bar{\mathcal{N}}_r$. It consists of the elements in Z which project to non-regular elements $[A, B] \in \mathbb{P}\bar{\mathcal{N}}_r$, so either $A = 0$ or $B = 0$. Therefore the projection of the branching set is equal to $(\{0\} \times \mathbb{CP}^1) \cup (\mathbb{CP}^1 \times \{0\}) \subset \mathbb{P}\bar{\mathcal{N}}_r$ and its components correspond to two skew lines in $\mathbb{CP}^3$ which are twistor fibres over $[0, 1]$ and $[1, 0] \in \mathbb{HP}^1$. As a result we can recover the diagram and the characterisation of harmonic spheres in S^4 as described in Section 3. A similar approach can be applied to quaternionic projective spaces $\mathbb{HP}^n$, $n \geq 2$. One can take the finite cover $\mathcal{N}' \to \mathcal{N}$ where $\mathcal{N}$ is the nilpotent variety in $\mathfrak{sp}(1,\mathbb{C}) \oplus \ldots \oplus \mathfrak{sp}(1,\mathbb{C})$ ($n+1$ summands). This corresponds to a generalisation of Section 3 where one projects $\mathbb{CP}^{2n+1}$ to a product of $n+1$ disjoint lines. But we can no longer use Gauss lifts and the condition on meromorphic functions on S^2 will be more complicated.

Real Grassmannians of oriented 4-planes. Let us consider the Grassmannian $Gr_4(\mathbb{R}^n)$, $n \geq 5$. According to the table on page 312 (cf. Example 7) we have a 2:1 branched covering $\sigma : Z_n \to \mathbb{P}\mathcal{N}$ where $Z_n = \mathbb{P}\mathcal{N}'$ is the twistor space of the Grassmannian. The orbit $\mathcal{N}$ has complex dimension $2n - 6$ and it fibres over the quadric $Q^{n-3} = \{[x] \in \mathbb{CP}^{n-2} \mid B(x, x) = 0\}$ where B is a $SO(n - 1, \mathbb{C})$-invariant nondegenerate symmetric bilinear form on $\mathbb{C}^{n-1}$. Since $\mathbb{P}\mathcal{N}_m$ is the quaternionic twistor space of $Gr_4(\mathbb{R}^{n-1})$, we have the following commutative diagram.

$$\begin{array}{ccccc}
Z_n \setminus Z_{n-1} & \xrightarrow{\phi}_{2:1} & \mathbb{P}\mathcal{N} & \xrightarrow{\tilde{\psi}} & \mathbb{P}T^*Q^{n-3} \\
\downarrow \pi & & \downarrow \psi & & \downarrow p \\
Gr_4(\mathbb{R}^n) \setminus Gr_4(\mathbb{R}^{n-1}) & & Q^{n-3} & = & Q^{n-3}
\end{array}$$

(9.1)

Let P be the stabiliser of a line $l \in Q^{n-3}$. Then P is a parabolic subgroup of $SO(n-1, \mathbb{C})$ and, according to Remark 3, it has a dense orbit in $\mathbb{P}T_l^* Q^{n-3}$ which is equal to the image of $\tilde{\psi}_l$. The holomorphic tangent space $T_l \mathbb{CP}^{n-2}$ can be identified with $\text{Hom}(l, \mathbb{C}^n/l)$ and then $T_l Q^{n-3} = \text{Hom}(l, l^\perp/l)$ (we take the polar $l^\perp$ with respect to B). As a result we have the identifications

$$\begin{aligned} \mathbb{P}T_l Q^{n-3} &\simeq \mathbb{P}(l^\perp/l) &\simeq \{\text{2-planes in } l^\perp \text{ which contain } l\} \\ \mathbb{P}T_l^* Q^{n-3} &\simeq \mathbb{P}(l^\perp/l)^* &\simeq \{\text{hyperplanes in } l^\perp \text{ which contain } l\}. \end{aligned} \quad (9.2)$$

The map $\mathbb{P}(l^\perp/l) \ni v \to v^\perp \in \mathbb{P}(l^\perp/l)^*$ gives a bijection between $\mathbb{P}T_l Q^{n-3}$ and $\mathbb{P}T_l^* Q^{n-3}$. Let us consider an orthogonal splitting $l^\perp = l \oplus l'$ and identify $l^\perp/l$ with l'. It is clear that $\mathbb{P}l'$ decomposes under the action of $O(n-3, \mathbb{C}) \subset P$ into two orbits: the orbit consisting of lines which are isotropic (i.e. null with respect to B) and the open dense orbit consisting of non-isotropic lines. The action of P preserves these orbits and we have the following lemma:

Lemma 4 *Let $\tilde{\psi} : \mathbb{P}\mathcal{N} \to \mathbb{P}T^* Q^{n-3}$ be as in Diagram 9.1. Then the image of $\tilde{\psi}$ consists of hyperplanes $v^\perp \subset TQ^{n-3}$ where $v \in \mathbb{P}TQ^{n-3}$ is a non-isotropic line.*

If we want to use Diagram 9.1 to characterise quaternionic superminimal spheres in $Gr_4(\mathbb{R}^n)$, we have to consider maps $\gamma : S^2 \to Gr_4(\mathbb{R}^n)$ which do not intersect $Gr_4(\mathbb{R}^{n-1})$. This is only a technical assumption as we have the following

Lemma 5 *Let us consider an inclusion $Gr_4(\mathbb{R}^{n-1}) \hookrightarrow Gr_4(\mathbb{R}^n)$, let Σ be a (real) surface and let $\gamma : \Sigma \to Gr_4(\mathbb{R}^n)$ be a smooth map. Then there exists $a \in SO(n)$ such that the image of $a \cdot \gamma$ is disjoint from $Gr_4(\mathbb{R}^{n-1})$.*

Proof. It will be easier to keep γ fixed and find a Grassmannian $Gr_4(\mathbb{R}^{n-1})$ which does not intersect the image of γ. The inclusions $Gr_4(\mathbb{R}^{n-1}) \hookrightarrow Gr_4(\mathbb{R}^n)$ can be parametrised by oriented hyperplanes in $\mathbb{R}^n$ i.e. by points of the unit sphere $S^{n-1} = SO(n)/SO(n-1)$. Let us call a vector $v \in S^{n-1}$ admissible if the corresponding Grassmannian $Gr_4(v^\perp)$ does not intersect the image of γ. The Grassmannians which contain a 4-plane $W \in Gr_4(\mathbb{R}^n)$ are parametrised by points of the $(n-5)$-sphere $\{w \in S^{n-1} : w \perp W\}$. Consequently, nonadmissible vectors in S^{n-1} are parametrised by points of the sphere bundle $S^{n-5}(\gamma^* V^\perp)$ where V is the tautological rank 4 vector bundle over $Gr_4(\mathbb{R}^n)$. More precisely, we have a map

$$\chi : S^{n-5}(\gamma^* V^\perp) \ni (z, v) \to v \in S^{n-1}$$

where $z \in \Sigma$ and v is a unit vector in $\gamma(z)^\perp$. Then the image of χ consists of nonadmissible vectors. The lemma follows since $\text{im}\,\chi$ has codimension 2 in S^{n-1}. □

Let us consider a harmonic map $\gamma : \Sigma \to Gr_4(\mathbb{R}^n)$ which does not intersect $Gr_4(\mathbb{R}^{n-1})$ and lifts horizontally to a holomorphic curve $\tilde{\gamma} : \Sigma \to Z_n$. Let us denote by W the curve $\tilde{\psi} \circ \phi \circ \tilde{\gamma} : \Sigma \to \mathbb{P}T^* Q^{n-3}$ and let $\sigma = p \circ W$. Since W is a contact curve, $T_z \sigma \subset W_z$ so W determines a codimension 1 subbundle of the normal bundle

$\nu_\sigma = TQ^{n-3}/T\sigma$. If we use the identifications in Formula 9.2, we can think of $T_z\sigma$ and $W_z^\perp$ as lines in $l^\perp/l$, where $l = \sigma(z)$. The condition $T_z\sigma \subset W_z$ can be then written as $W_z^\perp \subset (T_z\sigma)^\perp$. In other words $V = W^\perp$ is a line subbundle of the rank $n-4$ vector bundle $(T_z\sigma)^\perp \subset TQ^{n-3}$. It follows from Lemma 4 that the line bundle V is non-isotropic (the lemma also implies that if the curve W is nontrivial then its image cannot be contained in a fibre of the projection p, otherwise the curve V would intersect the quadric consisting of null lines in $T_l Q^{n-3}$). Conversely, if σ is a holomorphic curve in Q^{n-3} then a choice of a non-isotropic line subbundle $V \subset (T\sigma)^\perp$ gives a lifting $W = V^\perp : \Sigma \to \mathbb{P}T^* Q^{n-3}$ which takes values in im $\tilde{\psi}$. If $\Sigma = S^2$ then we can lift W to get two holomorphic horizontal curves in the twistor space Z_n and this gives a pair of superminimal spheres in $Gr_4(\mathbb{R}^n)$ (they differ by the action of σ from Example 7). As a result, we have the following theorem.

Theorem 8 *There is a 2:1 correspondence between quaternionic superminimal maps $\gamma : S^2 \to Gr_4(\mathbb{R}^n)$ which do not intersect $Gr_4(\mathbb{R}^{n-1})$ and pairs (σ, V) where $\sigma : S^2 \to Q^{n-3}$ is a holomorphic curve and V is a non-isotropic line subbundle of the rank $n-4$ vector bundle $(T\sigma)^\perp$.*

Remark 6 We get the simplest case when $n = 5$. The map
$$\mathbb{CP}^1 \times \mathbb{CP}^1 \ni ([a,b],[a',b']) \to (aa', ab', ba', bb') \in Q^2$$
shows that the 2-dimensional quadric $x_1 x_4 - x_2 x_3 = 0$ is just the product of two projective lines. There are two families of isotropic lines on the quadric (α and β lines) which under the above identification take the form $\{x\} \times \mathbb{CP}^1$ and $\mathbb{CP}^1 \times \{x\}$ respectively ($x \in \mathbb{CP}^1$). With the notation as in Theorem 8 the bundle $(T\sigma)^\perp$ has rank 1 so it is equal to V. The bundle V is nonisotropic if and only if $T\sigma$ is transversal to α and β lines. This means that the two maps given by projecting σ to each of the factors in $\mathbb{CP}^1 \times \mathbb{CP}^1$ have the same ramification divisor and we are back to the example in Section 3 since $Gr_4(\mathbb{R}^5) = S^4$.

The symmetric space $G_2/SO(4)$. The space $G_2/SO(4)$ can be identified with the set of subalgebras isomorphic to $\mathbb{H}$ in the nonassociative *Cayley algebra* $\mathbb{O}$. The algebra $\mathbb{O}$ (elements of which are called *octonions* or *Cayley numbers*) is equal to $\mathbb{H} \times \mathbb{H}$ with the multiplication defined along the lines of the multiplication in $\mathbb{C} = \mathbb{R} \times \mathbb{R}$ and $\mathbb{H} = \mathbb{C} \times \mathbb{C}$ (see [22], Appendix IV.A). The imaginary part of a product of two imaginary Cayley numbers gives a vector cross-product on $\mathbb{R}^7 = \text{im } \mathbb{O}$ and one can identify $G_2/SO(4)$ with the set of 3-planes in $\mathbb{R}^7$, closed under the cross-product (such 3-planes are called *associative*). $G_2/SO(4)$ is therefore a totally geodesic submanifold of the Wolf space $Gr_4(\mathbb{R}^7) \simeq Gr_3(\mathbb{R}^7)$.

The 3:1 branched covering $\mathbb{P}\mathfrak{g}^\mathbb{C} \supset \mathcal{N}' \to \tilde{\mathcal{N}}_r \subset \mathfrak{sl}(3,\mathbb{C})$ (Example 8) makes it possible to describe quaternionic superminimal spheres in $G_2/SO(4)$ by considering 3×3 trace-free matrices — a much more elementary object than the exceptional Lie algebra $\mathfrak{g}_2$ ([24, 25]). By combining this 3:1 covering with the contact map $\tilde{\psi} : \mathbb{P}\mathcal{N}_r \to \mathbb{P}T^* F_{12}(\mathbb{C}^3)$ induced by the canonical fibration $\mathbb{P}\mathcal{N}_r \to F_{12}(\mathbb{C}^3)$ (cf.

Example 3) we get the following commutative diagram (all maps in the diagram except for the twistor fibration π are holomorphic).

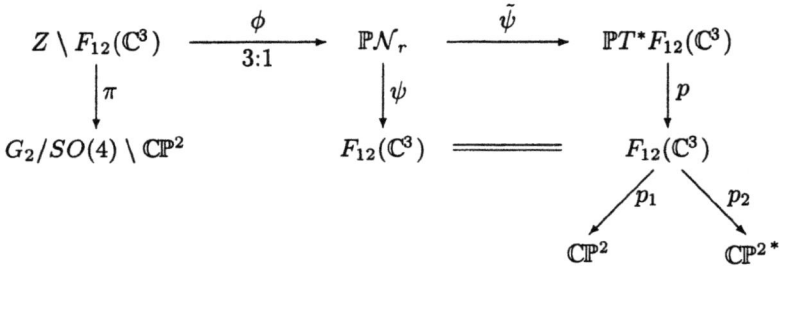

$$(9.3)$$

Remark 7 The submanifold $\mathbb{CP}^2 \hookrightarrow G_2/SO(4)$ in the above diagram was determined by a choice of $SU(3) \subset G_2$. Totally geodesic $\mathbb{CP}^2$'s in $G_2/SO(4)$ are parametrised by points of $S^6 = G_2/SU(3)$ in the following way. If $v \in \mathbb{R}^7$ is a unit vector then $v^\perp$ is naturally a complex 3-dimensional vector space with the complex structure given by vector multiplication by v. The intersections of associative 3-planes containing v with $v^\perp$ are complex lines in $v^\perp$ with respect to this complex structure so they form a $\mathbb{CP}^2$.

The map ϕ in the diagram above is not defined on $F_{12}(\mathbb{C}^3)$, so we will consider harmonic spheres in $G_2/SO(4)$ which do not intersect $\mathbb{CP}^2$. Similarly as before, we have the following

Lemma 6 Consider a totally geodesic submanifold $\mathbb{CP}^2 \hookrightarrow G_2/SO(4)$. If Σ is a (real) surface and the map $\gamma : \Sigma \to G_2/SO(4)$ is smooth then one can choose $a \in G_2$ so that the image of $a \cdot \gamma$ is disjoint from the $\mathbb{CP}^2$.

Proof. The proof is the same as in Lemma 5. This time we consider nonadmissible vectors in S^6. It follows from Remark 7 that they are parametrised by the 2-sphere bundle $S^2(\gamma^* V)$ where V is the tautological rank 3 vector bundle over $G_2/SO(4)$. Therefore the set of nonadmissible vectors has codimension 2 in S^6. □

We will now use the Diagram 9.3 to characterise quaternionic superminimal spheres in $G_2/SO(4)$ in terms of holomorphic data. So let $\gamma : S^2 \to G_2/SO(4)$ be a harmonic map which has a horizontal holomorphic lift $\tilde{\gamma} : S^2 \to Z$. If the image of γ does not intersect $\mathbb{CP}^2$ then im $\tilde{\gamma}$ does not intersect $F_{12}(\mathbb{C}^3) \hookrightarrow Z$. We can define holomorphic curves $W : S^2 \to \mathbb{P}T^* F_{12}(\mathbb{C}^3)$ and $\sigma : S^2 \to F_{12}(\mathbb{C}^3)$, where $W = \tilde{\psi} \circ \phi \circ \tilde{\gamma}$ and $\sigma = p \circ W$. Then W lies in the image of $\tilde{\psi}$ and it follows from Proposition 1 that it cannot be contained in a fibre of the projection p. Since W is a contact curve, it contains the tangent bundle $T\sigma$. Let L_i denote vertical bundles of the fibrations p_i. Their pull-backs by σ provide two holomorphic line bundles σ^*L_i over S^2. It follows from Proposition 1 that W is transversal to σ^*L_i. On the level of the normal bundle $\nu_\sigma = \sigma^*TF_{12}(\mathbb{C}^3)/T\sigma$ we have $\eta \oplus \sigma^*L_1 = \eta \oplus \sigma^*L_2$ where $\eta = W/T\sigma$. As a result $\sigma^*L_1 \simeq \sigma^*L_2$. Holomorphic curves $\sigma : S^2 \to F_{12}(\mathbb{C}^3)$ are determined by pairs of holomorphic maps $\alpha, \beta : S^2 \to \mathbb{CP}^2$ which satisfy the equation $\sum \alpha_i \beta_i = 0$ (we have $\alpha = p_1 \circ \sigma$, $\beta = p_2 \circ \sigma$, and we use the quadratic form $\sum x_i^2$ on $\mathbb{C}^3$ to identify $\mathbb{CP}^2$ with $\mathbb{CP}^{2*}$). Note that, since $T\sigma \subset W$ is transversal to σ^*L_i, the differentials of the curves α, β vanish to the same order i.e. α, β have the same ramification divisor. With the notation as above, this leads to the following theorem

Theorem 9 *There is a 3:1 correspondence between quaternionic superminimal maps $\gamma : S^2 \to G_2/SO(4)$ (which do not intersect a $\mathbb{CP}^2$) and triples (α, β, s) where*

1. *α, β are holomorphic curves $S^2 \to \mathbb{CP}^2$ with the same ramification divisor and $\sum \alpha_i \beta_i = 0$,*

2. *if $\sigma = (\alpha, \beta)$ denotes the corresponding holomorphic curve in $F_{12}(\mathbb{C}^3)$ then $\sigma^*L_1 \simeq \sigma^*L_2$,*

3. *s is a line subbundle of the normal bundle ν_σ, transversal to σ^*L_1 and σ^*L_2.*

The proof of the theorem will be complete if we show that the holomorphic data (α, β, s) gives rise to a triple of harmonic spheres in $G_2/SO(4)$. Since α, β have the same ramification divisor, the curve $\sigma = (\alpha, \beta) : S^2 \to F_{12}(\mathbb{C}^3)$ is transversal to the fibres of the projections p_1 and p_2. Then the line bundle s determines a rank 2 subbundle $W \hookrightarrow TF_{12}(\mathbb{C}^3)$ which contains $T\sigma$ and is transversal to the fibres of p_1 and p_2. As a result we get a contact curve $W : S^2 \to \mathbb{P}T^*F_{12}(\mathbb{C}^3)$ which lies in the image of the map $\tilde{\psi}$. Since S^2 is simply connected, $(\tilde{\psi})^{-1} \circ W$ lifts to three horizontal holomorphic curves in $Z \setminus F_{12}(\mathbb{C}^3)$. This gives three minimal spheres in $G_2/SO(4)$ which differ by the action of the centre of $SU(3)$ on $G_2/SO(4)$, described in Example 8.

Remark 8 Note that if $f : M \to F_{12}(\mathbb{C}^3)$ is a holomorphic map where M is a 2-dimensional complex manifold then at the points where TM is transversal to the fibres of the projections p_1, p_2 one can use the Gauss lift $Tf : M \to \mathbb{P}T^*F_{12}(\mathbb{C}^3)$ to construct a harmonic map $g : M \to G_2/SO(4)$. Being a projection of a J_2-holomorphic manifold in the twistor space, the map g must be $(1,1)$-*geodesic* (i.e. $(\nabla d\phi)^{1,1} = 0$, see [18, §4.43 and §7.9]). Note that holomorphic curves in M give rise to quaternionic superminimal surfaces in $G_2/SO(4)$.

Bibliography

[1] D. Alekseevskiĭ, *Compact quaternion spaces*, Funk. Anal. i Prilozh. **2** (1968), 11–20; English transl.: Funct. Anal. Appl. **2** (1968), 106–114.

[2] M.F. Atiyah, N.J. Hitchin, and I.M. Singer, *Self-duality in four-dimensional Riemannian geometry*, Proc. Roy. Soc. Lond. **A362** (1978), 425–461.

[3] A. Bahy-El-Dien and J.C. Wood, *The explicit construction of all harmonic twospheres in quaternionic projective spaces*, Proc. London Math. Soc. **62** (1991), 202–224.

[4] A. L. Besse, *Einstein manifolds*, Ergebnisse der Mathematik und ihrer Grenzgebiete, 3. Folge, vol. 10, Springer-Verlag, Berlin 1987.

[5] W. Borho and J. Brylinski, *Differential operators on homogeneous spaces. I*, Invent. Math. **69** (1982), 437–476.

[6] R.L. Bryant, *Conformal and minimal immersions of compact surfaces into the 4-sphere*, J. Diff. Geom. **17** (1982), 455–473.

[7] R.L. Bryant, *Lie groups and twistor spaces*, Duke Math. J. **52** (1985), 223–261.

[8] R. Brylinski and B. Kostant, *Nilpotent orbits, normality, and Hamiltonian group actions*, Bull. Amer. Math. Soc. (N.S.) **26** (1992), 269–275.

[9] D. Burns, *Harmonic maps from $\mathbb{CP}^1$ to $\mathbb{CP}^n$*, in: Harmonic Maps (Proceedings, New Orleans 1980), Lect. Notes in Math. 949, Springer-Verlag, Berlin 1982, 48–56.

[10] F.E. Burstall, *Recent developments in twistor methods for harmonic maps*, in: Harmonic Mappings, Twistors, and σ-models (Luminy, 1986), Adv. Ser. Math. Phys., vol. 4, World Scientific, Singapore, 1988, 158–176.

[11] F.E. Burstall, *Minimal surfaces in quaternionic symmetric spaces*, in: Geometry of Low-dimensional Manifolds, Cambridge University Press, 1990, 231–235.

[12] F.E. Burstall, *Nilpotent elements and parabolic subalgebras*, private communication, March 1990.

[13] F.E. Burstall and J.H. Rawnsley, *Twistor Theory for Riemannian Symmetric Spaces*, Lect. Notes in Math. 1424, Springer-Verlag, Berlin, 1990.

[14] F.E. Burstall and S.M. Salamon, *Tournaments, flags and harmonic maps*, Math. Ann. **277** (1987), 249–265.

[15] F.E. Burstall and J.C. Wood, *The construction of harmonic maps into complex Grassmannians*, J. Diff. Geom. **23** (1986), 255–297.

[16] A.M. Din and W.J. Zakrzewski, *General classical solutions in the $\mathbb{CP}^{n-1}$ model*, Nuclear Phys. **B174** (1980), 397–406.

[17] E.B. Dynkin, *Semisimple subalgebras of semisimple Lie algebras*, Mat. Sbornik N.S. **30** (1952), 349–462; English transl: A.M.S. Transl. Ser. 2, vol. 6 (1957), 111–244.

[18] J. Eells and L. Lemaire, *Another report on harmonic maps*, Bull. Lond. Math. Soc. **20** (1988), 385–524.

[19] J. Eells and S. Salamon, *Twistorial construction of harmonic maps of surfaces into four-manifolds*, Ann. Scuola Norm. Sup. Pisa **12** (1985), 589–640.

[20] J. Eells and J.C. Wood, *Harmonic maps from surfaces to complex projective spaces*, Advances in Math. **49** (1983), 217–263.

[21] A.G. Elashvili and A.N. Panov, *Polarizations in semisimple Lie algebras*, Bull. Acad. Sci. Georgian SSR **87** (1977), 25–28 (Russian).

[22] R. Harvey and H.B. Lawson, Jr, *Calibrated geometries*, Acta Math. **148** (1982), 47–157.

[23] W.H. Hesselink, *Polarizations in the classical groups*, Math. Zeitschrift **160** (1978), 217–234.

[24] P.Z. Kobak, *Quaternionic geometry and harmonic maps*, D. Phil. Thesis, University of Oxford, 1993.

[25] P.Z. Kobak and A.F. Swann, *Quaternionic geometry of a nilpotent variety*, Math. Ann. (to appear).

[26] B. Kostant, *The principal three-dimensional subgroup and the Betti numbers of a complex simple Lie group*, Amer. J. Math. **81** (1959), 973–1032.

[27] H.B. Lawson, Jr, *Surfaces minimales et la construction de Calabi-Penrose*, Sém. Bourbaki, 624 (1983/84), Astérisque **121–122** (Soc. Math. France, 1985), 197–211.

[28] T. Levasseur and S.P. Smith, *Primitive ideals and nilpotent orbits in type G_2*, Journal of Algebra **114** (1988), 81–105.

[29] B. Loo, *The space of harmonic maps of S^2 into S^4*, Trans. Amer. Math. Soc. **313** (1989), 81–102.

[30] J.H. Rawnsley, *Twistor theory for Riemannian manifolds*, in: New Developments in Lie Theory and Their Applications, J. Tirao and N. Wallach, eds., Progress in Mathematics 105, Birkhäuser, 1992, 115–128.

[31] R. Richardson, *Conjugacy classes in parabolic subgroups of semi-simple algebraic groups*, Bull. Lond. Math. Soc. **6** (1974), 21–24.

[32] S.M. Salamon, *Quaternionic Kähler manifolds*, Invent. Math. **67** (1982), 143–171.

[33] S.M. Salamon, *Harmonic and holomorphic maps*, in Geometry Seminar "L. Bianchi" II—1984, ed. E. Vesentini, Lect. Notes in Math. 1164, Springer-Verlag, Berlin 1985, 161–224.

[34] S.M. Salamon, *Riemannian Geometry and Holonomy Groups*, Pitman Research Notes in Math. 201, Longman, Harlow, 1989.

[35] T.A. Springer and R. Steinberg, *Conjugacy classes*, in: Seminar on Algebraic Groups and Related Finite Groups, Lect. Notes in Math. 131, Springer-Verlag, Berlin 1970, 167–266.

[36] A.F. Swann, *Hyperkähler and Quaternionic Kähler Geometry*, D. Phil. Thesis, University of Oxford, 1990.

[37] A.F. Swann, *Homogeneity of twistor spaces*, Twistor Newsl. **32** (1991), 8–9.

[38] A.F. Swann, *HyperKähler and quaternionic Kähler geometry*, Math. Ann. **289** (1991), 421–450.

[39] J. Wolf, *Complex homogeneous contact manifolds and quaternionic symmetric spaces*, J. Math. Mech. **14** (1965), 1033–1047.

[40] J.G. Wolfson, *Harmonic sequences and harmonic maps of surfaces into complex Grassmann manifolds*, J. Diff. Geom. **27** (1988), 161–178.

[41] J.C. Wood, *Twistor constructions for harmonic maps*, in: Differential Geometry and Differential Equations, Lect. Notes in Math. 1255, Springer-Verlag, Berlin 1987.

[42] J.C. Wood, *The explicit construction and parametrization of all harmonic maps from the two-sphere to a complex Grassmannian*, J. Reine Angew. Math. **386** (1988), 1–31.

[43] J.C. Wood, *Harmonic maps into symmetric spaces and integrable systems*, this volume.

INDEX

Index

(1, 1)-geodesic map, 317

action
    S^1, 44
    co-adjoint, 46
    group, 278–289
    integral, 30
    natural, 206, 211, 214, 216, 282–284
    Terng, 281
    Uhlenbeck, 284
Adler-Kostant-Symes scheme, 223–229
    and r-matrices, 239
    in infinite dimensions, 231–232
    on loop algebras, 232–234
almost complex
    curve, 61, 79
    torus, 79
associate minimal surfaces, 42
associated family, 139
associative 3-planes, 315, 316
asymptotic
    coordinates, 17, 133
    line, 111, 118, 129
        parametrisation, 116, 117
automorphism
    involutive, 189

Bäcklund transform, 17, 19, 24, 25, 36, 37, 41, 83, 139
Bäcklund-Darboux transform, 118
Baker function, 207, 208, 216–218
balanced diagram, 96–98
Bianchi
    diagram, 18
    surfaces, 84, 116–118, 121
Birkhoff
    decomposition, 278
    factorisation, 217, 218
Bogomolny bound, 194, 198, 200
Bonnet surface, 84
Bonnet's Theorem, 134
Borel algebra, 304, 307
branch point, 32

Bryant correspondence, 299, 304

canonical fibration, 306
    associated map of, 304, 309, 310
Cartan
    embedding, 41, 50, 264, 284
    immersion, 183
Casimir, 189
Cayley algebra, 315
Chebyshev net, 115
chiral, 183
    field, 149, 156, 159, 163, 165, 166, 169
    model, 151, 155, 159, 199
        $SU(2)$, 195, 196
Clifford torus, 76
cobordant, 124
Codazzi-Mainardi equation, 36, 113
Codazzi-Mainardi-Ricci equations, 50
compatibility condition, 36, 83, 86, 88, 108, 109, 113, 116, 117, 120, 134, 209, 211
complex analytic map, 31
composition law, 33
conformal, 43
    horizontally, 33
    invariance, 31, 33
    weakly, 31–33, 277, 297, 298
conservation laws, 9, 10, 31, 35
conserved density, 9
constant
    Gauss curvature, 83, 84, 111–115, 129, 134, 137, 138
    mean curvature, 32, 36, 37, 42, 50, 60, 84, 94–101, 129, 138, 139, 218
Coxeter-Killing automorphism, 215
CP^1 model, 194, 196–199
critical point, 29, 31
current, 176
    left, 149, 151, 160, 163, 165
curvature line, 93, 138
    coordinates, 133

deformation
    of a surface, 83, 84, 94, 105, 106, 109, 114, 117, 119, 120, 122, 123
    parameter, 94, 103, 105, 109, 114
Dirichlet integral, 30
Dodd-Bullough equation, 24, 85
dressing, 84, 147, 148, 153–156, 159, 170, 278–280
    orbit, 206
    transform, 42
    un-, 152
dual
    surface, 93
    symmetry, 178

Einstein manifold, 301
ellipse of curvature, 77
elliptic function, 35
energy
    density, 29
    functional, 188, 194
    integral, 29, 240
Enneper surface, 139
equiharmonic map, 70, 211, 213–215, 246
Euler-Lagrange equation, 130
extended
    framings, 257–259
    solutions, 43, 263, 274, 275

f-structure, 213
factorisation, 42, 276, 282
    of Lax operator, 17
    problem, 148, 152–156, 161
finite
    gap, 84, 118
    type, 50, 59, 68, 72, 76, 79, 243
flag
    canonical, 282
    manifold, 307
    manifolds, 298, 304
        as k-symmetric spaces, 250
        complete, 305, 310

maps into, 62
periodic, 212, 216
space, 207–208, 277
flatness condition, 131
flip number, 89, 91, 123
flux, 9
force, 97

gauge, 39
    group, 132
    theory, 273
    transformation, 25, 39
Gauss
    bundle, 277
    equation, 36, 96, 115, 116
    map, 32, 35, 37, 43, 60, 92, 95, 118, 133, 218
Gauss-Codazzi equations, 83, 86, 94, 114, 116, 119
Gauss-Weingarten equations, 36, 83, 86, 93
geodesic, 30
Grassmannian, 32, 41, 207–208

Hamiltonian
    system, 35, 135–138
        bi-, 129, 137
    theory, 47
    vector field, 46, 47, 185
harmonic
    $1/\sqrt{K}$, 120
    conjugate, 42
    function, 30
    inverse mean curvature, 84
    morphism, 33, 35
    sequence, 38, 61, 62, 206, 213, 214
harmonic maps, 29, 130, 150, 160, 165, 168, 169
    and minimal surfaces, 32
    between Kähler manifolds, 31
    conformal, 42, 50
    deformations of, 273, 287–289
    equivariant, 34–35, 221

examples, 30–33
existence problem, 34
framings of, 241
from a pseudo-Riemannian manifold, 30, 117
from a surface, 31, 38, 41, 62
from a torus, 34, 35, 50, 67–68, 72, 76, 79, 206, 217–218
from the 2-sphere, 31, 206
gauge-theoretic formulation of, 242
Lorentz-, 117
nonexistence, 34
pseudo-, 30, 176
reducible, 289
spaces of, 274, 276, 287–289
submersive, 32
to a complex projective space, 211–215
to a flag manifold, 206, 211–215
to a homogeneous space, 40
to a Lie group, 39–51, 260–263, 274–284
    of finite type, 261–263
    of finite uniton number, 267
to a quaternionic projective space, 289
to a reductive homogeneous space, 240–243
to a sphere, 34, 35, 50, 72, 253
to a symmetric space, 50, 130, 243–245, 284–287
    of G-finite type, 265–267
    of finite type, 243–245
to a unitary group, 41–42
to the 4-sphere, 298–300, 313, 315
to the 6-sphere, 78
twistorial constructions, 32
weakly conformal, 32
heat equation, 34
hidden symmetry, 273
higher fundamental form, 77
highest root nilpotent orbit, 302, 310

Hill's operator, 35
holomorphic
    map, 31, 44, 50, 277
    quadratic differential, 31, 37, 38
    section, 37
homogeneous space, 177
    reductive, 177
        naturally, 177
homotopic
    regularly, 123
homotopy class, 34
Hopf
    differential, 92
    problem, 138
horizontal
    distribution, 298, 300, 301
    holomorphic curve, 298, 300, 301, 314, 315, 317
    holomorphic map, 44, 66

inclusive map, 301
infinitesimal action, 273
integrability
    complete, 130, 136, 187, 205
    conditions, 39, 42, 45, 63, 67, 72, 131, 133, 275
    Liouville, 183
integrable, 83
    system, 273
isospectral flow, 13, 16
isothermal
    parametrisation, 93
isotopy, 123
isotropic, 277, 288–289
    complex, 38, 63
    real, 50, 298
    strongly, 277, 289
isotropy
    dimension, 253
    order, 38, 39, 64, 277
    total, 39
Iwasawa decomposition, 210–212, 214, 215, 218, 281
    of a semisimple Lie algebra, 228

of twisted loop algebras, 232
of twisted loop groups, 256–257

Jacobi identity
    violation of, 190
Jacobian, 136
    variety, 49, 218

k-symmetric space, 50, 69, 239–240
Kač-Moody
    algebra, 205, 273
    group, 274
Kähler
    angle, 68
    manifold, 31
KdV equation, 8, 23, 49, 206, 215
kink, 18
Koszul-Malgrange holomorphic structure, 31, 37

Lagrangian, 195
Laplace's equation, 30, 33
Laplacian, 30, 130
Lax
    equation, 187
    equations, 48, 49, 137, 234
    form, 187
    hierarchy, 15
    representation, 13, 83, 95, 96, 115, 116, 118
Lie
    algebra, 39
        metrisable, 177
    group, 39
    sphere geometry, 134
lines of curvature, 60
loop
    algebras, 154, 276
        based, 260
        twisted, 133, 232–234
    group, 161, 207, 212, 215, 274, 276
        Grassmannian model, 280–281
    of 1-forms, 43

of connections, 137
of maps, 42–44, 133
space, 44

Maurer-Cartan
    equations, 39, 229–231, 275
    form, 39, 40, 213
        of a Lie group, 229
        of a reductive homogeneous space, 240
metric
    Minkowski, 130, 133
    pseudo-Riemannian, 130, 133
    Sasaki, 184
minimal
    branched immersion, 32
    fibres, 33
    immersion, 35
    submanifold, 32
    surface, 61, 84, 92–93, 129
Miura map, 11
MKdV
    equation, 10, 23
    hierarchy, 17
moment map, 40
momentum, 97
    functional, 287
monodromy, 89, 96–98, 121–126, 217, 218
    matrices, 190
Morse-Bott function, 287

Neumann equation, 35
nilpotent
    orbits, 302
        characteristic numbers of, 308
        finite covers of, 311
        in $\mathfrak{g}_2^{\mathbb{C}}$, 308
        in $\mathfrak{sl}(n, \mathbb{C})$, 306
        minimal, 302, 305, 310
        regular, 302, 310
    variety, 302, 309
NLS equation, 22
Noether's Theorem, 40

nonlinear superposition formula, 18

octonions, 315
one-parameter family
   of harmonic maps, 42
   of surfaces, 94, 101, 105, 109, 111, 114, 117, 119, 120

pants, pair of, 123
parabolic algebra, 304, 307
parallel
   mean curvature, 32
   surfaces, 98–111
parity
   of the spin structure, 92, 126
   of twists, 90, 91, 93
Pauli matrices, 87
pendulum equation, 35
Peterson surface, 117
pluriharmonic maps, 49, 236
Pohlmeyer, 182
Poisson
   bracket, 47, 185, 186
   manifold, 47, 135
   structure, 46, 135
   structures, 224
      from r-matrices, 237
polar, 75
polarisation, 306
polynomial
   flows, 21
   Killing field, 138
primitive maps, 61, 62, 66, 68–72, 76, 213–215, 245–248
   and twistor lifts, 253–256
   are harmonic, 246
   have harmonic projections, 246
   of G-finite type, 265–267
   of finite type, 248
   zero-curvature equations for, 247
principal realisation, 212
pseudo-action, 279
   Birkhoff, 279
   Uhlenbeck, 279

pseudo-holomorphic, 38, 50, 63, 68, 277

quaternion-Kähler manifold, 301
quaternionic
   superminimal spheres
      in $Gr_4(\mathbb{R}^4)$, 314–315
      in $G_2/SO(4)$, 316–317
   superminimal surfaces, 298, 301
quaternions, 87, 130

r-matrices, 47, 237
   and the Adler-Kostant-Symes scheme, 239
   classical, 187
   on twisted loop algebras, 237
reality condition, 159–161, 166, 168, 210, 216
reduction theorems, 34–35
regularity
   of harmonic maps, 30
remarkable identity, 35
Richardson element, 306, 309
root space, 69

Schrödinger
   equation, 12
   operator, 35
second fundamental form, 30
self-dual 4-manifold, 297, 301
σ-model, 30, 176, 273
   $O(3)$, 194
sine-Gordon equation, 17, 19, 35, 36, 83, 115, 133, 139
   complex, 133
singularity
   higher order, 64, 67
sinh-Gordon equation, 24, 25, 36, 60, 96, 133, 218
Skyrmions, 198
$\mathfrak{sl}(2,\mathbb{C})$-triple, 303, 307
smoke ring, 134
solitary wave, 9
solitons, 9, 61, 83, 97, 129, 147, 161, 163, 168

multi-, 84
number of, 194
spectral
    curve, 49, 136
    parameter, 42–51, 83, 132, 135–139, 150, 214
    transform, 13
spectrum, 136
spin structure, 84, 89–93, 121–126
    of an immersion, 91, 96
spinor, 121–126
Springer resolution, 309, 310
stress energy, 32
superconformal, 50, 61, 62, 64, 67, 68, 72, 74, 76
superhorizontal, 286
    distribution, 298
superminimal, 38, 50, 63, 277, 298
Sym formula, 134
Symes method
    for based loop groups, 262–263
    for twisted loop groups, 257–259
symmetric spaces, 175, 239–240
    pseudo-Riemannian, 179
symplectic
    form, canonical, 184
    manifold, 135

τ-function, 14, 50, 213, 217
tension field, 30, 130
Theorem of permutability, 18
θ-function, 50, 206, 207, 215, 217
Toda
    equations
        for G_2, 78–80
        for $SO(2m+1)$, 72–78
        for $SU(n+1)$, 60, 62–68
        for a compact simple Lie group, 59, 68–72
    fields, 248–252
        and holomorphic maps, 252
        geometric interpretation of, 250
        Lagrangian for, 249

        zero-curvature equations for, 249
    frame, 71, 79, 250
    lattice, 24, 227–229
    system, 50, 134, 205
topological charge, 198, 201
torque, 97
torus, 49
    maps from, 122, 124
    with a hole, 124, 125
totally geodesic, 30, 33, 41, 50
transform, 38
    ∂-, 37, 206, 213
    flag, 41, 42, 44
    Gauss, 37, 206
    Riemann-Hilbert, 274
    scattering, 13
translation-holonomy strip, 90
trivialisation, 39
twistor, 33, 43
    fibration, 277, 297
    lifts, 253
    space, 273, 284, 285, 297
        of a symmetric space, 298
        quaternionic, 298, 301–303
    subfibration, 286
twistorial constructions
    of harmonic maps, 32

ultralocal, non, 186
uniton, 147, 170, 171, 276, 282
    number, 276
universal enveloping algebra, 188

variational principle, 97
Veronese sequence, 214

Weierstrass representation, 84, 92
Weingarten surface
    linear, 134
Wente tori, 138
Weyl group, 218
Willmore surface, 134
Wolf space, 301

Yang-Baxter equation, 136, 188, 237
Yang-Mills, self-dual, 200

Zakharov-Shabat representation, 83
zero curvature, 20, 147, 151, 158
    equation, 43, 131, 135–138
    representation, 178

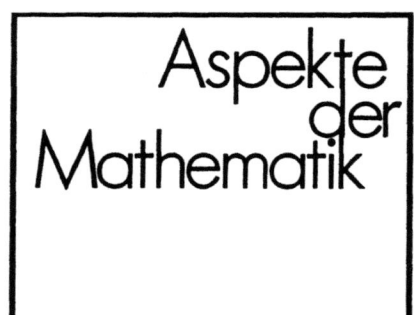

Edited by Klas Diederich

Band D 1: H. Kraft: Geometrische Methoden in der Invariantentheorie

Band D 2: J. Bingener: Lokale Modulräume in der analytischen Geometrie 1

Band D 3: J. Bingener: Lokale Modulräume in der analytischen Geometrie 2

Band D 4: G. Barthel/F. Hirzebruch/T. Höfer: Geradenkonfigurationen und Algebraische Flächen*

Band D 5: H. Stieber: Existenz semiuniverseller Deformationen in der komplexen Analysis

Band D 6: I. Kersten: Brauergruppen von Körpern

*A Publication of the Max-Planck-Institut für Mathematik, Bonn

If you have any concerns about our products,
you can contact us on
ProductSafety@springernature.com

In case Publisher is established outside the EU,
the EU authorized representative is:
**Springer Nature Customer Service Center GmbH
Europaplatz 3, 69115 Heidelberg, Germany**

Printed by Libri Plureos GmbH
in Hamburg, Germany